SYSTEMS AND COMPUTATIONAL BIOLOGY – MOLECULAR AND CELLULAR EXPERIMENTAL SYSTEMS

Edited by **Ning-Sun Yang**

Systems and Computational Biology – Molecular and Cellular Experimental Systems
Edited by Ning-Sun Yang

Published by InTech

Janeza Trdine 9, 51000 Rijeka, Croatia

Publishing Process Manager Davor Vidic

Technical Editor Teodora Smiljanic

Cover Designer InTech Design Team

Additional hard copies can be obtained from orders@intechopen.com

Systems and Computational Biology – Molecular and Cellular Experimental Systems,
Edited by Ning-Sun Yang
p. cm.
ISBN 978-953-307-280-7

Contents

Preface

Immediately after the first drafts of the human genome sequence were reported almost a decade ago, the importance of genomics and functional genomics studies became well recognized across the broad disciplines of biological sciences research. The initiatives of Leroy Hood and other pioneers on developing systems biology approaches for evaluating or addressing global and integrated biological activities, mechanisms, and network systems have motivated many of us, as bioscientists, to re-examine or revisit a whole spectrum of our previous experimental findings or observations in a much broader, link-seeking and cross-talk context. Soon thereafter, these lines of research efforts generated interesting, fancy and sometimes misleading new names for the now well-accepted "omics" research areas, including functional genomics, (functional) proteomics, metabolomics, transcriptomics, glycomics, lipidomics, and cellomics. It may be interesting for us to try to relate these "omics" approaches to one of the oldest omics studies that we all may be quite familiar with, and that is "economics", in a way that all "omics" indeed seemed to have meant to address the mechanisms/activities/constituents in a global, inter-connected and regulated way or manner.

The advancement of a spectrum of technological methodologies and assay systems for various omics studies has been literally astonishing, including next-generation DNA sequencing platforms, whole transcriptome microarrays, micro-RNA arrays, various protein chips, polysaccharide or glycomics arrays, advanced LC-MS/MS, GC-MS/MS, MALDI-TOF, 2D-NMR, FT-IR, and other systems for proteome and metabolome research and investigations on related molecular signaling and networking bioactivities. Even more excitingly and encouragingly, many outstanding researchers previously trained as mathematicians, information or computation scientists have courageously re-educated themselves and turned into a new generation of bioinformatics scientists. The collective achievements and breakthroughs made by our colleagues have created a number of wonderful database systems which are now routinely and extensively used by not only young but also "old" researchers. It is very difficult to miss the overwhelming feeling and excitement of this new era in systems biology and computational biology research.

It is now estimated, with good supporting evidence by omics information, that there are approximately 25,000 genes in the human genome, about 45,000 total proteins in the human proteome, and around 3000 species of primary and between 3000 and 6000 species of secondary metabolites, respectively, in the human body fluid/tissue metabolome. These numbers and their relative levels to each other are now helping us to construct a more comprehensive and realistic view of human biology systems. Likewise, but maybe to a lesser extent, various baseline omics databases on mouse, fruit fly, Arabidopsis plant, yeast, and E. coli systems are being built to serve as model systems for molecular, cellular and systems biology studies; these efforts are projected to result in very interesting and important research findings in the coming years.

Good findings in a new research area may not necessarily translate quickly into good or high-impact benefits pertaining to socio-economic needs, as may be witnessed now by many of us with regard to research and development in omics science/technology. To some of us, the new genes, novel protein functions, unique metabolite profiles or PCA clusters, and their signaling systems that we have so far revealed seemed to have yielded less than what we have previously (only some 5 to 10 years ago) expected, in terms of new targets or strategies for drug or therapeutics development in medical sciences, or for improvement of crop plants in agricultural science. Nonetheless, some useful new tools for diagnosis and personalized medicine have been developed as a result of genomics research. Recent reviews on this subject have helped us more realistically and still optimistically to address such issues in a socially responsible academic exercise. Therefore, whereas some "microarray" or "bioinformatics" scientists among us may have been criticized as doing "cataloging research", the majority of us believe that we are sincerely exploring new scientific and technological systems to benefit human health, human food and animal feed production, and environmental protections. Indeed, we are humbled by the complexity, extent and beauty of cross-talks in various biological systems; on the other hand, we are becoming more educated and are able to start addressing honestly and skillfully the various important issues concerning translational medicine, global agriculture, and the environment.

I am very honored to serve as the editor of these two volumes on Systems and Computational Biology: (I) Molecular and Cellular Experimental Systems, and (II) Bioinformatics and Computational Modeling. I believe that we have collectively contributed a series of high-quality research or review articles in a timely fashion to this emerging research field of our scientific community.

I sincerely hope that our colleagues and readers worldwide will help us in future similar efforts, by providing us feedback in the form of critical comments, interdisciplinary ideas and innovative suggestions on our book chapters, as a way to pay our high respect to the biological genomes on planet earth.

Dr. Ning-Sun Yang
Agricultural Biotechnology Research Center, Academia Sinica
Taiwan, R.O.C

Part 1

Approaches for Studying Genomes, Transcriptomes, and Proteomes

1

Gene Expression Analysis Using RNA-Seq from Organisms Lacking Substantial Genomic Resources

Yingjia Shen, Tzintzuni Garcia and Ronald B. Walter
Texas State University, USA

1. Introduction

Development of massively parallel "next generation" sequencing technology (NGS) has dramatically revolutionized biological studies. Among the many applications of NGS, RNA-Seq is one of the most important uses of this technology. RNA-Seq enables investigators to accurately probe the current state of a transcriptome and assess many biologically important issues, such as; gene expression levels, differential splicing events, and allele-specific gene expression. Compared with previous technologies (e.g., microarrays, etc.) NGS has the clear advantage of not being limited to experimental systems having well characterized genomes or transcript sequence libraries. This positions RNA-seq approaches as important and versatile techniques for experimental systems and species where specific genetic information may be limited or altogether lacking.

A major goal of most transcriptomic studies is the identification and characterization of all transcripts within a developmental stage or specific tissue. NGS techniques have made the massive amount of data required to carry out such studies both inexpensive and available to an unprecedented extent. Clever computer algorithms have made the assembly of these massive data sets the work of one or two people with reasonably powerful workstations or a moderate analytical server.

Once a reference transcriptome has been assembled, analyses can be carried out that involve several steps, such as; mapping short sequence reads to transcriptome, quantifying the abundance of genes or gene sets, and comparing differential expression patterns among all samples. Herein we outline the processes from obtaining raw short read data to advanced comparative gene expression analysis and we review bioinformatic programs currently available, such as Tophat, Cufflinks, DESeq, that are specifically designed to address each of the above steps. We will discuss both accuracy and ease of use of these tools by biologists beginning to pursue these types of analyses. In addition to individual programs, we will also discuss integration of multiple programs into pipelines for more rapid and complete expression analyses. Overall, the future applications of RNA-Seq will open new avenues for transcriptome analyses of less well-studied and/or wild caught species that could not have previously been approached. This will yield a wealth of new comparative data highlighting the many ways plants and animals have developed to survive in this rapidly changing environment.

2. *De Novo* sequence assembly and expression analysis with NGS data

There are many phases to an NGS research project where the end goal is expression analysis in a non-model organism. This chapter is dedicated to the many phases and options available to the researcher. In general however, bioinformatic analyses at some point begin with gathering raw sequence data from a biological sample of interest and having it sequenced. The raw data will often need to be filtered for quality. If any pre-existing sequences are available from a closely related species, their use as a reference should be considered, but is not necessary. Assembling the short reads derived from one or more of the NGS platforms comes next, but should not be considered a definitive, terminal process. Most frequently assembly of short read data entails an iterative refinement phase in which a wide range of parameters are modified in the search for a sufficiently contiguous and complete assembly. Analyzing the assembly can entail searching for signatures of assembly errors and trying to identify the assembled contigs. Once a satisfactory group of transcripts is produced they are locked for the expression level analysis. Mapping the short reads to the assembled transcripts is the first step in assessing gene expression levels. The next is determining the expression levels of each contig based on the number of short reads mapped to it. Generally a comparative gene expression analysis will follow in which two or more samples are compared and alternate regulation patterns or profiles determined. We end the chapter with a specialized comparative expression study in F1 hybrid organisms in which differential expression may reveal evolutionary divergence in gene regulation mechanisms.

2.1 Next-generation sequencing

Next-generation sequencing (NGS) techniques produce millions of reads per run but each read may be as short as 25 bp. Using NGS allows one to apply complex samples (i.e., total DNA or RNA libraries) on the NGS instrument. These mixed samples contain fragments of larger molecule targets sheared to some pre-set fragment length distribution. NGS techniques allow the sequencing of completely unknown samples in a massively parallel fashion. In order to perform massively parallel sequencing most NGS instruments require a run time of days to weeks in carefully controlled conditions for complete data acquisition. There are many competing technologies, and new challengers are in constant development to increase both the speed and quantity of NGS per sample run. It is beyond the scope of this chapter to examine all of the current and upcoming techniques so we will briefly focus on two most common NGS instruments currently in use: the Illumina Genome Analyzer and ABI SOLiD platforms. Each of these platforms has its strengths and weaknesses that are very important to understand when designing research strategies.

2.1.1 The ABI SOLiD platform

The SOLiD system produces short sequencing lengths (i.e., termed "reads") ranging from 35 to 75 bp and has run times of between one and seven days depending on the amount and type of reads desired. Typically, instruments will have 6 or 12 independent lanes available per run and samples can be multiplexed in each of those for up to 96 unique barcodes. Product literature states the daily sequencing throughput is between 10-30 Gbp.

The SOLiD sequencing process begins by fragmenting high molecular weight DNA into smaller fragments to be sequenced (Fig 1A). Fragments are size selected in a narrow range, typically around 200 bp, and primers are ligated to both ends of the fragments (Fig 1A).

Glass beads coated with complimentary primers are mixed with the fragments (Fig 1A) and emulsified in such manner that an aqueous droplet will contain a single bead and a single fragment along with the biochemistry necessary for PCR (Fig 1B). Several rounds of emulsion PCR later each bead is coated with sequences identical to the original fragment (Fig 1B). The DNA coated beads are then released from the emulsion, and washed into tiny wells in a plate sized to admit a single bead per well (Fig 1B). Finally the cyclic sequencing phase begins during which each position is iteratively read (Fig 1C).

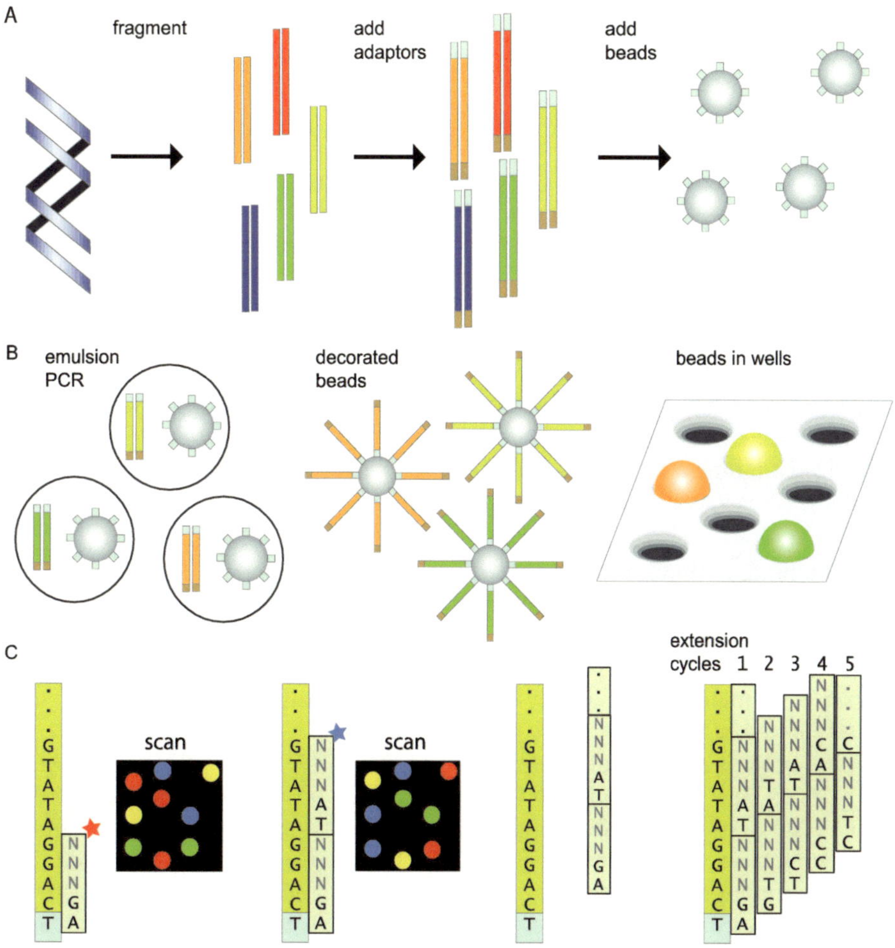

Fig. 1. A simplified outline of the ABI SOLiD sequencing procedure. A) Sample preparation and the addition of glass beads decorated with primers. B) Emulsion PCR amplifies a single template so that its copies are primed by primers bound to a glass bead. C) The sequencing reaction repeats through five extension cycles where the primers are offset by one position in each cycle so that each position in the template is interrogated twice.

The most distinctive feature of SOLiD data is the fact that during the sequencing phase nucleotides are added in dinucleotide probes. In each cycle a nucleotide pentamer is added in which the two 5′ bases are determined by the attached dye (Fig 1C). Once the plate is imaged, the dye is removed leaving the newly added pentamer (Fig 1C). Each cycle therafter interrogates two more bases offset by three positions from the previous cycle (Fig 1C). As the growing fragment reaches the desired length the entire fragment is washed off and a new primer bound at an offset of one so that a different set of bases are interrogated as this new strand grows (Fig 1C). This process is repeated five times, each one offset by one base from the last so that each position is ultimately interrogated twice (Fig 1C).

Four fluorescent dyes are used but each dye can be carried by one of four nucleotide dimers. As each color is read, the recorded data corresponds to a sequence of colors coded by 0, 1, 2, or 3; this is called color-space. This arrangement means that for any given string of numbers there are four possible nucleotide sequences that it may encode. Given knowledge of the first base it is possible to determine the most likely nucleotide sequence encoded by the entire read. However, to do this prior to assembly of the reads into contiguous sequences (i.e., contigs) for comparison or alignment to a reference genome would result in losing the advantage of SOliD's built-in error checking (afforded by reading each base twice). For example, If a read was determined to possess a position that does not match a consensus reference sequence, it would be ambiguous in other technology platforms whether it were a real difference or sequencing error. With the double-coverage afforded by SOLiD color-space the same "error" is not likely to be made twice in subsequent cycles and it is much more likely that a real variation has been identified instead of a sequencing error.

It should be noted the SOLiD color-space, in which short reads are reported, can be difficult to work with for some assembly applications. Most assembly programs are initially designed to work with nucleotides and require special pre- and post-processing programs to properly assemble color-space reads and these are not always available. Many, but not all, of the specialized alignment programs that can align short reads to a reference library are also able to handle color-space reads but require special options to be enabled.

2.1.2 The illumina genome analyzer platform

The Genome Analyzer (GA) platform typically produces read lengths in the range of 35-150 bp and requires 2 to 14 days for a sequencing run depending on the amount of data desired. Each flow cell contains 8 lanes each of which can produce 80 million reads or more. Daily throughput is estimated at 6.5 Gb for a run in which both ends of fragments (i.e., paired end) are sequenced to 100bp.

The Illumina process also begins by shearing sample DNA (or cDNA) into fragments that are size selected in a target range, often around 200 bp (Fig 2A). These fragments then have short adaptors ligated to both ends of the sample fragments such that unique primer sequences are ligated to either end (Fig 2A). The fragments are then washed onto the flow cell that has sequences complimentary to the two unique primers bound to its surface (Fig 2A). The concentration of fragments on the flow cell is controlled such that they bind sparsely enough on the surface to be optically distinguished from neighboring fragments. Template sequences are only bound by base-paring to primers covalently bound to the flow cell. An initial PCR step produces a complimentary copy of the template now covalently bound to the flow cell, and following this the original template is removed by washing.

The next steps (Fig 2B) are repeated several times to produce a colony of copies of the sequence via 'bridge' PCR. The free end of the template pairs with one of the primers covalently bound to the flow cell and a PCR cycle produces a new copy bound by one end to the flow cell a short distance from the first. After several bridge PCR cycles, a cluster of copies is built up around the originally bound sequence.

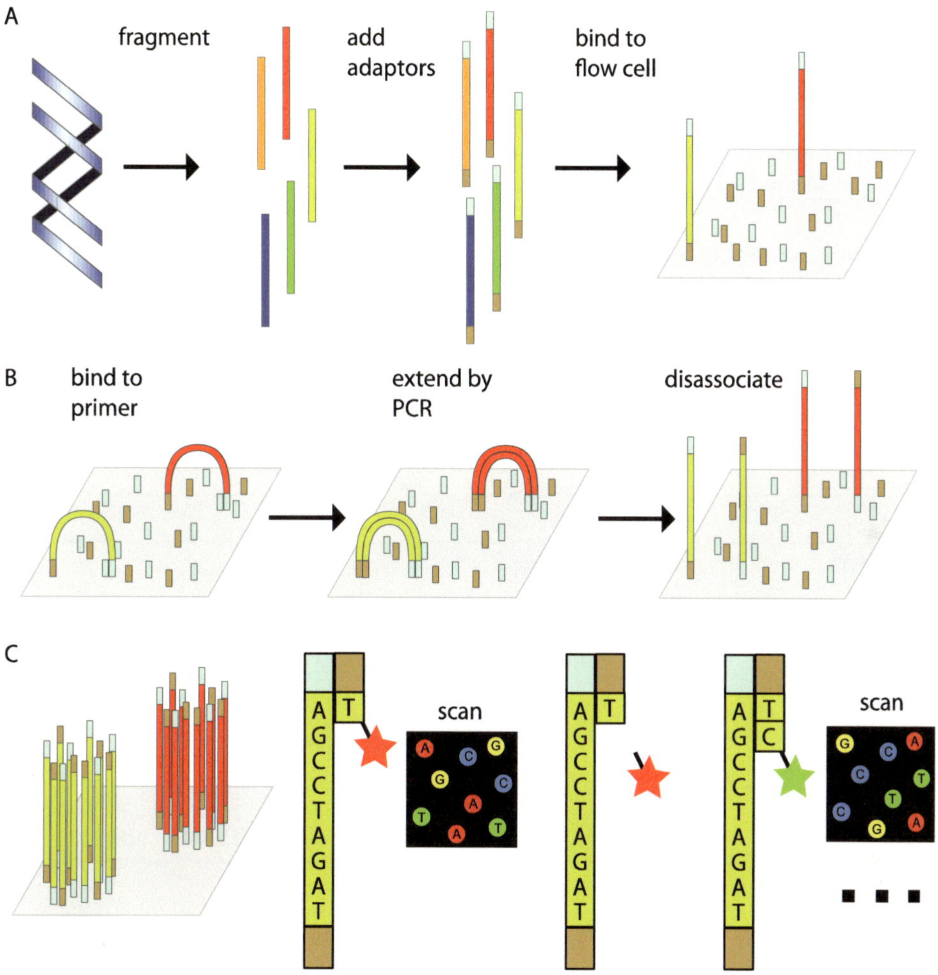

Fig. 2. Simplified outline of the Illumina Genome Analyzer process A) Sample preparation and attachment to the flow cell. B) Bridge PCR amplifies each bound fragment producing a cluster of copies. C) The sequencing reaction extends the growing strand by one nucleotide, excites attached fluorophores which are read optically, and removes the terminator and fluorescent dye before repeating with the next nucleotide.

When spots have reached sufficient density to produce clear signals (Fig 2C) the cyclic sequencing reaction can begin. One of the two unique primers is attached to the free ends and nucleotide addition cycles commence. Each nucleotide contains a different fluorescent reporter tag and a reversible terminator. During each cycle all four bases are flowed onto the reaction chamber, but since each contains a replication terminator only a single one can be incorporated into any elongating sequence (Fig 2C). Laser sources excite the fluorescent reporter of the added nucleotide and an optical sensor detects the wavelength of light emitted. The color of each spot is tracked with each cycle and interpreted directly as a nucleotide base (Fig 2C). This cycle is repeated until the reads reach the desired length and the entire sequencing process is then repeated using the other unique primer to sequence the complementary copies of the DNA.

We have briefly covered two popular NGS sequencing techniques to introduce the capabilities of the technologies and what types of data are produced. There are several other sequencing technologies and many recent reviews covering them are available (Metzker, 2009; Voelkerding et al., 2009; Bräutigam and Gowik, 2010; Nowrousian, 2010). The reader is encouraged to seek out the latest reviews as these technologies are advancing with immense speed and published information quickly becomes outdated.

2.2 Sequence assembly algorithms

When the human genome project first began capillary sequencing base on Sanger technology was the primary sequencing tool employed (Lander et al., 2001). It was extremely labor intensive yet at the time an amazing amount of sequence data was being produced. The Sanger reads produced where about 700 bp in length. Some current NGS techniques are now able to produce reads close to this length, while others hold the promise of producing several hundreds to thousands of base pair length reads.

Package	Availability
phrap	www.phrap.org
wgs-assembler (celera)	sourceforge.net/apps/mediawiki/wgs-assembler/
ARACHNE	ftp.broadinstitute.org/pub/crd/ARACHNE/
Phusion	www.sanger.ac.uk/resources/software/phusion/
RePS	Contact authors at: reps@genomics.org.cn
PCAP	seq.cs.iastate.edu/pcap.html
Atlas	www.hgsc.bcm.tmc.edu/cascade-tech-software_atlas-ti.hgsc

Table 1. Several overlap assembly programs.

The basic strategy for assembling sequences of this length is to use an overlap graph. In an overlap graph nodes represent whole reads and connections represent overlap between the reads. In this case the reads are large and a significant amount of unique information is held in each overlap. Many repetitive features and similar sequence properties that would stymie a short read assembler are easily resolved by long reads and an overlap strategy. Still, assembly problems are not trivial and many packages have continued to mature and acquire a variety of tools. A listing of overlap-based assemblers is given in Table 1.

2.2.1 De Bruijn graph assemblers

As NGS data became available it was quickly apparent that new algorithms were needed to assemble the very short sequences being produced. This problem was addressed by application of discoveries made independently by both De Bruijn and Good in 1946 (de Bruijn, 1946; Good, 1946). All of the most successful short sequence assembly programs in use today utilize the De Bruijn graph as a central data structure and then leverage other aspects of the data to improve upon the assembly process.

The first step in a De Bruijn based assembler is to build the graph. To do so, each short read is broken into k-mers where k is a pre-defined integer length; each k-mer will be a node in the graph (Fig 3A). The k-mers are defined by recording the sequence in a window of size k and sliding that window down by one position for the length of the short read - producing a new k-mer at each position (Fig 3A). If a short read has a length of L, it will contribute L-k+1 k-mers to the graph. The number of occurrences of each k-mer is also counted and will come into play in a subsequent step.

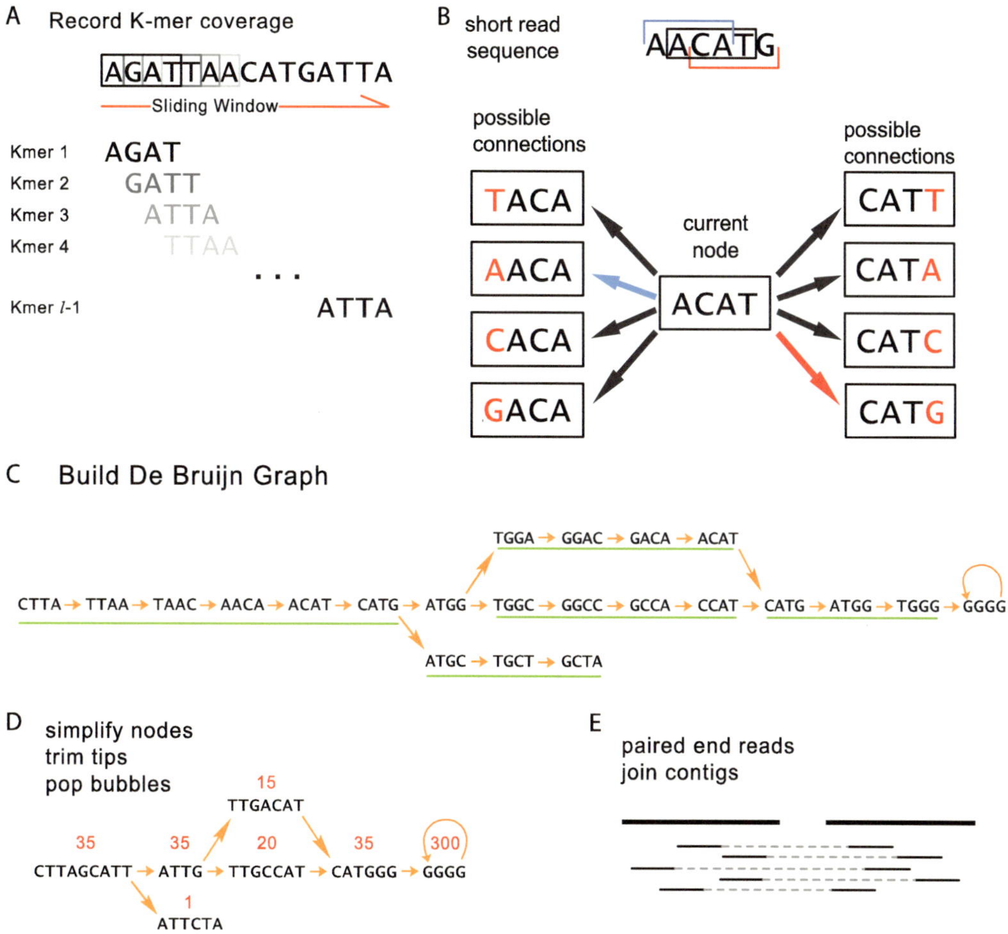

Fig. 3. Outline of a De Bruijn graph based assembler

The edges (or connections between nodes) represent a k-1 overlap between the connected nodes. Thus, we see that each node can have 8 possible connections (Fig 3B). Connections are recorded as they are observed in the raw read data. As reads are passed into the graph building algorithm discrete seed graphs begin to expand and are joined as the reads connecting them are found. In the end several thousand discrete, internally connected graphs exist in the working memory of the computer. An idealized example of one is given in Fig 3C. This is a very simple example but several complicating features are represented here. At this stage a simplified graph can be constructed in which linear stretches (underlined in green in Fig 3C) are condensed into nodes and edges are still k-1 overlaps. The resulting simplified graph is given in Fig 3D, and some of the problems can begin to be addressed. The leftmost is a 'tip'; a dead end likely caused by a sequencing error near the beginning or end of a short read. A bubble is also present in the center of the graph indicating two alternative possible paths through the k-mer space are present in the short read data. This also could be the result of a sequencing error or a genuine sequence variant. The depth of coverage for each simplified node is indicated by a red number above each node. This information can be used to trim off any tips and remove bubbles with low coverage. Higher coverage anomalies may merit incorporation into alternately assembled contigs depending on the application.

The right-most feature in this graph is a cyclic node. This creates a problem for short read assemblers when repetitive sequence regions are encountered. It could be the sequence has only 4 guanines in a row, or 40, it is impossible to tell from the information generated. This sort of assembly problem is more difficult to resolve by addressing coverage alone and usually results in breaks in contigs. Paired-end information can rescue some of these repetitive situations but scaffold contigs may be broken for many other reasons as well. However, if sufficient paired-end sequences are available that join two contigs it is possible to estimate the size of the gap between them given the expected fragment size (Fig 3E).

One major practical drawback of De Bruijn graph based assemblers is the amount of memory (RAM) required to build, and traverse the graph during an assembly. For example, the Velvet assembler package may require use of 70-100 GB of physical memory to build a vertebrate transcriptome assembly from 100 million reads. Although single machines with such large amounts of memory are not as rare and expensive as they once were, they remain somewhat difficult to find and gain access to. There are several assembler packages that have attempted to address this requirement for large memory. For example, a distributed approach has been implemented in the Abyss assembler and this spreads the workload across several nodes in a computer cluster. Optimizations in the SOAPdenovo package first seek to reduce the amount of memory required by attempting to correct erroneous k-mers produced by sequencing errors. In one study, this approach allowed the number of 25-mers in an assembly of the human genome to be reduced from 14.6 billion to 5.0 billion (Li et al., 2010a).

An alternative to purchasing computer capability with very large memory is use of a cloud computing services, such as the Amazon Elastic Compute Cloud (aws.amazon.com/ec2/). For a fee, computer time is available in dynamically generated computing environments. Several instance types are available including some with up to 68.4 GB of memory and two cluster instance types optimized for traditional compute nodes or GPU nodes. While no assembly process has yet been reported as having used this resource several similarly complex analyses have reported favorable experiences (Afgan et al., 2010; Di Tommaso et al., 2010; Wall et al., 2010).

De Bruijn Assemblers	Availability
EULER-SR	euler-assembler.ucsd.edu/portal/
Velvet	www.ebi.ac.uk/~zerbino/velvet/
ALLPATHS-LG	ftp.broadinstitute.org/pub/crd/ALLPATHS/Release-LG/
Abyss	www.bcgsc.ca/platform/bioinfo/software/abyss
SOAPdenovo	soap.genomics.org.cn/soapdenovo.html

Table 2. Several De Bruijn graph based assemblers

2.3 Overview of sequence assembly process

We have discussed the basic workings of assembly algorithms in order to provide a foundation for further discussion of assembly and the effects that different choices can have on the outcome. We now turn to a larger view of the practical assembly process. At each step we will give recommendations based on our experience and mention other sources for information and help.

2.3.1 Sequence filtration

Prior to NGS read assembly it can be beneficial to remove reads that are more likely to carry erroneous sequences. This is most important for De Bruijn graph based assemblers because each erroneous base call creates up to k erroneous nodes in memory. Thus, large data sets can very quickly exceed even very large memory systems. There are many types of sequencing errors that may need to be removed and some are unique to certain types of techniques. For example, sample DNA can become contaminated by bacterial or vector sequences and so screening reads against appropriate libraries can help to remove some of these contaminants. Short reads produced by the Illumina GA platform tend to decrease in quality as they are lengthened as well as have an increased error rate in the first few bases. To deal with this some tools (built in options in BWA and Bowtie short read alignment programs) will allow one to trim all reads by a certain length from either end in a set after measuring average quality scores across a read set. Other tools such as the FASTX-Toolkit (hannonlab.cshl.edu/fastx_toolkit/) are more adaptive and deal with each read individually. Another strategy that attempts to correct short reads is to enumerate all the k-mers defined by a set and modify those with very low occurrence frequencies (Schröder et al., 2009; Li et al., 2010b; Shi et al., 2010). Few papers primarily address this issue but the quality of the final assembly can only be as good as data you begin with.

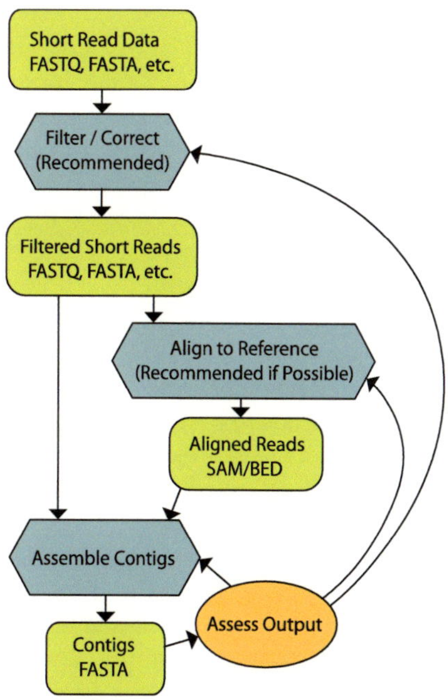

Fig. 4. Outline of an assembly process

2.3.2 Use of a reference library

Reference sequences can be used in many ways to aid in assembly. The most straight forward is to map reads onto a set of closely related reference sequences (using a tool such as BWA, Bowtie, Tophat, etc. section 2.4), then derive a consensus sequence from the reads aligned to each reference sequence. Among others, the samtools pileup or mpileup tools can aid in this approach. This limits the resulting sequences to the set of previously known reference sequences but that is not necessarily a problem. The Cufflinks tool is a unique take on reference based assembly. It is specifically designed to find exons and intron-exon junctions by mapping transcript sequence to a reference genome.

Reference sequences can also be used to guide a de novo assembly. It is possible there are other tools which enable this procedure but here we describe the use of the Columbus extension in the Velvet package. In this case short reads are again aligned with a separate tool to reference sequences which may be genome or transcript seqeunces. The resulting alignment file and reference sequences are then given as input to Velvet which will initially carry out its de novo assembly process as normal. The reference sequences are treated in a sense as long reads and are used to scaffold together appropriate contigs that resulted from the initial assembly process. This technique uses known sequences to extend assembled contigs while also allowing for the discovery of novel sequences.

2.3.3 Sequence assembly

While many NGS assembly packages utilize the De Bruijn graph to represent k-mer connectivity, each has a slightly different algorithm to traverse the graph, prune it, and extract contigs. Most of the freely-available, academically-developed assembly packages have extensive manuals and, more importantly, active communities of users and developers. An extensive listing of the settings and options that can be modified in even one of these packages is far beyond the scope of this discussion. Some considerations however transcend all of these software packages and are discussed here.

The selection of k (k-mer size or hash length) will have a huge impact on assembly. Short k-mers allow for the assembly of low coverage regions since for any two reads to be linked in k-mer space they must overlap by at least k-1. Conversely a too-short k-mer size could allow contigs to be linked in k-mer-space which are not truly linked; thus leading to a chimeric assembly. Very high k-mer sizes significantly cut down on chimeric contigs but impair the assembly of low expression level transcripts and reduce the contiguity overall. A good approach is to scan a range of k-mer sizes and compare the results of several assemblies to determine a k-mer size that gives the best balance.

Another important parameter to consider is how the assembler uses the coverage levels to assemble contigs. In Velvet, for example, the minimum coverage cutoff and expected coverage parameters define a range of coverage levels to consider. This is fine for genomic sequences where coverage levels should be much more consistent, but is extremely problematic for transcript assembly. The Oases extension in Velvet is designed to adapt to varying coverage depth levels and is allowed to report alternative contigs instead of selecting only high coverage paths through the graph.

The diverse range of De Bruijn graph-based assemblers each take different approaches to traversing the graph and pre- and post-processing the data. Software documentation is an excellent place to begin to understand the various assembly parameter modifications and settings allowed. As previously mentioned most of the academically developed packages have an associated community that communicate via e-mail listserv (many of which are archived online) or internet forum.

2.3.4 Assessing assembly quality

This is likely to be the most challenging step in an assembly. A set of basic statistics that are often seen in literature are the N50 value, overall length, number of contigs, and largest contig. The N50 is the length-weighted median length. Another way to think about it is to say that at the N50 length, half of the length in the set of assembled contigs is in contigs equal to or shorter than this value. It is a measure of contiguity since the N50 length increases as sequence length is shifted into longer contigs. The overall length is simply the sum of the lengths of all contigs, and the number of contigs and largest contig are self-explanatory. These are basic statistics often seen in literature but they are fairly limited in assessing assembly quality.

It is generally desirable to quantify correctly assembled contigs, but this is a very tricky thing to do especially with novel transcriptomes. There is no one good approach to assess this easily so we will present several and discuss advantages and disadvantages of each. One approach is to use BLAST or other similarity search tool to compare the assembly to a well-annotated transcriptome of a closely related species if available or a large curated set like the non-redundant (nr) database maintained by the NCBI. A tool like Blast2Go (Conesa et al., 2005; Conesa and Götz, 2008; Götz et al., 2008) can be used to analyze the BLAST results and select a good match for each contig. Trying to maximize unique hits may be a useful indicator but the annotation by BLAST may give different results for alternate splice forms.

Another useful metric is to measure how completely a reference transcriptome from a closely related species is covered by the assembled contigs. This depends heavily on the quality of the reference transcriptome and may not tell very much about the contiguity of the assembled contigs.

A third indication that contigs have been properly assembled is their ability to map to a reference genome. A tool like gmap (Wu and Watanabe, 2005; Wu and Nacu, 2010) can quickly map a large set of contigs to a large genome and report its results in a variety of formats including some basic statistics for each mapping. This would seem like the best method but some software development may be necessary to extract full meaning from such an alignment.

Analyzing the assembly often leads to another round of refinement possibly reaching all the way back to doing more sequencing. More stringent or different filtering, replacing the reference with the assembled contigs, or modifying assembler settings can all help to refine an assembly. Usually this process continues until a 'good enough' transcriptome is reached and that is defined by each researcher for their specific needs.

2.4 RNA short read mapping

After a reference transcriptome or background genome sequence has been efficiently assembled, the next step in many experimental designs is to accurately map RNA-seq reads derived from specific cell or organisms states to it as a method to profile global gene expression (Fig 5). Generally speaking, programs designed for EST mapping [i.e., MUMmer and BLAT(Kent, 2002; Kurtz et al., 2004)] are suitable for reads generated from Roche 454 platfoms, but are not nearly efficient enough for use with short reads generated by Illumina Gene Analyzer or ABI SOLiD NGS platforms. Alignment algorithms designed specifically for NGS short reads are necessary to map reads from latter two platforms. Over the past two years, a wide variety of different programs have been developed to meet the challenge of efficiently mapping millions of short reads and the number of available programs seems to be continuously growing. The challenge for biological scientists is how to choose the best program that is optimally suited to their specific project. Table 3 shows five currently popular programs available for short read mapping that will be evaluated herein.

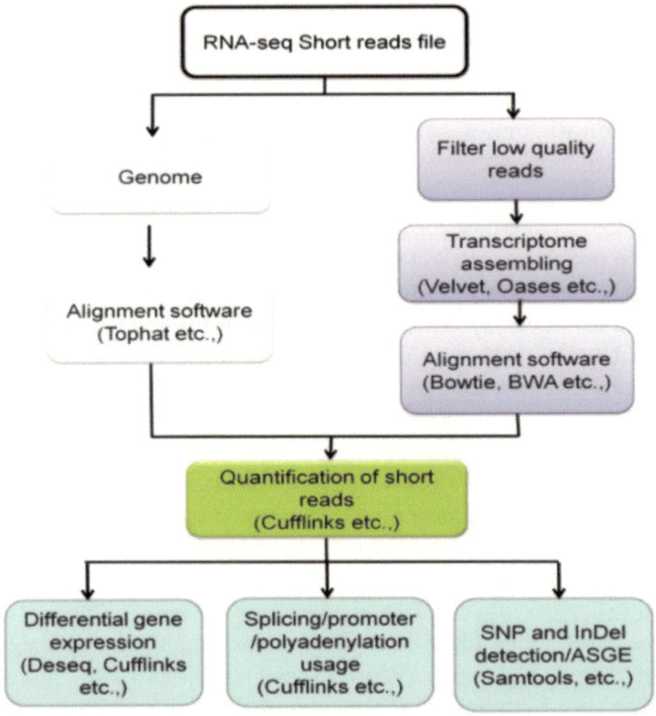

Fig. 5. RNA-Seq project pipeline and commonly used programs. (see; http://en.wikipedia.org/wiki/List_of_sequence_alignment_software#Short-Read_Sequence_Alignment)

To evaluate these programs, we used an *X. maculatus* reference transcriptome [for description of the species see section 2.7 and (Walter and Kazianis, 2001; Kallman and Kazianis, 2006; Meierjohann and Schartl, 2006)] built from over 200 million paired-end reads sequenced from the brain, heart, and liver tissues of mature individuals using the Illumina GAIIx platform (Expression Analysis® Inc. Durham, NC). We used the Velvet assembly package (Zerbino and Birney, 2008) to integrate the combined read set with a hash length (k-mer size) of 43. Oases (http://www.ebi.ac.uk/~zerbino/oases/) was used to perform the final assembly and resulted in a final transcriptome having 110,604 transcripts with an average length of 2,197 bp, and a total size of 243 Mb. In addition, we employed 34 million 60 bp paired-end reads (GAIIx, no custom filtration) sequenced from RNA isolated from *X. maculatus* liver tissue and mapped them back to the reference transcriptome described above to test and compare all five programs in terms of RAM usage, computing time and mapping sensitivity (e.g., the percent of mapped reads).

As shown in the Table 3, the five different programs required different amount of RAM and produced different mapping efficiencies. Bowtie is currently one of the most popular mapping programs and has a reputation for very rapid mapping speeds. It employs a Burrows-Wheeler Transform (BWT) and full-text minute-space index (for review of alignment algorithms, see (Li and Homer, 2010), which greatly reduces both the memory usage and computational time. In our test, Bowtie proved to be the fastest mapping program and also used a modest amount of RAM. The small RAM usage and speed of Bowtie allows it to run on a standard desktop computer. However, Bowtie's fast performance speed is not without cost. Bowtie only allows non-gapped alignments between reads and references, thus sacrificing some sensitivity for faster mapping speed. Therefore, it was not surprising that Bowtie had the lowest mapping percentage of all tested programs. In addition, using genomic sequences as the reference for mapping RNA-seq reads with Bowtie might not be appropriate since reads spanning two exons cannot be mapped without the support of gap alignment.

Program	Maximum RAM Usage	Time	%of mapped reads	Feature	Reference
Bowtie	2.6G	40 min	42.51	Ultra fast aligner	(Langmead et al., 2009)
BWA	1.2G	64 min	52.05	Support gap alignment	(Li and Durbin, 2009)
Novoalign	1.4G	41 hr[a]	59.81	High sensitivity and allows up to 8 mismatches	www.novocraft.com
SHRiMP	7.0G	14 days	53.08	A collection of mapping tools	(David et al., 2011)
Tophat[b]	63G[b]	5.5hr[b]	52.92[b]	Splice junction reads aligner	(Trapnell et al., 2009)

[a]Only one thread is used for free version. Licensed user can use multi-threads feature of Novoalign.
[b]Transcriptome is used as the reference in this case. Tophat is designed for using genome sequence as reference so the actual time and mapping efficiency may vary when genome is used.

Table 3. Popular short-read alignment software.

An alternative to Bowtie is BWA (Li and Durbin, 2009), which also uses a full-text minute-space index based algorithm but supports gapped alignments. In our test, BWA used least amount of RAM and was comparable to Bowtie in computing time. The gapped alignment feature of BWA makes it more suitable should variations (i.e., small insertion/deletion or InDels) exist between the reference genome or transcriptome and the RNA-seq reads being mapped. This serves to increase the mapping sensitivity of alignments. In our test, BWA reported more reads properly mapped than Bowtie, suggesting BWA is more sensitive in identifying possible alignments between short reads and reference sequences.

Two other programs tested were Novoalign and SHRiMP. They were both noticeably slower than Bowtie or BWA programs. Both Novoalign and SHRiMP programs use a hashing reference based algorithm, which can be traced back to BLAST searching but is optimized for alignment of short reads. For Novoalign, we tested the free version and thus only one thread was used while in all other cases four threads were used during the read mapping trials. Therefore it is likely the licensed version of Novoalign, employing fully multi-thread functions, will exhibit greatly reduced computation time. Novoalign showed the highest mapping percentage in all tested program, indicating the excellent sensitivity of the hashing reference based algorithm. Unlike Novoalign, SHRiMP employs a k-mer hashing index and Smith-Waterman algorithm which gives it robust mapping sensitivity and specificity (David et al., 2011). However, SHRiMP requires large amounts of RAM and was the slowest program tested. The increased computational time and RAM requirements make SHRiMP less attractive for projects needing high-throughput data analyses.

The final program we tested is Tophat (Trapnell et al., 2009). Tophat is a splice junction mapping program quite different from the previous four programs. Tophat is designed to align RNA-seq reads to a reference genome. Using Tophat, RNA-seq reads can be analyzed to identify novel splice variants of genes. Tophat first employs iterative rounds of Bowtie mapping to identify genomic regions with RNA-seq read mapping, and then to generate potential splice donor/acceptor sites flanking the sequence. Unmatched reads are then mapped to these splice junction sequences again by Bowtie to confirm possible splice junctions. Tophat prefers a genome sequence as a reference and mapping results may not be reliable if only a transcriptome reference is used. Of all five programs tested, Tophat required the most RAM for alignment processing. Thus, Tophat may be best used in a high-performance computing environment.

Overall, the choice of which alignment program to use should be based on both the available computer resources and experimental design. If the alignment process is to be performed on a standard desktop computer (e.g., about 4G RAM), SHRiMP and Tophat should be avoided due to memory constraints. However, Bowtie, BWA, and Novoalign can map reads efficiently on standard office computers. On the other hand, if a genome sequence is available for a reference, or the purpose of study is to identify InDel's between a reference and reads, Bowtie may not be the best choice since it lacks gap alignment capabilities. Tophat is preferred when a genome sequence is present because it fully considers potential donor/acceptor sites in the genome and allows the alignment to cross splice junctions accurately, compared to the other programs. However, should a transcriptome be used as mapping reference, Tophat should be avoided as it is designed for use with genome sequence data. Finally, although all programs tested herein fully support both Illumina and SOLiD single or paired ends reads, SHRiMP and BWA (through its BWA-SW module) also support mapping of mixed RNA-seq short reads with longer Sanger or 454 Roche based reads. In such situations, where mixed reads are to be used, employing a single program saves both time and effort in the subsequent analyses.

Overall, with the continuous increase in throughput for recently developed sequencing technologies, new algorithms are becoming available almost monthly; while older programs are continually refined to reduce computational time and memory demands. However, there is not a perfect program suited for all experimental designs and hardware requirements. The choice of programs will need to be reviewed and evaluated on a case-by-case basis.

2.5 Quantification of gene expression level

Currently most short read alignment programs adopt SAM (or its binary version, BAM) as the alignment output format. SAM (Sequence Alignment/Map format) is a tab-delimited text format designed for recording short read alignment information. Although it is human readable, a typical SAM file will consist of millions of lines of mapping information that is required for downstream analyses. In the next steps of data processing, most RNA-Seq projects aim to utilize read mapping as a means to quantify gene expression levels across entire reference transcriptomes or genomes (Fig 5).

An early approach of using RNA-seq to quantify gene expression relied on simply counting the total number of reads mapping to each transcript in sample. However, since the total number of reads varied between each sample, read counts could not be use for direct comparison or determination of differential expression between samples. In addition to total read count numbers between samples, the length of transcripts within each sample may vary and longer transcripts are generally more likely to have more reads mapped to them than shorter ones. Thus, performing tasks such as finding the highest expressed genes in a sample via direct read counting proved to be inaccurate. In an effort to normalize the sample size and transcript lengths for head-to-head read count comparisons, Mortazavi and coworkers (2008) developed the term Reads Per Kilobase per Million of mapped reads (RPKM) as a standard to compare different genes within or across different samples (Mortazavi et al., 2008). RPKM and its derived term FPKM (Fragments Per Kilobase per Million of mapped reads) for paired end reads, have been widely adopted in RNAseq studies employing various experimental systems.

Since RPKM is easy to calculate and understand, it provides a platform to facilitate comparison of transcript levels both within and between samples. However, since the purpose of most studies involving RPKM is to compare differential gene expression, one must be aware that RPKM values may be affected by both experimental and computational issues. Experimental issues such as the quality of RNA, contamination of ribosomal RNA and length of output reads (Pepke et al., 2009; Costa et al., 2010) and computational influences including accuracy of gene modeling, and inclusion/exclusion of multiple mapped reads, may all affect the results obtained. One issue deserving special attention is the diminished statistical power one accepts when using RPKM to detect differential expression of longer transcripts (Oshlack and Wakefield, 2009). Employing RPKM, where the number of reads from a given transcript is divided by the length of the transcript, serves to deflate statistical power by producing a large sample size (more reads). To illustrate this, assume a 1000 bp gene (gene A) has 5 and 10 mapped reads in sample 1 and sample 2, respectively. In the same samples, a 10,000 bp gene (gene B) has 50 and 100 mapped reads, respectively. By definition of RPKM, since gene B is 10 times longer and has 10 times more reads mapped, both genes have identical RPKM values and fold changes in the two samples. Thus one would assume the confidence of gene A and gene B being differentially expressed is exactly same. However, since gene A has a much smaller sample size (15 reads in total) compared with gene B (150 reads), gene A is more prone to statistical error when trying to identify a 2 fold-change in expression between samples 1 and 2. Therefore, although RPKM is widely used to provide a scalable value to quantify gene expression levels, it is affected by variation in a transcript length dependent manner and should not be used to directly compare gene expression.

2.6 Comparison of differential expression

One common goal of many large-scale transcriptome studies is to identify differentially expressed genes between two or more samples. While microarrays have been widely used for over a decade to assess transcriptome-wide gene expression levels, RNA-seq technologies have displayed several advantages over microarrays, such as the ability to identify novel transcripts and to assess quantitative allele-specific gene expression. However, it is still debatable which tool is better to accurately assess gene expression values. In a recent study (Bloom et al., 2009), microarray and RNA-seq results were compared using quantitative RT-PCR (qRT-PCR) assays and it was determined that both methods performed similarly in measuring differential gene expression. The microarray had an advantage over RNAseq in better measure of low-abundance transcripts (Bloom et al., 2009); however, when results of microarray and RNA-seq were further assessed with 2D LC-MS/MS the expression values estimated by RNA-Seq appeared to be better correlated with the proteomics data (Fu et al., 2009). Overall, these studies prove that RNA-Seq may serve as a reliable method to accurately estimate absolute transcript levels.

Since both microarray and RNA-seq are used to quantify expression levels of transcripts, statistical methods developed for microarrays have been adopted to compare gene expression using RNA-seq. However, there are notable differences between the two technologies and methods successfully used for microarray analysis might not be appropriate for RNA-seq data (Costa et al., 2010). First, the gold standard for any microarray studies is to have at least three replicates in each condition while many RNA-seq projects lack the luxury of replicates due to the relatively expensive cost of sequencing RNA-seq libraries. Methods that have been used in microarray analysis range from simple t-testing to more complicated statistical modeling; but all these techniques rely on having multiple replicates to identify differentially expressed genes. The absence of multiple replicates greatly reduces the statistical power of RNAseq methods. Secondly, for microarray analysis, fluorescence intensity is utilized as the measurement of transcript levels and these data may be treated as continuous data. However, RNA-seq studies utilizing read counts (or RPKM) to gauge the expression of a particular transcript generate discrete data. Thus, statistical models developed for continuous data might not be effective when applied to data generated from an RNA-seq experiment.

Many studies have utilized different statistical tools to identify differentially expressed transcripts in RNA-seq experiments. Simple approaches such as classical Z-test and Fishers exact test have been employed for this purpose (Bloom et al., 2009; Hashimoto et al., 2009). Although these methods are appropriate for hypothesis testing of discrete data, they do not consider the global variations of all genes, thus less robust than more advanced approaches discussed below. There are several studies reported where more sophisticated microarray based methods have been modified and made suitable for RNA-seq projects. One of the pioneering reports involved RNAs extracted from liver and kidney of the same individual that were separated into seven aliquots for each sample and sequenced in individual lanes of a Illumina genome analyzer (Marioni et al., 2008). The variations of these technological replicates were then calculated and were found to fit the variance predicted by a Poisson model. Using the Poisson model allowed the authors to identify 30% more differentially expressed genes than a standard statistic analysis and employing microarrays with the same samples (Marioni et al., 2008). Based on the notion that a Poisson distribution can predict the variations in RNA-seq data, DEGseq, a Bioconductor software package, has been developed for examining differential expression of RNA-seq read count data (Wang et al., 2010). DEGseq modeled the number of reads derived from a gene into a Poisson distribution and used the Fisher's exact test and likelihood ratio test to identify differentially expressed genes (Wang et al., 2010). However, it has been argued the Poisson distribution will underestimate actual variations in replicated samples and tends to predict smaller variations than are actually present in the data (Nagalakshmi et al., 2008). As a result, methods based on the Poisson distribution do not control false discoveries very well. In addition to Poisson distributions, two other Bioconductor packages, DESeq and EdgeR, both take read counts as input and use negative binomial distributions to estimate variations of RNA-seq data (Anders and Huber, 2010; Robinson et al., 2010). EdgeR employs negative binomial distributions to account for variability and assesses differential expression based on Empirical Bayes methods (Robinson et al., 2010). The DESeq package models distributions of read count data by negative binomial distribution, with variance and mean linked by local regression (Anders and Huber, 2010). Compared with previous Poisson based program, both DESeq and EdgeR control the probability of false discoveries and produce good fits when the number of replicates is small (Anders and Huber, 2010).

In addition to the Bioconductor packages discussed above, another standalone tool termed "Cufflinks", written by same research group that developed Bowtie and Tophat, may be used to read SAM files produced from Tophat directly and compare differential expression in pair-wise manner (Trapnell et al., 2010). The program extracts read count information from SAM files and computes the entropy of the average distribution minus the average of the individual entropies [Jensen-Shannon divergence; see (Menendez et al., 1997)] and the difference between abundances of transcripts in two conditions may be calculated as the square root of this divergence. Cufflinks can be easily integrated with Bowtie/Tophat workflow and outputs FPKM values for two samples and the significance level of the statistics tests. In addition to transcript expression, Cufflinks may also be used to find significant changes in transcript splicing and promoter usage between two samples.

2.7 SNP identification and allele specific gene expression

One major advantage of RNA-seq technology over microarray based approaches is that one may quantify not only total gene expression, but also allele specific gene expression (ASGE) at same time. To study allele specific gene expression using microarrays, one must have very detailed characterization of genome polymorphisms and then specifically design probes to assess the abundance of each allele independently on the array. Therefore, it is difficult to study ASGE in less well-characterized species or genetic models that possess little information of known polymorphisms. With rapid progress in next generation sequencing technologies (NGS), RNA-Seq has been shown to provide single-base resolution and quantitative information for thousands of genes simultaneously (Pastinen, 2010). Notably, this approach does not rely on previous knowledge of known variations and can be used for both identifying polymorphisms and quantifying ASGE. Using both 454 and Illumina sequencing platforms respectively, allelic expression imbalances have been assessed in *Drosophila* hybrids, *Xiphophorus* fishes, and in humans (Serre et al., 2008; Daelemans et al., 2010; Fontanillas et al., 2010; Shen et al., 2011).

Here we demonstrate our recent ASGE study using *Xiphophorus* interspecies hybrid fishes. The genus *Xiphophorus* has at least 27 species of live-bearing fishes found from northern Mexico south into Belize and Guatemala (Kallman and Kazianis, 2006). The *Xiphophorus* genus couples extreme genetic variability among *Xiphophorus* species with the capability of producing fertile interspecies hybrids that have allowed chromosomal inheritance of complex traits to be followed into individual F_1 and backcross hybrid progeny (Kazianis et al., 2001; Walter and Kazianis, 2001; Meierjohann and Schartl, 2006). Using interspecies hybrids provides a unique opportunity to reveal underlying mechanisms of genetic variation.

We have assembled the transcriptome of *X. maculatus* Jp163 A, a highly inbred line species of *Xiphophorus* (Fig 6) using RNA-seq sequencing from brain, heart, and liver tissues (see section 2.5). We first investigate transcriptome-wide SNP polymorphisms between two highly inbred *Xiphophorus* species: *X. maculatus* Jp 163 B and *X. couchianus*. To do this RNA-seq reads sequenced from *X. maculatus* Jp163 B were mapped to the reference transcriptome of *X. maculatus* Jp163 A by Bowtie (Langmead et al., 2009) and SNPs were called by

Samtools (Li et al., 2009). The density of intraspecific SNPs was about 1 SNP/49 kb of transcriptome [Figure 7; (Shen et al., 2011)].

Fig. 6. Fishes used in this study. *X. maculatus* Jp 163 A carrying the Sd pigment pattern is the species utilized for deep transcriptome development and eventual assembly of the reference transcriptome. F1 interspecies hybrids utilized in these studies were produced by crossing the *X. maculatus* Jp 163 B (Sp pigment pattern) and *X. couchianus* parental species. RNA-seq reads analyzed in this study were sequenced from *X. maculatus* Jp 163 B, *X. couchianus* and their F_1 interspecies hybrids respectively.

We wished to ascertain ASGE between *X. maculatus* Jp 163 B, *X. couchianus* and an F_1 hybrid produced from crossing these two species (Fig 6). Thus, we first determined that the 90,788 SNPs, identified between the *X. maculatus* reference transcriptome and *X. couchianus* were also polymorphic between the *X. maculatus* Jp 163 B strain and *X. couchianus*. To improve the accuracy of ASGE analysis in the hybrid, we scored only genes that exhibited greater than 20 SNP supporting reads. These constraints resulted in 38,746 SNPs between *X. maculatus* Jp 163 B and *X. couchianus* that could be clearly assigned to one or the other parental alleles and were unambiguously mapped to 6,524 *Xiphophorus* transcripts in the reference transcriptome.

After identification of SNPs, ASGE can be calculated as number of reads mapped to each allele in the F_1 hybrid (for a diagrammatic illustration of the process, see Fig 7). Since most short alignment programs only allow a limited number of base mismatches (i.e., 2 in case of Bowtie) between reads and reference sequences, the reads representing the *X. couchianus* alleles possessed natural disadvantages in mapping efficiency since they carried an extra mismatch (i.e., the SNP) compared with *X. maculatus* reads. In the F_1 hybrid, we found many transcripts showed more *X. maculatus* mapped reads than *X. couchianus* ones when the mapping was back to the *X. maculatus* reference transcriptome. To eliminate this read mapping bias and create an environment where reads from both X.

maculatus and *X. couchianus* alleles had equal chances of mapping to the transcriptome, we first duplicated the *X. maculatus* reference and then introduced all *X. couchianus* specific SNP's into it to produce an *in silico X. couchianus* reference transcriptome (based on *X. maculatus* transcriptome with masked SNPs). The induction of *X. couchianus* reference transcriptome allowed reads with *X. couchianus* alleles to have comparable likelihood of being mapped in ASGE study.

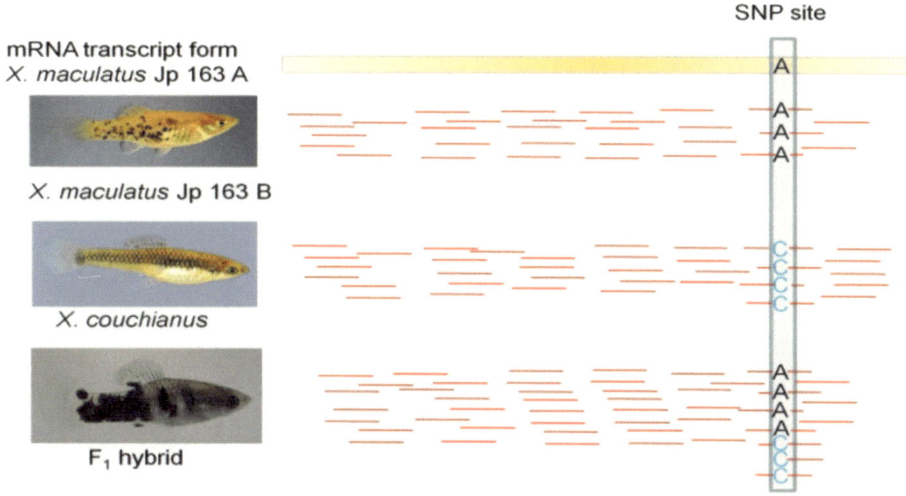

Fig. 7. A diagrammatic example of identification of SNPs and measurement of ASGE in F_1 interspecies hybrids. Red bars represent RNA-Seq reads mapped to the reference transcriptome. Most reads from *X. maculatus* Jp 163 B match perfectly to the Jp 163 A reference transcriptome. RNA-seq reads from *X. couchianus* were also mapped to *X. maculatus* Jp 163 A reference transcriptome and SNPs sites were identified by comparing consensus bases of RNA-seq reads (C in this case) to the corresponding base in the reference transcriptome (A in this case). In the hybrid, reads mapped to SNPs sites are classified by the bases they carry and counted separately as the measurement of ASGE. In this SNP, 4 *X. maculatus* allele reads and 3 *X. couchianus* allele reads were counted in the hybrid.

As shown in Fig 8, using the corrected reference transcriptome allowed both *X. maculatus* and *X. couchianus* alleles to exhibit a more balanced expression pattern (Fig 8b) in the hybrid genetic background than without normalization (Fig 8a). Without proper normalization, we found over 84% of genes in the transcriptome were biased toward over-representation of the *X. maculatus* allele (fraction > 0.5, Fig 8a). After production of the *in silico* reference transcriptome and tolerating 5 mismatches, analyses of the distribution of ASGE in F_1 hybrids indicate that most genes (5,980 of 6,524 genes or 92%) exhibit relatively balanced allele expression in the hybrid genetic background (<70% of preference of one particular allele, those between 0.3 and 0.7 in Fig 8b). Overall, employment of high throughput sequencing technology and proper normalization approaches allow direct and accurate assessment of ASGE in the interspecies hybrids.

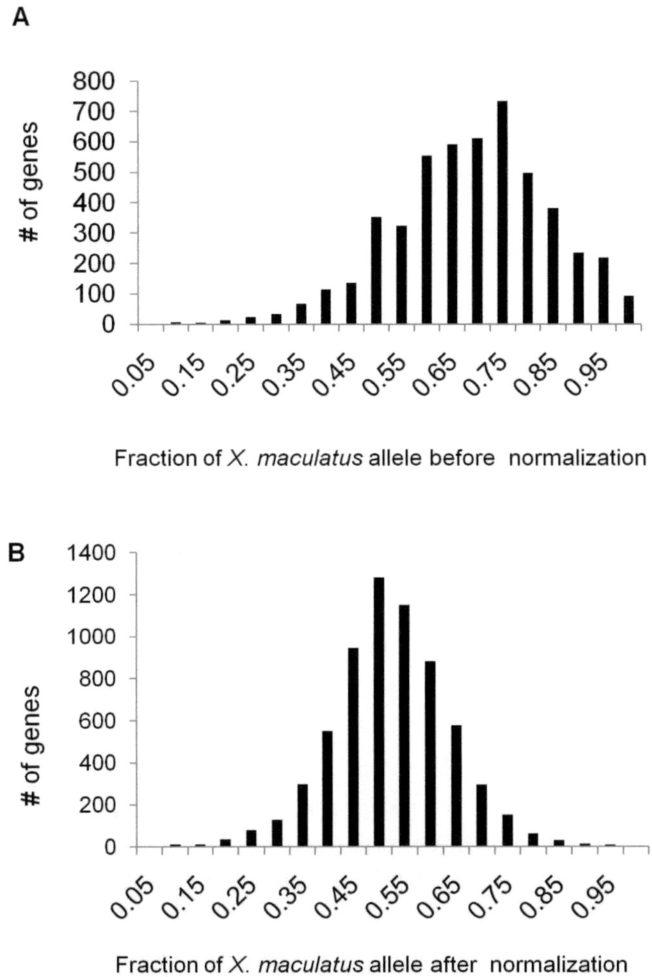

Fig. 8. Allele distribution in F1 hybrid background. A: A histogram shows the distribution of F1 transcripts carrying different parental alleles before normalization. X axis is the fraction of reads carrying *X. maculatus* allele. 0.5 means in that gene, half of F1 hybrid reads can be identified from *X. couchianus* and another half are from *X. maculatus*. 1.0 and 0.0 means reads exclusively carrying *X. maculatus* and *X. couchianus* alleles, respectively. B: Fraction of *X. maculatus in* hybrid background after normalization. We masked *X. maculatus* reference with consensus bases from *X. couchianus* and allowing five mapping mismatches.

3. Conclusion

The bottleneck of large-scale NGS projects has shifted from obtaining experimental data to downstream bioinformatic analyses. With the continuous development of software infrastructure to suit the needs of RNA-Seq analyses, there are several competent programs in each of the analysis step; such as transcriptome assembly, read mapping, and identification of differential gene expression. The real challenge facing many biologists is to find the right tool to use and carefully weighing the strength and weakness of each tool. The constant advance in sequencing technology will continue to increase the amount of data produced, urging the use of the most efficient tool within the capacity of available computer resources. The combination of the carefully designed experiment and right methodology utilizing NGS data will open a new era for studying species with little historical background genetic information available.

4. References

Afgan, E., Baker, D., Coraor, N., Chapman, B., Nekrutenko, A., and Taylor, J. (2010). Galaxy CloudMan: delivering cloud compute clusters. *BMC Bioinformatics,* Vol.11, (2010), pp. S4, ISSN

Anders, S., and Huber, W. (2010). Differential expression analysis for sequence count data. *Genome Biol,* Vol.11, No.10, (October 2010), pp. R106, ISSN 1465-6914

Bloom, J.S., Khan, Z., Kruglyak, L., Singh, M., and Caudy, A.A. (2009). Measuring differential gene expression by short read sequencing: quantitative comparison to 2-channel gene expression microarrays. *BMC Genomics,* Vol.10, (May 2009), pp. 221, ISSN 1471-2164

Bräutigam, A., and Gowik, U. (2010). What can next generation sequencing do for you? Next generation sequencing as a valuable tool in plant research. *Plant biology (Stuttgart, Germany),* Vol.12, (2010), pp. 831-841, ISSN

Conesa, A., and Götz, S. (2008). Blast2GO: A Comprehensive Suite for Functional Analysis in Plant Genomics. *International journal of plant genomics,* Vol.2008, (2008), pp. 619832, ISSN

Conesa, A., Götz, S., García-Gómez, J.M., Terol, J., Talón, M., and Robles, M. (2005). Blast2GO: a universal tool for annotation, visualization and analysis in functional genomics research. *Bioinformatics (Oxford, England),* Vol.21, (2005), pp. 3674-3676, ISSN

Costa, V., Angelini, C., De Feis, I., and Ciccodicola, A. (2010). Uncovering the complexity of transcriptomes with RNA-Seq. *J Biomed Biotechnol,* Vol.2010, (June 2010), pp. 853916, ISSN 1110-7251

Daelemans, C., Ritchie, M.E., Smits, G., Abu-Amero, S., Sudbery, I.M., Forrest, M.S., Campino, S., Clark, T.G., Stanier, P., Kwiatkowski, D., *et al.* (2010). High-throughput analysis of candidate imprinted genes and allele-specific gene expression in the human term placenta. *BMC Genet,* Vol.11, (June 2010), pp. 25, ISSN 1471-2156

David, M., Dzamba, M., Lister, D., Ilie, L., and Brudno, M. (2011). SHRiMP2: Sensitive yet Practical Short Read Mapping. *Bioinformatics,* Vol.27, No.7, (January 2011), pp. 1011-1012, ISSN 1367-4811

de Bruijn, N.G. (1946). A Combinatorial Problem. *Koninklijke nederlandse Akademie v Wetenschappen,* Vol.49, (1946), pp. 758-764, ISSN

Di Tommaso, P., Orobitg, M., Guirado, F., Cores, F., Espinosa, T., and Notredame, C. (2010). Cloud-Coffee: implementation of a parallel consistency-based multiple alignment algorithm in the T-Coffee package and its benchMarchking on the Amazon Elastic-Cloud. *Bioinformatics (Oxford, England),* Vol.26, (2010), pp. 1903-1904, ISSN

Fontanillas, P., Landry, C.R., Wittkopp, P.J., Russ, C., Gruber, J.D., Nusbaum, C., and Hartl, D.L. (2010). Key considerations for measuring allelic expression on a genomic scale using high-throughput sequencing. *Mol Ecol,* Vol.19 Suppl 1, (March 2010), pp. 212-227, ISSN 1365-294X

Fu, X., Fu, N., Guo, S., Yan, Z., Xu, Y., Hu, H., Menzel, C., Chen, W., Li, Y., Zeng, R., *et al.* (2009). Estimating accuracy of RNA-Seq and microarrays with proteomics. *BMC Genomics,* Vol.10, (April 2009), pp. 161, ISSN 1471-2164

Good, I.J. (1946). Normal recurring Decemberimals. *Journal of the London Mathematical Society,* Vol.21, (1946), pp. 167-169, ISSN

Götz, S., García-Gómez, J.M., Terol, J., Williams, T.D., Nagaraj, S.H., Nueda, M.J., Robles, M., Talón, M., Dopazo, J., and Conesa, A. (2008). High-throughput functional annotation and data mining with the Blast2GO suite. *Nucleic acids research,* Vol.36, (2008), pp. 3420-3435, ISSN

Hashimoto, T., de Hoon, M.J., Grimmond, S.M., Daub, C.O., Hayashizaki, Y., and Faulkner, G.J. (2009). Probabilistic resolution of multi-mapping reads in massively parallel sequencing data using MuMRescueLite. *Bioinformatics,* Vol.25, No.19, (October 2009), pp. 2613-2614, ISSN 1367-4811

Kallman, K.D., and Kazianis, S. (2006). The genus Xiphophorus in Mexico and central america. *Zebrafish,* Vol.3, No.3, (April 2006), pp. 271-285, ISSN 1557-8542

Kazianis, S., Gimenez-Conti, I., Trono, D., Pedroza, A., Chovanec, L.B., Morizot, D.C., Nairn, R.S., and Walter, R.B. (2001). Genetic analysis of neoplasia induced by N-nitroso-N-methylurea in Xiphophorus hybrid fish. *March Biotechnol (NY),* Vol.3, No.Supplement 1, (June 2001), pp. S37-43, ISSN 1436-2228

Kent, W.J. (2002). BLAT--the BLAST-like alignment tool. *Genome Res,* Vol.12, No.4, (April 2002), pp. 656-664, ISSN 1088-9051

Kurtz, S., Phillippy, A., Delcher, A.L., Smoot, M., Shumway, M., Antonescu, C., and Salzberg, S.L. (2004). Versatile and open software for comparing large genomes. *Genome Biol,* Vol.5, No.2, (February 2004), pp. R12, ISSN 1465-6914

Lander, E.S., Linton, L.M., Birren, B., Nusbaum, C., Zody, M.C., Baldwin, J., Devon, K., Dewar, K., Doyle, M., FitzHugh, W., *et al.* (2001). Initial sequencing and analysis of the human genome. *Nature,* Vol.409, (2001), pp. 860-921, ISSN

Langmead, B., Trapnell, C., Pop, M., and Salzberg, S.L. (2009). Ultrafast and memory-efficient alignment of short DNA sequences to the human genome. *Genome Biol,* Vol.10, No.3, (March 2009), pp. R25, ISSN 1465-6914

Li, H., and Durbin, R. (2009). Fast and accurate short read alignment with Burrows-Wheeler transform. *Bioinformatics,* Vol.25, No.14, (July 2009), pp. 1754-1760, ISSN 1367-4811

Li, H., Handsaker, B., Wysoker, A., Fennell, T., Ruan, J., Homer, N., Marchth, G., Abecasis, G., and Durbin, R. (2009). The Sequence Alignment/Map format and SAMtools. *Bioinformatics,* Vol.25, No.16, (August 2009), pp. 2078-2079, ISSN 1367-4811

Li, H., and Homer, N. (2010). A survey of sequence alignment algorithms for next-generation sequencing. *Brief Bioinform,* Vol.11, No.5, (September 2010), pp. 473-483, ISSN 1477-4054

Li, R., Fan, W., Tian, G., Zhu, H., He, L., Cai, J., Huang, Q., Cai, Q., Li, B., Bai, Y., *et al.* (2010a). The sequence and de Novembero assembly of the giant panda genome. *Nature,* Vol.463, (2010a), pp. 311-317, ISSN

Li, R., Zhu, H., Ruan, J., Qian, W., Fang, X., Shi, Z., Li, Y., Li, S., Shan, G., Kristiansen, K., *et al.* (2010b). De Novembero assembly of human genomes with massively parallel short read sequencing. *Genome research,* Vol.20, (2010b), pp. 265-272, ISSN

Marchioni, J.C., Mason, C.E., Mane, S.M., Stephens, M., and Gilad, Y. (2008). RNA-seq: an assessment of technical reproducibility and comparison with gene expression arrays. *Genome Res,* Vol.18, No.9, (September 2008), pp. 1509-1517, ISSN 1088-9051

Meierjohann, S., and Schartl, M. (2006). From Mendelian to molecular genetics: the Xiphophorus melanoma model. *Trends Genet,* Vol.22, No.12, (December 2006), pp. 654-661, ISSN 0168-9525

Menendez, M.L., Pardo, J.A., Pardo, L., and Pardo, M.C. (1997). The Jensen-Shannon divergence. *J Franklin I,* Vol.334B, No.2, (March 1997), pp. 307-318, ISSN 0016-0032

Metzker, M.L. (2009). Sequencing technologies — the next generation. *Nature Reviews Genetics,* Vol.11, (2009), pp. 31-46, ISSN

Mortazavi, A., Williams, B.A., McCue, K., Schaeffer, L., and Wold, B. (2008). Mapping and quantifying mammalian transcriptomes by RNA-Seq. *Nat Methods,* Vol.5, No.7, (July 2008), pp. 621-628, ISSN 1548-7105

Nagalakshmi, U., Wang, Z., Waern, K., Shou, C., Raha, D., Gerstein, M., and Snyder, M. (2008). The transcriptional landscape of the yeast genome defined by RNA sequencing. *Science,* Vol.320, No.5881, (June 6 2008), pp. 1344-1349, ISSN 1095-9203

Nowrousian, M. (2010). Next-generation sequencing techniques for eukaryotic microorganisms: sequencing-based solutions to biological problems. *Eukaryotic cell,* Vol.9, (2010), pp. 1300-1310, ISSN

Oshlack, A., and Wakefield, M.J. (2009). Transcript length bias in RNA-seq data confounds systems biology. *Biol Direct,* Vol.4, (April 2009), pp. 14, ISSN 1745-6150

Pastinen, T. (2010). Genome-wide allele-specific analysis: insights into regulatory variation. *Nat Rev Genet,* Vol.11, No.8, (June 2010), pp. 533-538, ISSN 1471-0064

Pepke, S., Wold, B., and Mortazavi, A. (2009). Computation for ChIP-seq and RNA-seq studies. *Nat Methods,* Vol.6, No.11 Suppl, (November 2009), pp. S22-32, ISSN 1548-7105

Robinson, M.D., McCarthy, D.J., and Smyth, G.K. (2010). edgeR: a Bioconductor package for differential expression analysis of digital gene expression data. *Bioinformatics,* Vol.26, No.1, (January 2010), pp. 139-140, ISSN 1367-4811

Schröder, J., Schröder, H., Puglisi, S.J., Sinha, R., and Schmidt, B. (2009). SHREC: a short-read error correction method. *Bioinformatics (Oxford, England),* Vol.25, (2009), pp. 2157-2163, ISSN

Serre, D., Gurd, S., Ge, B., Sladek, R., Sinnett, D., Harmsen, E., Bibikova, M., Chudin, E., Barker, D.L., Dickinson, T., *et al.* (2008). Differential allelic expression in the human genome: a robust approach to identify genetic and epigenetic cis-acting mechanisms regulating gene expression. *PLoS Genet,* Vol.4, No.2, (February 2008), pp. e1000006, ISSN 1553-7404

Shen, Y., Catchen, J., Garcia, T., Amores, A., Beldorth, I., Wagner, J.R., Zhang, Z., Postlethwait, J., Warren, W., Schartl, M., *et al.* (2011). Identification of transcriptome wide SNPs between Xiphophorus lines and species for assessment of allele specific gene expression within F1 interspecies hybrids. *Comparative Biochemistry and Physiology, Part C,* Vol.In press, (2011), ISSN 1532-0456

Shi, H., Schmidt, B., Liu, W., and Müller-Wittig, W. (2010). Quality-score guided error correction for short-read sequencing data using CUDA. *Procedia Computer Science,* Vol.1, (2010), pp. 1129-1138, ISSN

Trapnell, C., Pachter, L., and Salzberg, S.L. (2009). TopHat: discovering splice Junections with RNA-Seq. *Bioinformatics,* Vol.25, No.9, (May 2009), pp. 1105-1111, ISSN 1367-4811

Trapnell, C., Williams, B.A., Pertea, G., Mortazavi, A., Kwan, G., van Baren, M.J., Salzberg, S.L., Wold, B.J., and Pachter, L. (2010). Transcript assembly and quantification by RNA-Seq reveals unannotated transcripts and isoform switching during cell differentiation. *Nat Biotechnol,* Vol.28, No.5, (May 2010), pp. 511-515, ISSN 1546-1696

Voelkerding, K.V., Dames, S.A., and Durtschi, J.D. (2009). Next-generation sequencing: from basic research to diagnostics. *Clinical chemistry,* Vol.55, (2009), pp. 641-658, ISSN

Wall, D.P., Kudtarkar, P., Fusaro, V.A., Pivovarov, R., Patil, P., and Tonellato, P.J. (2010). Cloud computing for comparative genomics. *BMC bioinformatics,* Vol.11, (2010), pp. 259, ISSN

Walter, R.B., and Kazianis, S. (2001). Xiphophorus interspecies hybrids as genetic models of induced neoplasia. *ILAR J,* Vol.42, No.4, (October 2001), pp. 299-321, ISSN 1084-2020

Wang, L., Feng, Z., Wang, X., and Zhang, X. (2010). DEGseq: an R package for identifying differentially expressed genes from RNA-seq data. *Bioinformatics,* Vol.26, No.1, (January 2010), pp. 136-138, ISSN 1367-4811

Wu, T.D., and Nacu, S. (2010). Fast and SNP-tolerant detection of complex variants and splicing in short reads. *Bioinformatics (Oxford, England),* Vol.26, (2010), pp. 873-881, ISSN

Wu, T.D., and Watanabe, C.K. (2005). GMAP: a genomic mapping and alignment program for mRNA and EST sequences. *Bioinformatics (Oxford, England),* Vol.21, (2005), pp. 1859-1875, ISSN

Zerbino, D.R., and Birney, E. (2008). Velvet: algorithms for de Novembero short read assembly using de Bruijn graphs. *Genome Res,* Vol.18, No.5, (May 2008), pp. 821-829, ISSN 1088-9051

2

Linguistic Approaches for Annotation, Visualization and Comparison of Prokaryotic Genomes and Environmental Sequences

Oliver Bezuidt, Hamilton Ganesan, Phillip Labuschange, Warren Emmett, Rian Pierneef and Oleg N. Reva
University of Pretoria, Dep. Biochemistry, Bioinformatics and Computational Biology Unit, Pretoria, South Africa

1. Introduction

Sequencing of bacterial genomes has become a common technique of the present day microbiology. Thereafter, data mining in complete genome sequence is an essential step to uncover the uniqueness and evolutionary success of microorganisms. Oligonucleotide usage (OU or k-mer) statistics provides invaluable tools to get insight into genome organization and functionality.

The study of genome OU signatures has a long history dating back to early publications by Karlin et al. 1995, 1997, 1998, who focused mainly on dinucleotide compositional biases and their evolutionary implications. Statistical approaches of OU comparison were further advanced by Deschavanne et al., 1999, who applied chaos game algorithms; and by Pride *et al.*, 2003, who extended the analysis to tetranucleotides using Markov Chain Model simulations. Later, a number of practical tools for phylogenetic comparison of bacterial genomes (Coenye & Vandamme, 2004; van Passel et al., 2006); identification of horizontally transferred genomic islands (Mrázek & Karlin, 1999; Pried & Blaser, 2002; Nakamura et al., 2004; Azad & Lawrence, 2005; Dufraign et al., 2005; Becq et al., 2007) and assignment of unknown genomic sequences (Abe et al., 2003; Teeling et al., 2004) based on OU statistics became publicly available. These approaches exploited the notion that genomic OU composition was less variable within genomes rather than between them, regardless of which genomic regions had been taken into consideration (Jernigan & Baran, 2002). A general belief was that if a significant compositional difference was discovered in genomic fragments relative to the core genome, these loci most likely can be assigned to horizontally transferred genetic elements (transposons, prophages or integrated plasmids). This approach was criticized by several researchers (Koski et al., 2001; Wang 2001), who pointed out that codon bias and base composition are poor indicators of horizontal gene transfer. Therefore, there is a need for more informative parameters which also take into account higher order DNA variation. An overview of the current OU statistical methods based on di-, tetra- and hexanucleotides has been published recently (Bohlin et al., 2008). The conclusion of the review was that all methods were context dependent and, though being efficient and powerful, none of them were superior in all applications. Thus, the major motivation of our work was to develop more flexible and informative algorithms seamlessly integrating di- to heptanucleotides OU analysis for reliable identification of divergent genomic regions.

2. Linguistic approaches for genomics and metagenomics

Genome linguistics is respectively known as the analysis of frequencies of k-mers in genome wide DNA sequences. The basic hypothesis is that biased distribution of oligonucleotides in bacterial genome is genome specific and may serve as a signature. Each OU pattern may be characterized by a number of OU statistical parameters, namely: local pattern deviation (D), pattern skew (PS), relative variance (RV) and several others that will be explained below. The requirements for the OU statistics are as follows: i) distances between patterns of different word length (from di- through to heptanucleotides) calculated for the same sequence must be comparable; i.e. one may use longer word patterns to perform a large scale analysis and then switch to shorter word patterns for a more detailed view; ii) OU patterns calculated for sequences of different lengths must be comparable provided that the length of the sequence is longer than the specified thresholds; iii) alterations of OU patterns may be analyzed by different non-redundant parameters (D, PS and RV with different schemes of normalization by frequencies of shorter constituent words). Superimposition of these OU characteristics allows better discrimination of divergent genomic regions.

2.1 Oligonucleotide usage pattern concept

OU pattern was denoted as a matrix of deviations $\Delta_{[\xi 1 \ldots \xi N]}$ of observed from expected counts for all possible words of length N. Oligonucleotides or words are distributed in sequences logarithmically and deviations of their frequencies from expectations may be found as follows:

$$\Delta_w = \Delta_{|\xi 1 \ldots \xi N|} = 6 \times \frac{\ln\left(\dfrac{C^2_{|\xi 1 \ldots \xi N||obs} \sqrt{C^2_{|\xi 1 \ldots \xi N||e} + C^2_{|\xi 1 \ldots \xi N||0}}}{C^2_{|\xi 1 \ldots \xi N||e} \sqrt{C^2_{|\xi 1 \ldots \xi N||obs} + C^2_{|\xi 1 \ldots \xi N||0}}}\right)}{\sqrt{\ln\left(\left[C^2_{|\xi 1 \ldots \xi N||0} \Big/ C^2_{|\xi 1 \ldots \xi N||e}\right] + 1\right)}} \qquad (1)$$

where ξ_n is any nucleotide A, T, G or C in the N-long word; $C_{[\xi 1 \ldots \xi N]\,|\,obs}$ is the observed count of a word $[\xi_1 \ldots \xi_N]$; $C_{[\xi 1 \ldots \xi N]\,|\,e}$ is its expected count and $C_{[\xi 1 \ldots \xi N]\,|\,0}$ is a standard count estimated from the assumption of an equal distribution of words in the sequence: ($C_{[\xi 1 \ldots \xi N]\,|\,0} = L_{seq} \times 4^{-N}$).

Expected counts of words $C_{[\xi 1 \ldots \xi N]\,|\,e}$ were calculated in accordance to the applied normalization scheme. For instance, $C_{[\xi 1 \ldots \xi N]\,|\,e} = C_{[\xi 1 \ldots \xi N]\,|\,0}$ if OU is not normalized, and $C_{[\xi 1 \ldots \xi N]\,|\,e} = C_{[\xi 1 \ldots \xi N]\,|\,n}$ if OU is normalized by empirical frequencies of shorter constituent words of length n. The expected count of a word $C_{[\xi 1 \ldots \xi N]\,|\,e}$ of the length N in a L_{seq} long sequence normalized by frequencies of n-mers ($n < N$) is calculated as follows:

$$C_{[\xi_1 \ldots \xi_{l_w}]|n} = L_{seq} \times F_{[\xi_1 \ldots \xi_n]} \times \prod_{i=2}^{N-n+1} \left(\frac{F_{[\xi_i \ldots \xi_{i+n-1}]\xi_{i+n}}}{\sum_{\xi}^{A,T,G,C} F_{[\xi_i \ldots \xi_{i+n-1}]\xi}} \right) \qquad (2)$$

Where the $F_{[\xi 1 \ldots \xi n]}$ values are the observed frequencies of a particular word of length n in the sequence and ξ is any nucleotide A, T, G or C. For instance, the expected count of a word ATGC in a sequence of L_{seq} nucleotides normalized by frequencies of trinucleotides would be determined as follows:

$$C_{ATGC} = L_{seq} \times F_{ATG} \times \frac{F_{TGC}}{F_{TGA} + F_{TGT} + F_{TGG} + F_{TGC}} \tag{3}$$

Two approaches of normalization have been exploited where the F values are calculated for the complete genome (generalized normalization) or for a given sliding window (local normalization).
The distance D between two patterns was calculated as the sum of absolute distances between ranks of identical words (w, in a total 4^N different words) after ordering of words by $\Delta_{[\xi 1 \ldots \xi N]}$ values (equation 1) in patterns i and j as follows:

$$D(\%) = 100 \times \frac{\sum_{w}^{4^N} \left| rank_{w,i} - rank_{w,j} \right| - D_{\min}}{D_{\max} - D_{\min}} \tag{4}$$

Application of ranks instead of relative oligonucleotide frequency statistics made the comparison of OU patterns less biased to the sequence length provided that the sequences are longer than the limits of 0.3, 1.2, 5, 18.5, 74 and 295 kbp for di-, tri-, tetra-, penta-, hexa- and heptanucleotides, respectively (Reva and Tümmler, 2004).
PS is a particular case of D where patterns i and j are calculated for the same DNA but for direct and reversed strands, respectively. $D_{max} = 4^N \times (4^N - 1)/2$ and $D_{min} = 0$ when calculating a D, or, in a case of PS calculation, $D_{min} = 4^N$ if N is an odd number, or $D_{min} = 4^N - 2^N$ if N is an even number due to the presence of palindromic words. Normalization of D-values by D_{max} ensures that the distances between two sequences are comparable regardless of the word length.
Relative variance of an OU pattern was calculated by the following equation:

$$RV = \frac{\sum_{w}^{4^N} \Delta_w^2}{\left(4^N - 1\right) \times \sigma_0} \tag{5}$$

where N is word length; Δ^2_w is the square of a word w count deviation (see equation 1); and σ_0 is the expected standard deviation of the word distribution in a randomly generated sequence which depends on the sequence length (L_{seq}) and the word length (N):

$$\sigma_0 = \sqrt{0.02 + \frac{4^N}{L_{seq}}} \tag{6}$$

2.2 Compositional polymorphism of bacterial genomes

Biased distribution of k-mers may be explained by selective forces of DNA reparation enzymes of microorganisms, which may sense stereochemical properties of DNA fragments. A strong correlation was discovered between frequencies of oligonucleotides and their physicochemical properties such as base stacking energy, propeller twist angle, bendability, protein deformability and position preference in the DNA helical repeats (Fig. 1) calculated by the additive scale approach proposed by Baldi & Baisnée, 2000. It looks plausible that proteins of the replication-reparation system may sense the stereochemical properties of the DNA molecule and allow higher mutation rates in atypical regions; however, it has not yet been proved experimentally. The latter may explain the pervasive properties of genomic signatures that are reported for bacterial genomes (Jernigan & Baran, 2002). Despite a significant conservation of the OU pattern in genomic core DNA sequences, every bacterial genome contains loci of DNA which differ significantly from the core sequence. These loci usually contain gene clusters for ribosomal RNA and ribosomal proteins, horizontally transferred genomic islands, DNA fragments with multiple repeats and some other features. Superimposition of different OU parameters allows discrimination of divergent genomic regions. Briefly: rRNA operons are characterized by extremely high PS and low RV; giant genes with multiple repeated elements have high or moderate PS and high RV; horizontally transferred genetic elements are characterized by increased divergence between RV and GRV accompanied by high D; and genes for ribosomal proteins show a moderate increase of D, PS and RV above genomic averages. In the examples given above D denotes the distance between a local pattern calculated for a sliding window and the global pattern determined for the complete genome; PS is local pattern skew; and RV and GRV are variances of local OU patterns normalized by GC-content of the sliding window and the complete genome, respectively.

A Web-based applet SeqWord Genome Browser (SWGB) was developed and available on-line at www.bi.up.ac.za/SeqWord/ to visualize DNA compositional variations in pre-calculated bacterial and viral genomes. The SWGB is basically comprised of two views, denoted by the 'Gene Map' and 'Diagram' tabs. The 'Gene Map' tab offers a simple view of an entire genome at a glance and gives users access to a number of important pre-calculated OU statistics superimposed on the gene map (Fig. 2). The 'Diagram' tab allows flexible filtering of the underlying data based on the criteria chosen by users. Although the underlying data is pre-calculated, the user may, by simply changing selected parameters, generate many alternative plots, which give different insights into the natural genomic variation. On the dot-plot diagram, each genomic fragment selected by the sliding window is represented by a dot with X and Y coordinates, which correspond to values of OU parameters chosen from X and Y drop-down lists, respectively. The Z axis parameter may be set as well. In this case, the dots are coloured by values of OU parameters selected for the Z axis, and the colour range is displayed on the vertical colour bar on the left of the plot area (Fig. 3).

Several routines have been developed to identify horizontally transferred genomic islands, genes for ribosomal RNA and proteins, non-functional pseudogenes and genes of other functional categories. All these routines are described in detail with illustrations in supplementary web-pages (use the 'Help' link in the applet window). Take for example the genome of *Pseudomonas putida* KT2440, a known mosaic genome with 105 genomic islands above 4000 bp in length (Weinel et al., 2002). Many of these features can be visualized at a glance using the SWGB without any in depth analysis (see Fig. 2). On the 'Diagram' view the parameters n1_4mer:RV, n1_4mer:GRV and n0_4mer:D were selected for the X, Y and Z axes, respectively, as we showed previously (see Fig. 3).

Plotting local relative OU variance (RV) against global relative variance (GRV) basically shows the effect of normalization by global mononucleotide content. The core genome is then represented on the dot plot as the positive linear correlation line where RV ≈ GRV (Fig. 3). In other words, these fragments exhibit such compositional closeness to the core genome that normalizing by local mononucleotide content does not have any effect compared to normalizing by the global content. These genomic fragments also exhibit compositional similarity to the genomic average; and are therefore coloured blue. Scattered dots lying peripheral to the expected strong linear correlation do not belong to the core genome and also have a higher distance from the genomic average and are hence coloured green and red.

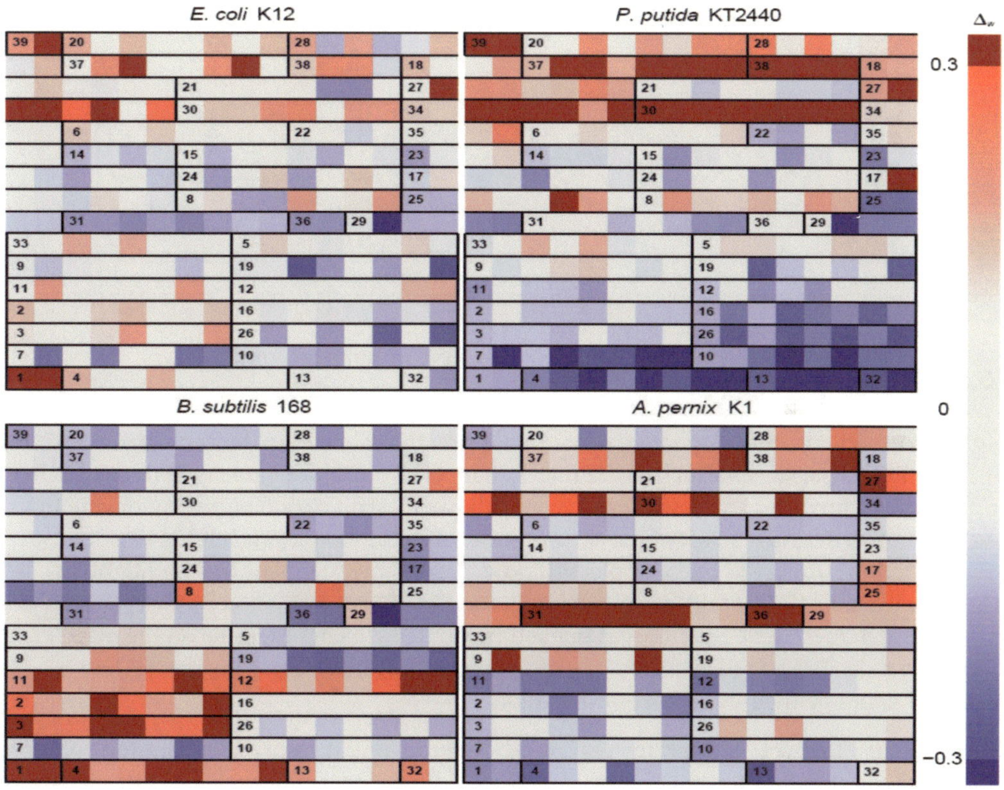

Fig. 1. Tetranucleotide usage patterns calculated for genomes of four different organisms. The deviations Δ_w of observed from expected counts are shown for all 256 tetranucleotide permutations (16×16 cells) by a colour code (right bar) depicting overrepresented (red) and rare (blue) words. The words are grouped into 39 equivalence classes and ordered by decreasing base stacking energy row-by-row starting from the upper corner (class 39).

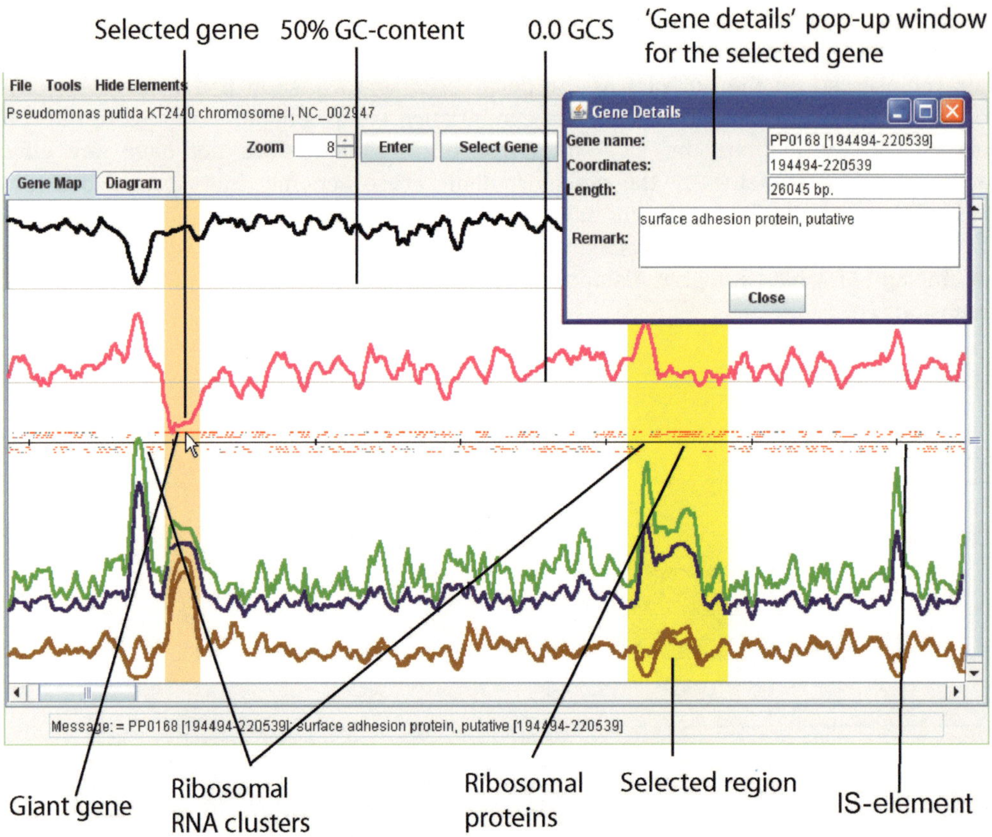

Fig. 2. Identification of divergent genomic regions on the 'Gene Map' view. Superimposition of different OU parameters such as GC (black line), GCS (pink), PS (green), D (blue), GRV (upper brown line) and RV (lower brown line) allows discrimination of divergent genomic regions. In this example a part of the chromosome of *Pseudomonas putida* KT2440 (127-774 kbp) is displayed in the applet window. A genomic fragment was highlighted using the function 'Select region' and a giant gene, PP0168, was selected by 'Select gene'. A pop-up window 'Gene Details' was opened by double-clicking the gene on the map. Genes are indicated by red and grey (for hypothetical) bars. The black horizontal line separates genes by their direction of translation.

Changing of the set of parameters as shown on Fig. 4 allows separation of core housekeeping genes from clusters of genes encoding ribosomal proteins and ribosomal RNA, vestigial regions with pseudogenes and giant genes with multiple repeats.
SWGB is linked to a database of pre-calculated OU patterns of bacterial genomes (2243 complete sequences, including bacterial chromosomes, plasmids and some viruses were available at the time of writing of this chapter and new sequences are regularly being added). The SWGB allows tentative annotation of the various divergent regions and provides overviews for use in comparative genomics. Users may download the command line version of the OligoWords program to analyze locally their own sequences. A packaged version of the SWGB allows users to view and manipulate their OligoWords results locally using a compatible web-browser.

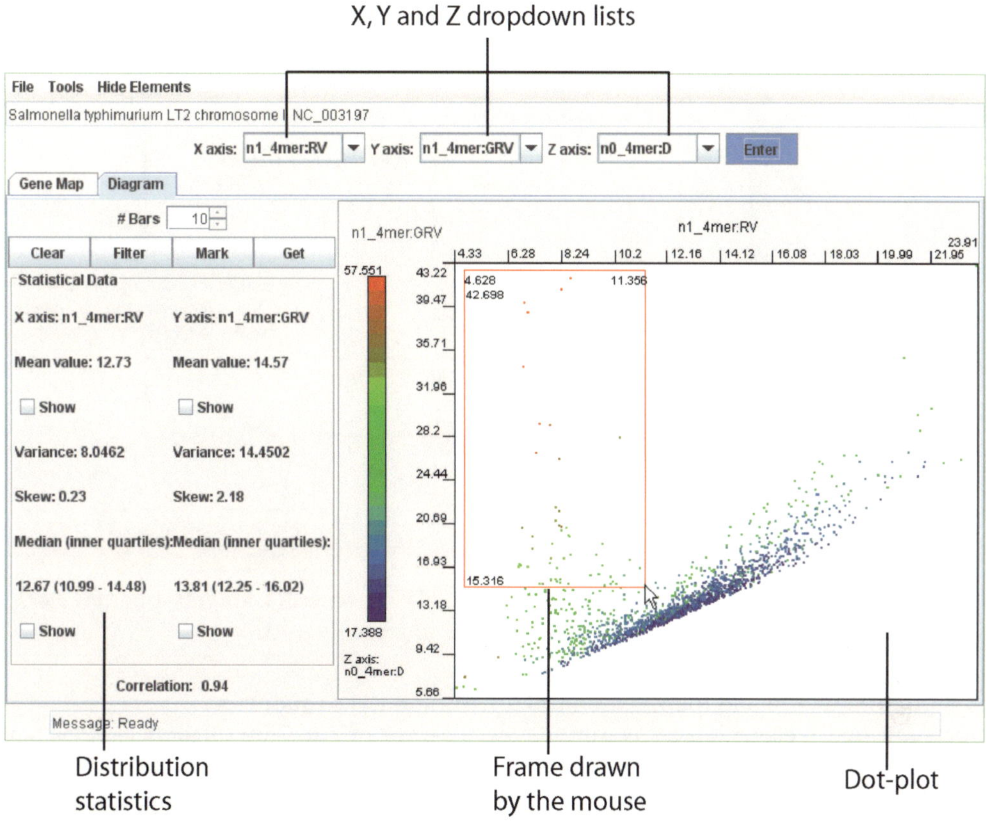

Fig. 3. The 'Diagram' view. In this example n1_4mer:RV, n1_4mer:GRV and n0_4mer:D were selected for the X, Y and Z axes, respectively. Every dot on the dot-plot corresponds to a genomic fragment selected by the sliding window. Dots are spread and coloured in accordance with their values of the selected statistical OU parameters. Information for each dot may be found by one of the following methods: i) information for a dot pointed by the mouse is shown in the 'Message' bar; ii) double clicking a dot returns us to the 'Gene map' tab with the corresponding genomic fragment highlighted; iii) framing the dots and clicking the 'Get' button opens a new applet window with the information about all selected regions. In this example the genomic regions of *Salmonella typhimurium* LT2 (NC_003197) which correspond to horizontally transferred genetic elements were selected.

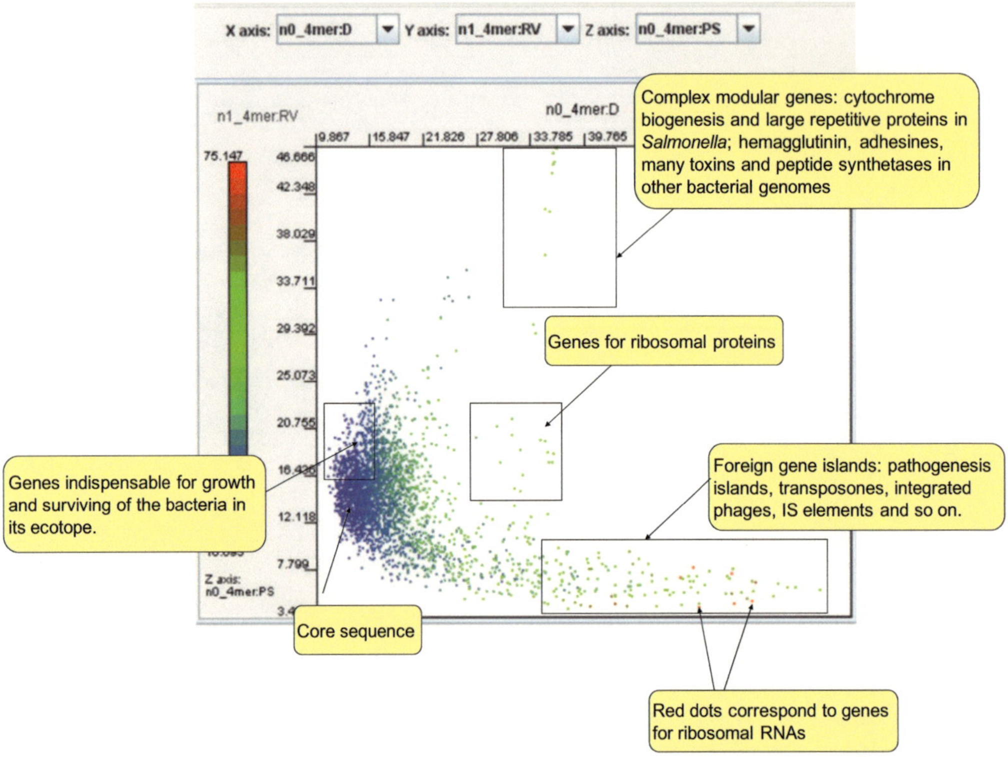

Fig. 4. The 'Diagram' view. In this example n0_4mer:D, n1_4mer:RV and n0_4mer:PS were selected for the X, Y and Z axes to identify genomic areas of interest.

2.3 Signature words and identification of environmental sequences

In genomic and metagenomic literature the occurrences of 2 to 7 bp oligonucleotides have been studied extensively. Patterns of short oligomers (words) have been used successfully for DNA read clustering (Chatterji et al., 2008; Kislyuk et al., 2009; Saeed & Halgamuge, 2009). However, short oligonucleotide patterns usually do not provide enough information for binning DNA reads to bacterial species or higher taxonomic units. Longer words of 8 to 14 nucleotides generally are more specific. Nevertheless, it was illustrated that the approach based on the analysis of frequencies of all the permutations of oligonucleotides of a given length such as discussed above is not effective for analysis of 8 to 14 letter words (Bohlin et al., 2008). Furthermore, an analysis of all possible permutations of 8 to 14 bp words would be computationally expensive because the total number of possible permutations of words is 4^L where L is the word length. For words of length 8 to 14 bp, this quantity becomes very large. Additionally, the random changes in the frequencies of such a large number of words obscure the genome specific information present in a few signature words. According to Kirzhner et al., 2005, less than 1% of 10-mers are informative in a large-scale comparison of bacterial genomes.

A first investigation into exploiting the information present in 8 to 14 letter words in 13 strains from the genus *Pseudomonas* was made by Davenport et al., 2009. It has been shown as well that certain profiles of signature words may help to distinguish DNA fragments that originated from different genomes (Saeed & Halgamuge, 2009).

In this work an attempt was made to standardize the linguistic approaches of binning metagenomic DNA reads by creating a database of signature words of Eubacteria and Archaebacteria represented in GenBank. The first step was to develop methods for summarizing the large amounts of data associated with these words. To avoid using the words that do not provide any taxonomic information, the most divergent words were selected and stored in the database. Currently, the frequencies of 172,636 signature words calculated in 768 bacterial chromosomes are stored in a binary database file available for download from the SeqWord project Web-site at www.bi.up.ac.za/SeqWord/oligodb/. Furthermore, scoring functions were designed, which measure the likelihood that a given DNA fragment originated from a given taxonomic group.

This tool also may be used to identify the origin of DNA sequences or whole clusters of DNA sequences. There are a number of programs such as LikelyBin (Kislyuk et al., 2009), CompostBin (Chatterji et al., 2008) and some others that cluster DNA sequences, but there is no default methodology for inferring the taxonomic affinity of these clusters. Typically, BLAST is used to compare these clusters to the databases of DNA sequences. Frequently these clusters consist of several short sequences that cannot be easily assembled, which makes using BLAST complicated. TETRA identifies long unknown DNA sequences by comparison of the whole patterns of frequencies of tetranucleotides (Teeling et al., 2004). A tool based on occurrences of 8 to 14 letter words is expected to work equally well both on clusters of sequences and single long sequences.

2.3.1 Statistical background of selection of signature words

To identify prospective signature words, the distribution score coefficients were calculated for each 8 to 14-mer permutation as follows:

$$DS = \frac{100000}{\sqrt{\mu^2 + \sigma^2}} \qquad (7)$$

where μ indicates the average length of spans (in base pairs) between the repeated words in the sequence (i.e., μ = sequence_length/number_of_words); and σ is the standard deviation of the span lengths. The DS for a word increases in value when there is a high frequency of occurrence (μ is minimal) and the words are evenly distributed (σ tends to 0). The DS coefficient assigns low scores to infrequent words and to local repeats while giving higher scores to words occurring frequently and evenly distributed throughout the genome. The words that have DS above the threshold value of 0.3 in at least one genome were included in the template of signature words. Furthermore, their frequencies were recalculated for all genomes. The threshold value was empirically determined to ensure that the template contained similar numbers of words for each different word length and that an appropriate ratio between template size and word specificity is obtained. The final template contains 172,636 signature words; that is approximately 0.1% of the total number of all possible permutations of 8 bp to 14 bp oligonucleotides. Note that in this work each oligonucleotide and its reverse complement were considered as the same word so that the two different strands of the DNA molecule will be assigned identical scores.

To improve maintenance and operational flexibility of the database, the numeric frequencies of words may be replaced by percentile values without any significant loss of information. The empirical cumulative distribution of the frequency of occurrence of the words in the template was studied and the following non-linear regression model was fitted to the data:

$$f = \frac{\exp(3p + 9)}{L^{4.5}} \tag{8}$$

where f is the frequency of a word per 100 Kbp, L is the word length and p is the probability that the word occurs at a frequency less than or equal to f. For example, according to equation 8 for 50% of words of the length 8 bp ($p = 0.5$; $L = 8$) the frequency f is in the range from 0 to 3.13 words per 100 Kbp of the given sequence; and 90% ($p = 0.9$) of 8-mers have frequencies from 0 to 10.41. Four categories were designated for rare ($p < 0.1$), common ($0.1 \le p < 0.5$), frequent ($0.5 \le p < 0.9$) and abundant ($0.9 \le p$) words. The borders of the percentile categories calculated by equation 8 are shown in Table 1.

The performance of a signature word to separate DNA reads of different origins or to bin a cluster of reads to a taxonomic unit depends on the set of taxonomic units to be differentiated and the task formulation. Several scoring algorithms were used in this study. All the scores were normalized to a range from 0 to 10. The scores were used to order the words in the database and to select the ones with the highest scores.

Word divergence is scored by the variance of percentile values (see Table 1) in the selected genomes normalized by the maximum possible variance. The most diverse word would be rare in one half of the selected genomes and abundant in the other half of the genomes.

To select the words, which are rare or abundant in all selected genomes, the following score was used:

$$\text{Score} = 10 \times (\text{Av} - 0.05)/0.9 \tag{9}$$

where Av is the average of the percentile values calculated for a word in selected genomes. To select rare words ($10 - Score$) was used.

The perfect word to distinguish between taxa is one that is similarly distributed in genomes belonging to the same taxon but is differently distributed in different taxa. The scores were assigned in the spirit of ANOVA by computing the ratio of the sums of square deviations over the average values between taxa and within every taxon.

Another practical task may consist in distinguishing one taxon (outgroup) from a number of other taxa (counterparts) by diverse, abundant or rare words. In our study this approach was termed as *confronted comparison*. Three scoring algorithms were used:

$$DiversityScore = \left|Av_0 - Av_g\right| / \sqrt{1 + Var_0/n} \tag{10}$$

$$AbundanceScore = \left(10 + Av_0 - Av_g\right) / 2\sqrt{1 + Var_0/n} \tag{11}$$

$$ScarceScore = \left(10 + Av_g - Av_0\right) / 2\sqrt{1 + Var_0/n} \tag{12}$$

where Av_0 and Av_g are average frequencies of the word in genomes of the outgroup and counterpart taxonomic units, correspondingly; Var_0 is the variance of the word frequencies in the outgroup genomes and n is the number of genomes in the outgroup taxonomic unit.

Computer simulation of metagenomic datasets was done by the MetaSim program (Richter et al., 2008). DNA reads were clustered by the LikelyBin algorithm (Kislyuk et al., 2009). The database of signature words and the OligoDBViewer program are available for download from www.bi.up.ac.za/SeqWord/oligodb/.

Word length	Percentiles			
	Rare - (0.0)*	Common + (0.25)	Frequent ++ (0.75)	Abundant +++ (1.0)
8 bp	< 0.94†	≥ 0.94 and < 3.13	≥ 3.13 and < 10.4	≥ 10.4
9 bp	< 0.56	≥ 0.56 and < 1.85	≥ 1.85 and < 6.13	≥ 6.13
10 bp	< 0.35	≥ 0.35 and < 1.15	≥ 1.15 and < 3.81	≥ 3.81
11 bp	< 0.23	≥ 0.23 and < 0.75	≥ 0.75 and < 2.48	≥ 2.48
12 bp	< 0.15	≥ 0.15 and < 0.51	≥ 0.51 and < 1.68	≥ 1.68
13 bp	< 0.11	≥ 0.11 and < 0.35	≥ 0.35 and < 1.17	≥ 1.17
14 bp	< 0.08	≥ 0.08 and < 0.25	≥ 0.25 and < 0.84	≥ 0.84

*Rare, common, frequent and abundant words are marked as –, +, ++ and +++, respectively. The numeric values representing each percentile category are used for score calculations.
† These are *f*-values calculated by equation 8 for the cumulated likelihoods (p) 0.1, 0.5 and 0.9, respectively.

Table 1. The percentile border frequencies calculated for the words of different length.

2.3.2 OligoDBViewer and the database of signature words

The main window of OligoDBViewer is shown in Fig. 5. The functionality of the OligoDBViewer is described in detail on the project Web site. The program allows selecting genomes or taxonomic units from the list and searching for the best discriminative words by using the program functions accessible through the toolbar and the 'Command' menu. The resulting list of ordered words will be shown in the pop-up panel on the right hand side as in Fig. 6. In this example the diverse words were searched with the goal of separating the genomes of *Mycobacterium avium* k10 [NC_002944] and *M. tuberculosis* F11 [NC_009565]. The divergence scores are shown in column 'C'. The number of times an oligomer falls into the categories of rare, common, frequent or abundant words is shown respectively from left to right in column 'Stat'.

The list of the words returned by OligoDBViewer may be highly redundant. For example, words that only differ by a single nucleotide may be expected to have similar distributions in genomes. To reduce the redundancy of the selected words, several filter options may be set. The filter settings in Fig. 6 removes all the words that differ from the words with the highest scores by less than 30 % similarity (another option is available to set the minimal number of mismatches) as well as the words that are left or right shifted sub-words or shorter constituents of longer words that have higher scores. Then the list is cut off so that only the top 10 words remain. Additionally, the list of selected words may be filtered by the word length and the score threshold. All word filtering settings is reversible.

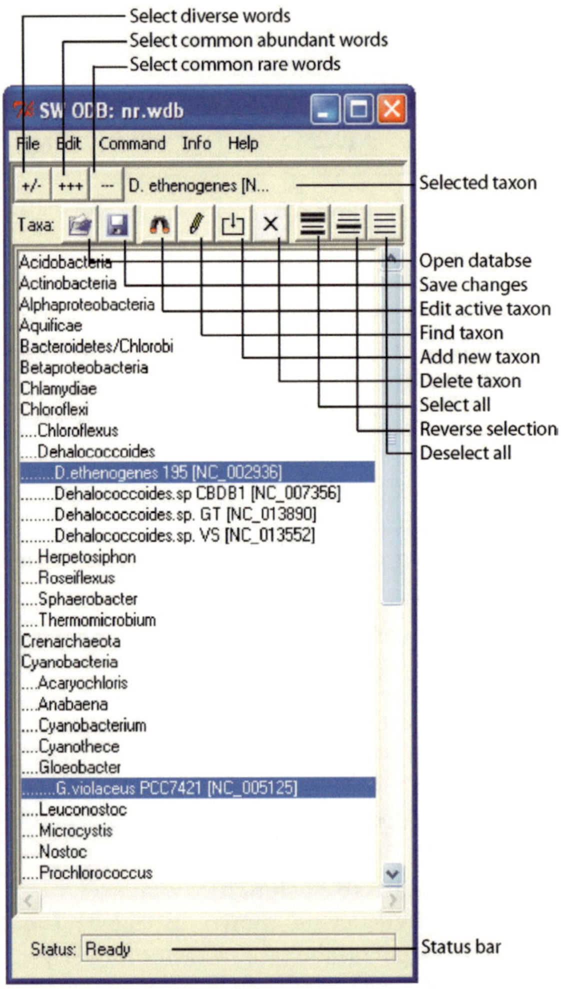

Fig. 5. The main window of OligoDBViewer.

To facilitate large scale calculations and the database updates on remote servers, several command line utilities are available for download. They are fully described on the project Web site.

2.3.3 Algorithms of binning of clusters of DNA reads to taxonomic units

To estimate the similarity of a cluster of DNA reads to bacterial taxonomic units the percentile values were used (Table 1). All DNA reads of the cluster were concatenated in an artificial sequence and the frequency of the words normalized per 100 Kbp were counted (*f*-value). Then the *f*-values were converted to percentiles:

$$p = \frac{\ln(f) + 4.5\ln(L) - 9}{3} \tag{13}$$

Note that equation 13 is the inverse of the equation 8. The meanings of the coefficients were explained above.

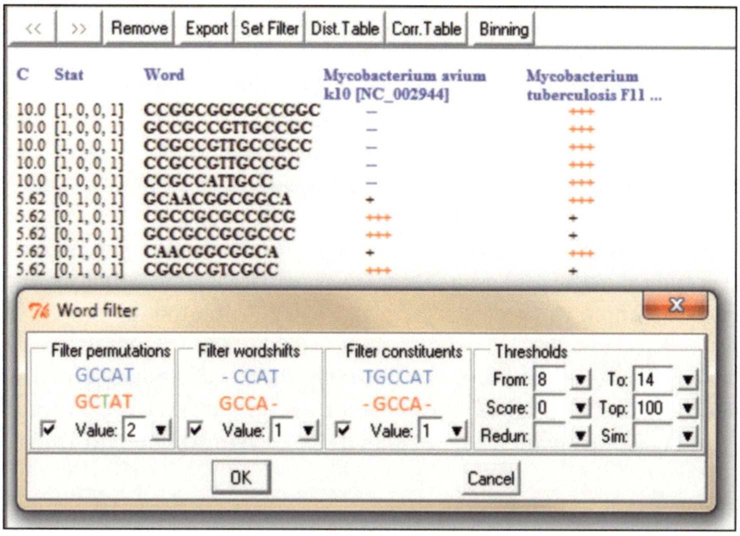

Fig. 6. The top 10 most diverse words which separate *M. tuberculosis* from *M. avium*. The filter was set to reduce the redundancy of the selected words.

Next, the distance values D between an unknown sequence and the taxonomic units were calculated as follows:

$$D = 10 \times \sqrt{\frac{(p_i{'} - p_i)^2 \times \sum\left(1 + \frac{p_i{'} - p_i}{2 - p_i}\right)}{(m_i - p_i)^2 \times \sum\left(1 + \frac{m_i - p_i}{2 - p_i}\right)}} \tag{14}$$

where p_i is the percentile value from the OligoDB database for the word i in a bacterial genome, or the average of the percentile values for the genomes of a taxonomic unit; $p_i{'}$ is the percentile value of this word in the query sequence as calculated by equation 13; and m_i is an indicator variable equal to 0 if $p_i \geq 0.5$ and equal to 1 if $p_i < 0.5$. Thus, the denominator is the maximum possible distance. D values fall in the range from 0 to 10.

Consider the following example: let the 10-mer TTAAAGAAAA be distributed in the concatenated cluster sequence with the frequency 2.81 words per 100 Kbp and let the 8-mer TCTTTTAA occur 6.35 times per 100 Kbp. According to equation 13, the percentile value of the word TTAAAGAAAA is:

$$p = \frac{\ln(2.81) + 4.5\ln(10) - 9}{3} = .79 \tag{15}$$

and for the word TCTTTTAA the percentile value is:

$$p = \frac{\ln(6.35) + 4.5\ln(8) - 9}{3} = .74 \tag{16}$$

Next D is calculated by equation 14. The motivation for using D values rather than Euclidian distances is based on the fact that the observed frequency of occurrence of words in the clustered reads is frequently lower than in the original genome due to asymmetric distribution of word frequencies. Another factor that contributes to this observation is that the clusters of metagenomic DNA reads often contain fragments from more than one organism. This leads to false similarity of metagenomic sequences to taxa where the signature words are uncommon. To remove this bias, the equation 14 was constructed so that the difference between p_i' and p_i is given less weight if p_i' is smaller than p_i than if p_i' is larger than p_i.

To evaluate the discriminative power of the algorithms, several simulated metagenomic datasets were prepared using MetaSim. Then DNA reads were clustered by LikelyBin.

The first set was a simple random selection of 50 DNA fragments of the chromosome of *Bacillus subtilis* 168 [NC_000964]. The total length of all the fragments was 691 Kbp. The OligoDB program was used to compare the compositional similarity of these randomly selected sequences with the original chromosome and the closely related organisms of the genus *Bacillus* and class Firmicutes. The obtained distances are shown in Table 2.

Genome	D*
B. subtilis [NC_000964]	1.74
B. amyloliquefaciens [NC_009725]	2.22
B. licheniformis [NC_006270]	2.23
B. pumilus [NC_009848]	3.43
B. clausii [NC_006582]	3.70
Lactobacillus brevis [NC_008497]	3.83
B. halodurans [NC_002570]	4.05
B. pseudofirmus [NC_013791]	4.49
B. anthracis [NC_003997]	5.60
B. cereus [NC_004722]	5.68

*In this and the following tables the filter settings for the program were as follows: only the top 100 words of 8 to 12-mers with the sequence similarity ≤ 30 % were considered. Further, one nucleotide shifted words and one nucleotide shorter constituent words were filtered out as in the filter setting window in Fig. 6.

Table 2. Identification of DNA fragments generated from the *B. subtilis* chromosome.

For the next test, two quite distant organisms were selected: *Burkholderia cenocepacia* AU1054 [NC_008062] and *Psychrobacter arcticus* 273-4 [NC_007204]. 552 genomic fragments with an average length of 500 bp were generated randomly by MetaSim from the *B. cenocepacia* chromosome and 448 fragments of the same average length were obtained from the *P. arcticus* chromosome. All these fragments were mixed together and used as the input for LikelyBin. These randomly generated genomic fragments were then grouped by DNA composition similarity into 13 clusters. The two biggest clusters contained DNA fragments that were generated exclusively from one origin: 347 of the fragments generated from *B. cenocepacia* were grouped into cluster A and 437 of the fragments from the *P. arcticus* chromosome were in cluster B. Now, the OligoDB algorithm was used to identify the organisms most similar to the cluster. For the comparative analysis several representatives of β- and γ-Proteobacteria were selected (Table 3).

Cluster A (172 Kbp)	D	Cluster B (220 Kbp)	D
B. cenocepacia [NC_008062]	2.07	*P. arcticus* [NC_007204]	1.59
B.ambifaria [NC_010557]	3.05	*P. cryohalolentis* [NC_007969]	1.66
B. mallei [NC_006348]	3.51	*P. haloplanktis* [NC_007481]	1.92
B. phymatum [NC_010622]	6.69	*P. atlantica* [NC_008228]	2.25
B. xenovorans [NC_007952]	6.73	*P. ingrahamii* [NC_008709]	2.28
R. solanacearum [NC_003295]	7.67	*S. baltica* [NC_009052]	2.72
R. eutropha [NC_007347]	7.94	*S. enterica* [NC_003198]	7.06
C. metallidurans [NC_007974]	8.59	*E. pyrifoliae* [NC_012214]	7.72
P. arcticus [NC_007204]	8.59	*B. cenocepacia* [NC_008062]	8.67
R. pickettii [NC_010682]	8.66	*P. putida* [NC_002947]	8.90

Table 3. Identification of DNA fragments generated from *B. cenocepacia* (cluster A) and *P. arcticus* (cluster B).

Cluster A (87 Kbp)	D	Cluster B (99 Kbp)	D
P. haloplanktis [NC_007481]	3.00	*S. enterica* [NC_003198]	2.66
P. cryohalolentis [NC_007969]	3.06	*E. pyrifoliae* [NC_012214]	3.21
P. ingrahamii [NC_008709]	3.06	*P. putida* [NC_002947]	3.50
P. mirabilis [NC_010554]	3.10	*S. baltica* [NC_009052]	7.00
P. arcticus [NC_007204]	3.29	*P. atlantica* [NC_008228]	7.57
P. atlantica [NC_008228]	4.25	*P. arcticus* [NC_007204]	7.92
S. baltica [NC_009052]	4.80	*P. cryohalolentis* [NC_007969]	7.99
S. enterica [NC_003198]	7.49	*P. ingrahamii* [NC_008709]	8.03
E. pyrifoliae [NC_012214]	7.57	*P. mirabilis* [NC_010554]	8.09
P. putida [NC_002947]	8.04	*P. haloplanktis* [NC_007481]	8.17

Table 4. Identification of a chimerical cluster A that contains DNA fragments from *P. haloplanktis* and *S. enterica*, and a monophyletic cluster B containing fragments of the *S. enterica* genome.

The clusters were identified correctly; however, the separation of genomic fragments of *P. arcticus* (Cluster B) from other close relative organisms of genera *Psychrobacter*, *Psychromonas* and *Pseudoalteromonas* was not reliable. An additional round of identification is needed where the signature words are selected specifically to distinguish between these organisms.

The next set of DNA fragments was generated from two genomes of γ-Proteobacteria: *Pseudoalteromonas haloplanktis* TAC125 [NC_007481] and *Salmonella enterica* CT18 [NC_003198]. LikelyBin clustered the fragments into 49 clusters. Half of the clusters contained sequences generated from both chromosomes. The two biggest clusters were selected for analysis by the OligoDB algorithm. Cluster A contains 166 fragments of the *P. haloplanktis* chromosome and 9 sequences originating from *S. enterica*. Cluster B contains 195 DNA fragments generated from *S. enterica* only. Results of the identification are shown in Table 4.

Both of these clusters were identified correctly. The mix of two organisms in Cluster A yields D values that are higher than all other examples in this paper. D values calculated for more complex chimerical clusters were around 5; indicating that it will be difficult to associate such a set of sequences with a specific taxonomic unit.

2.4 Stratigraphic analysis of bacterial genomes

DNA molecules encoding functional enzymes, transcriptional regulators and virulence factors are fluxing through the bacterial taxonomic walls. They endow environmental and clinical strains of bacteria with new unexpected properties. Lateral genetic exchange, particularly of drug tolerance genes has been recognized for a long time; however the ontology of genomic islands and their donor-recipient relations remain generally obscure because of methodological problems. Horizontally transferred genes are highly mutable and the mobilome entities having been inserted into host chromosomes undergo multiple events of fragmentation, partial duplications and deletions. Even prediction of insertion sites in host chromosomes remained to be a challenge.

Genome linguistics methods are applicable to study and visualize intrinsic relationships between mobile genetic elements in bacterial genomes. *Mycobacterium tuberculosis*, a bacterial pathogen which is a leading cause of human death worldwide, was selected as a subject for this study. Emergence and evolution of this deadly pathogen are still ambiguous and not fully understood even after having done the sequencing and comparative studies on multiple strains of this genus.

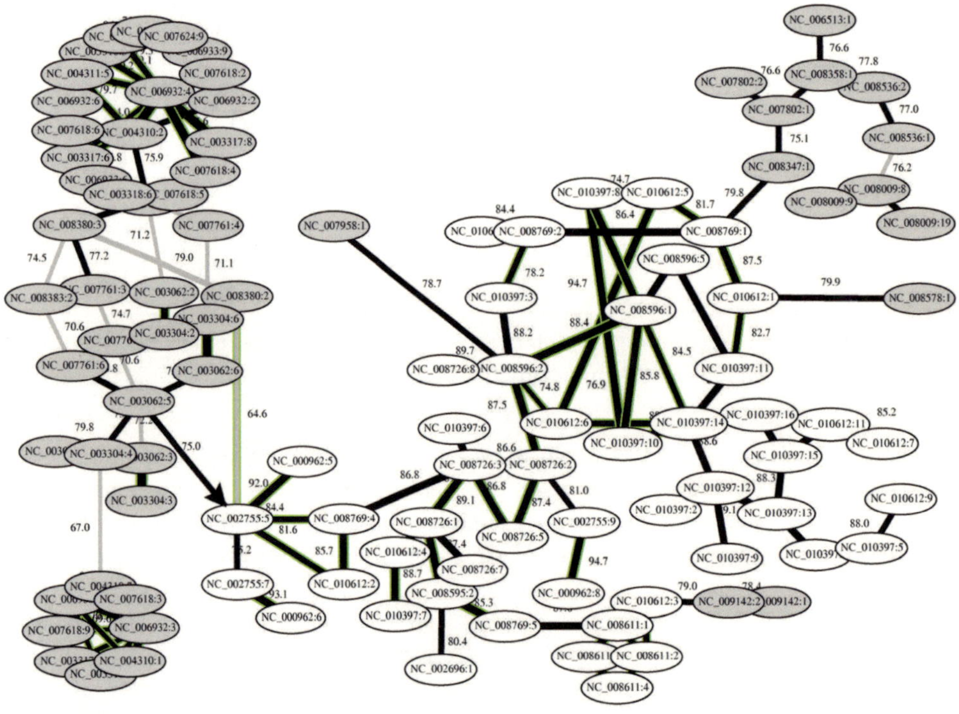

Fig. 7. GIs identified in *Mycobacterium* genomes and other organisms share compositional similarity. GIs identified in Mycobacteria are represented by white nodes and species of other genera by grey nodes. Each node represents one GI tagged by NC number of the host organism as in NCBI followed by the reference number of GIs as in GEI-DB. The edges depicted by green halo link GIs sharing similar DNA sequences longer than 100 bp identified by blast2seq. The layout was created by an in-house Python program that incorporates executable files of Graphviz 2.26.3 for Windows.

2.4.1 Identification and grouping of mycobacterial genomic islands

Linguistic methods were applied to study the distribution of genomic islands (GIs) in complete genome sequences of *Mycobacterium*. GIs were identified by SeqWord Gene Island Sniffer (SWGIS available at www.bi.up.ac.za/SeqWord/sniffer/). The identified GIs were grouped by compositional similarity of oligonucleotide usage (OU) patterns (Fig. 7). They were further pair-wise compared by blast2seq and the proteins encoded by GIs' genes were searched by BLASTp through the local databases of bacterial, plasmid and phage proteins. The latter analysis was performed to check if the GIs that cluster together share syntenic genes and to also deduce the types of genes that are most frequently transferred horizontally across species and genus borders.
In genomes of virulent and environmental *Mycobacterium* multiple genomic islands were identified which share both sequence and OU similarity (Fig. 7). An exception is *M. leprae* which genomic islands were unrelated to GIs of other Mycobacteria (data not shown but check http://anjie.bi.up.ac.za/geidb/geidb-home.php). In Fig. 7 GIs identified in *M. tuberculosis, M. bovis, M. marinum, M. vanbaalenii, M. abscessus* and *M. smegmatis* are represented by white nodes and those of species of other genera by grey nodes. Each node represents one GI tagged by NC accession number of the host organism as in NCBI followed by colons and reference numbers of GIs as in GEI-DB (http://anjie.bi.up.ac.za/geidb/geidb-home.php). Furthermore, six GIs identified in *M. tuberculosis, M. bovis* and *M. marinum* (framed in Fig. 2) share similarity in both DNA sequence and OU with GIs distributed among α-Proteobacteria, particularly to those of *Rhizobium* and *Agrobacterium*.

2.4.2 Stratigraphic analysis of genomic inserts

To determine the relative time of GI insertions, the similarity in OU patterns of GIs and corresponding host chromosomes was calculated for all organisms. The results are depicted by grey gradient colors in Fig. 8. GIs that significantly deviate from their hosts (recent inserts) are shown dark grey; and those that already underwent genomic amelioration (Lawrence & Ochman, 1997) are shown light grey. Most mycobacterial GIs revealed to be ancient inserts that is in consistence with the fact that they are shared by different species. Few of the GIs that showed to be in possession of OU patterns similar to GIs of *Rhizobium* and *Agrobacterium* are relatively recent acquisitions. Comparison of the patterns of the GIs and host genomes was revised in order to determine donor-recipient relationships between these organisms (Fig. 9). The analysis revealed that these mycobacterial GIs are compositionally more similar to the chromosomes and mobilomes of *Agrobacterium* and that they are most likely originated from this source as indicated in Fig. 9.
43 Mycobacterial GIs (unframed in Fig. 2) contain 910 annotated genes among which 386 were hypothetical or unknown. Functional genes are listed in Table 5. Predominance of phage related genes suggests that these GIs are mostly prophages. Genes that are harboured by the GIs of the α-Proteobacteria origin (framed in Fig. 2) encode several transferases, esterases, mmcH proteins and hypothetical proteins organized into operon structures (Fig. 10), which may be involved in the biosynthesis of some yet unknown compounds. Shaded areas in Fig. 10 link regions sharing DNA sequence similarity determined by blast2seq. The compared genomes are NC_000962 (*M. tuberculosis* H37Rv); NC_002755 (*M. tuberculosis* CDC1551); NC_008769 (*M. bovis* BCG str. Pasteur 1173P2); and NC_010612 (*M. marinum* M). Lengths of GIs are also indicated.

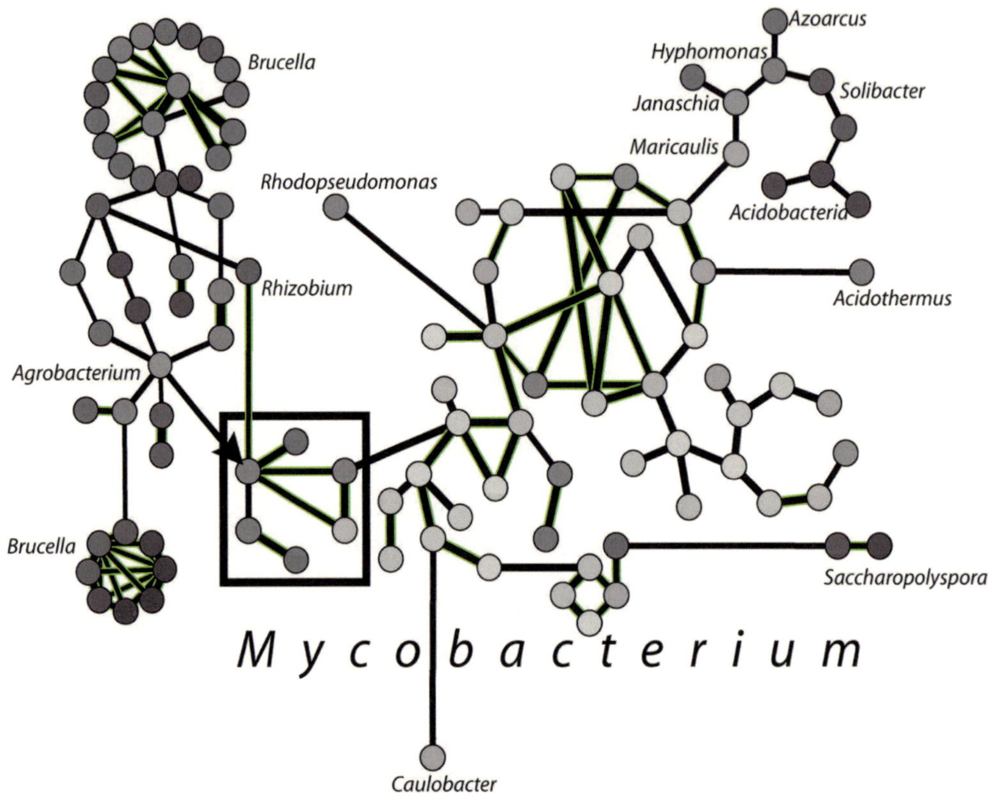

Fig. 8. Stratigraphic analysis of GIs. The edges depicted by green halo link GIs sharing similar DNA sequences longer than 100 bp identified by blast2seq. The layout of nodes is the same as in Fig. 7.

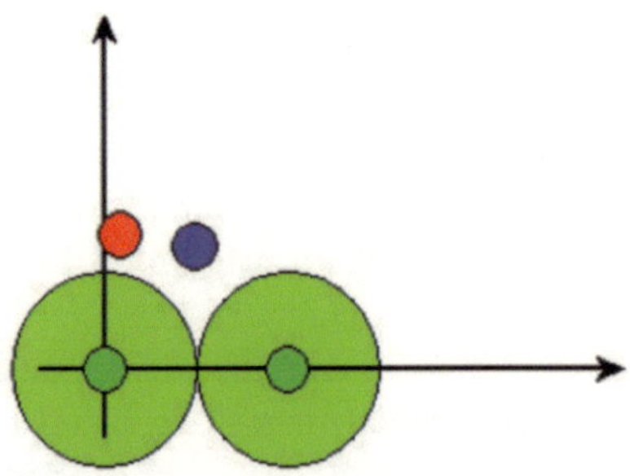

Fig. 9. Donor-recipient relationships between GIs and host organisms of *Agrobacterium* and *Mycobacterium*. Dark green circles indicate OU patterns of the host organisms. Light green shaded areas represent half-distances between chromosomal OU patterns. OU patterns of genomic islands of *M. tuberculosis* NC_002755 and *A. tumefaciens* NC_003062 (blue and red circles respectively) were plotted according to the calculated distances between them and OU patterns of the chromosomes. Plotting was done by an in-house Python program.

Gene categories	Number of genes
Phage related proteins, integrases and transposases	91
Dehydrogenases	31
Transcriptional regulator	23
Peptide synthetase and polyketide	13
Membrane proteins	23
Monooxygenase	11
Glycosyl transferases	11
Oxidoreductase	10
Dioxygenase	9
PE-PGRS proteins	7
Esterases	5

Table 5. Proteins encoded by genes in ancient GIs of *Mycobacterium*.

Protein BLAST analysis of Mycobacterial GIs retrieved similarities in proteins shared with a great variety of bacterial plasmids and phages, particularly in the plasmid pSOL1 from *Clostridium acetobutylicum* ATCC 824. Acquisition of genetic materials from intracellular parasitic and symbiotic species of α-Proteobacteria by an ancestral strain of *Mycobacterium* may be an event that had triggered the evolution of former saprophytic organisms towards the parasitic lifestyle.

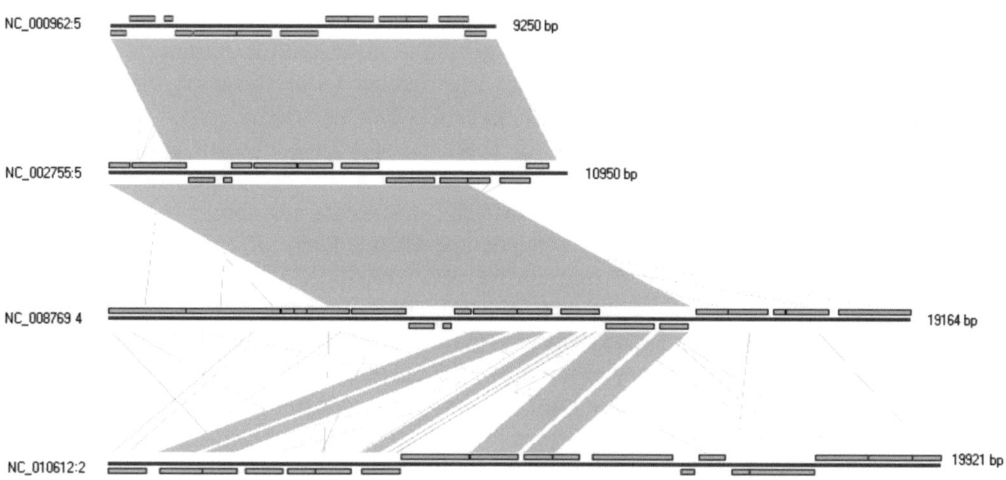

Fig. 10. Homologous genes and operons in GIs shared by *Mycobacterium*. GIs are referred by NC number of the host organism in the NCBI database followed by the reference number of GIs in GEI-DB.

2.4.3 Overview of the horizontal gene transfer in the bacterial world

The exchange of genetic material was found to have occurred in different domains of life: Archaea, Bacteria, and Eukarya (Choi and Kim, 2007). Horizontal gene transfer, defined as a mechanism that promotes the transfer of foreign genomic segments between lineages was found to be relatively common in prokaryotes and less common in higher-order organisms. The transfer of operational genes is a continual process and is far more important in prokaryotic diversity of different sources (Jain et al., 1999; Ochman, 2000). For horizontal gene transfer to become a success, the acquisition of foreign DNA segments must be counterbalanced by DNA loss. Acquired DNA providing functions that are beneficial to the host may be maintained, while DNA providing less beneficial functions may be lost (Lawrence, 1999). Mobile genetic elements possess genes that contribute to bacterial speciation and adaptation to different niches, but also carry with them factors that contribute to the bacteria's fitness traits, secondary metabolism, antibiotic resistance and symbiotic interactions (Dobrindt et al., 2004; Mantri & Williams, 2004) that are of medical and agricultural importance.

The transfer of GIs occurs through three mechanisms: transformation, conjugation and transduction. These mechanisms mediate the movement and transfer of DNA segments intercellularly. Conjugation and transduction are the common players in genetic transfer. They require mobile elements such as plasmids and bacteriophages to transfer genetic elements along with the sequence features of their donor to recipient cells (Hacker & Carniel, 2001). Upon transfer, these genetic elements get established into the recipient cell either as self replicating elements or by getting integrated into the chromosome either by homologous or illegitimate recombination techniques (Dutta and Pan, 2002; Beiko et al., 2005). Transformation, unlike conjugation and transduction does not require any form of a vector to transport genomic elements between bacteria. It is mediated by the uptake of a naked DNA in the environment. The uptake usually takes place upon the release of DNA from decomposing and disrupted cell, or viral particles, or even excretions from living cells (Thomas & Nielsen, 2005).

DNA composition comparisons between lineages have uncovered that genes acquired by the above mechanisms display features that are distinct from those of their recipient genomes (Hacker and Carniel, 2001; van Passel et al., 2006). Genes acquired by horizontal transfer can often display atypical sequence characteristics and a restricted phylogenetic distribution among related strains, thereby producing a scattered phylogenetic distribution (Ochman et al., 2000; Dutta and Pan, 2002). Bacterial species are variable in their overall GC content but the genes in genomes of particular species are fairly uniform with respect to their base composition patterns and frequencies of oligonucleotides (Ochman et al., 2000). The phylogenetic aspect of similarity in base composition among closely related species arises from their common origin. Similarity is also influenced by genome specific mutational pressures that act upon their genes to promote the maintenance of composition stability. Native or core genes in a given organism exhibit homogeneous OU content and codon usage, while foreign genes display atypical characteristic features shared with their mobilomes (phages and conjugative plasmids) or previous host organisms for the genetic segments which were mobilized and integrated by mobilomes (Davenport et al., 2009). Compositional specificity of GIs allows their precise identification by the SWGIS program (see above in this chapter). In this work SWGIS was used to search each prokaryotic genome for foreign inserts based on the comparisons of tetranucleotide usage patterns, whereby the frequencies of particular tetramers are compared with expected occurrences of the same tetramers throughout the whole genome. Identified GIs were stored to GEI-DB (http://anjie.bi.up.ac.za/geidb/geidb-home.php) that contains a set of 3518 precalculated GIs identified in 637 prokaryotic genomes. All these GIs were clustered by the compositional OU pattern similarity that is believed to represent their common ancestry. Similarity between GIs was calculated as 100 - D(%), where D(%) was found by the equation 4. GIs which share more than 75% of similarity were grouped together. Groups of GIs and their distribution among bacteria are shown in Fig. 11.

GIs were identified in all bacterial classes. There are more GIs from *E. coli* and other Enterobacteria and γ-Proteobacteria that partially may be explained by a biased overrepresentation of these microorganisms among other sequenced genomes in the GenBank database. *E. coli*, *Shigella* and *Salmonella* share GIs of one common origin but GIs found in other species often showed to have originated from several different origins. For example, GIs from *Pseudomonas* form several separate clusters associated with either other γ-Proteobacteria or α-Proteobacteria. GIs of α-Proteobacteria and Firmicutes show extreme diversity. *Brucella*, *Agrobacterium* and *Rhizobium* share several unrelated pools of their mobilomes. Relations which were found between GIs of *Mycobacterium* and those of *Agrobacterium* and *Rhizobium* have been discussed above in detail. GIs of *Prochlorococcus* and *Nostoc* cyanobacteria most likely originated from marine γ-Proteobacteria, but GIs of *Synechococcus* are very specific and share no similarity with any other microorganisms.

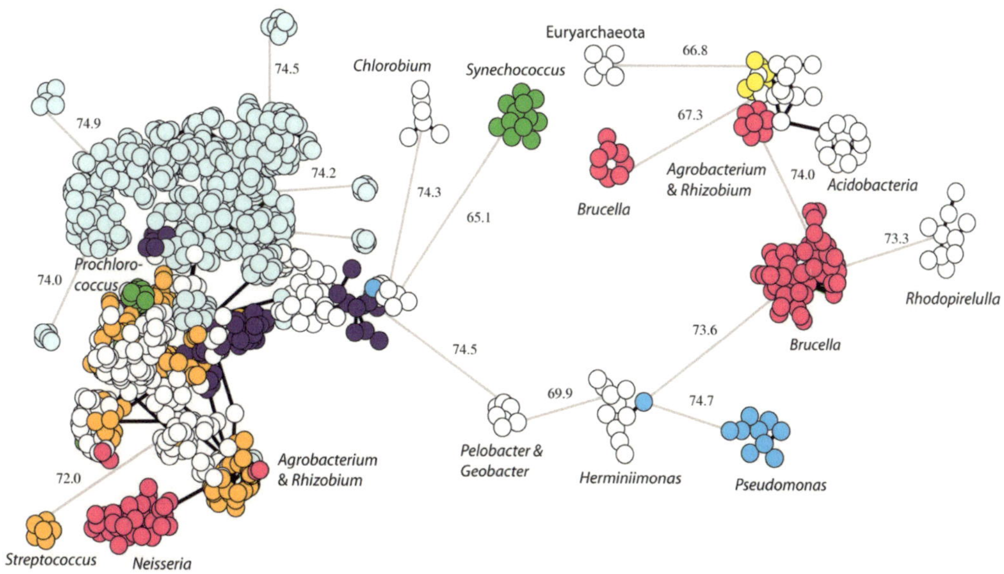

Fig. 11. Groups of GIs joined by compositional OU pattern similarity. Each node represents one GI. Genera of γ-Proteobacteria are shown in light blue (enterobacteria *Escherichia*, *Shigella* and *Salmonella*), cyan (*Pseudomonas*) and dark blue (marine bacteria *Shewanella*, *Hahella*, *Pseudoalteromonas* and *Alcanivorax*); α-Proteobacteria *Agrobacterium*, *Brucella*, *Neisseria*, *Rhizobium* and *Synorhizobium* are depicted by magenta nodes; Firmicutes (*Bacillus*, *Clostridium*, *Geobacillus* and *Streptococcus*) - orange; Actinobacteria (*Corynebacterium* and *Mycobacterium*) - yellow; Cyanobacteria (*Prochlorococcus*, *Nostoc* and *Synechococcus*) - green. Nodes representing other organisms are white. Black edges link nodes which share strong OU similarity above 75% and grey edges represent weaker similarity below 75%.

It may be concluded that GIs indeed may flux through the bacterial taxonomic walls but not in a random fashion. Several species and genera share pools of horizontally transferred genetic elements, which include pathogenicity, antibiotic resistance, O-antigen synthesis and catabolic GIs, whereas the genetic exchange between other groups of microorganisms is seemed very unlikely. Detailed analysis of gene exchange pathways among microorganisms will shed light on the roles played by the horizontal gene transfer in the evolution and pathogenicity of bacteria.

26732 proteins encoded by GIs' genes used in this study were pair-wise compared by BLASTp. The bit-score results were used to produce clusters of proteins by Markov clustering algorithm (MCL) (Vlasblom & Wodak, 2009). MCL with an inflation parameter of 1.8 produced 10837 clusters, however, many of them were of a single hypothetical protein. Due to the large amount of hypothetical and unknown genes in the database not all of these clusters would present biologically significant data. Top 24 clusters containing more than 50 proteins were chosen as significant to represent categories of proteins which are most often mobilized and transferred horizontally among bacteria. Besides phage related proteins which are in a majority, the most frequently bacteria acquire ABC-transporters, transcriptional regulators including GGDEF diguanylates, polysaccharide and O-antigen biosynthesis proteins, dehydrogenases and outer membrane proteins (Table 6).

Functional group	Nr of proteins identified in 3518 GIs
Phage related proteins, IS-elements, transposases;	792
Transcriptional regulators;	599
Polysaccharide and O-antigen biosynthesis proteins;	352
ABC-transporter;	252
Outer membrane proteins;	241
Dehydrogenases;	67
RHS-family proteins;	64

Table 6. Predominant categories of horizontally transferred proteins.

3. Conclusion

Comparative genomics exploits the methods of two major categories based on the analysis of composition and sequence similarity. Having been developed at the beginning of the genomics era, sequence similarity comparison by BLAST (Altschul, 1990) and FASTA (Pearson, 1995), and sequence composition simulation by Markov Chain Models (Schbath, 2000) remain the algorithms of first choice. The algorithms for sequence similarity comparison are widely used because of speed, more straightforward statistics and a clearer biological relevance of sequence alignment based considerations. However, a number of practical tools based on OU statistics have become publicly available. Several novel OU analytical tools of the SeqWord project for genome visualization, genomic island detection and identification of unknown sequences have been presented in this chapter.

Composition based methods are termed genome linguistics as they deal with frequencies of words written as chains of given alphabets of nucleotides or amino acids of variable lengths. Genome linguistic approaches may complement or even outperform the sequence similarity comparison in clustering DNA reads (Kislyuk et al., 2009) and detecting inserts of genomic islands (Hsiao et al., 2003). These approaches have also been shown to be instrumental in viral metagenomics (Delwart, 2007). During composition based analysis, longer DNA sequences are rather preferred to shorter ones for the word distribution statistics to be reliable.

DNA similarity vanishes much faster in phylogenetically distant organisms than the OU composition does, especially in highly variable virus, phage, plasmids and genomic islands. Protein similarity may mislead binning or identification of unknown sequences for it mostly reflects the functional conservation of protein domains rather than the taxonomic unity. Another common limitation of the similarity based methods is that the sequence identification is possible only if a homologous DNA or protein sequence is present in the searched database. On the contrary, the genome specific OU pattern is a pervasive property of the whole genome (Jernigan & Baran, 2002) that allows binning of DNA reads to their putative origin even if they do not share any significant sequence similarity.

The advancement in genome sequencing technologies made large scale sequencing affordable for many laboratories. An attractive approach of alignment-independent phylogenetic studies based on the comparison of OU patterns was discussed in several publications and a number of web-based services were proposed (Chapus et al., 2005). We suggest rather a cautious use of these methods as a significant convergence of OU patterns was observed between unrelated organisms. For instance, *Pseudomonas* and *Mycobacterium* share similar OU patterns. Furthermore, a wider application of OU patterns is hindered by the absence of any noteworthy mathematical models simulating the evolutionary changes in OU patterns between organisms in contrast to sequence similarity methods which provide plenty of models of nucleotide and amino acid substitutions. Development and testing of such models is the task that urgently needs to be looked into to advance applicability of genome linguistic approaches.

4. Acknowledgment

Funding for this research was provided by the MetaLingvo grant 71261 of the National Research Foundation (NRF) of South Africa.

5. References

Abe, T., Kanaya, S., Kinouchi, M., Ichiba, Y., Kozuki, T. & Ikemura, T. (2003). Informatics for unveiling hidden genome signatures. *Genome Res.*, Vol. 13, No. 4, pp. 693-702.

Altschul, S. F., Gish, W., Miller, W., Myers, E. W. & Lipman, D. J. (1990). Basic local alignment search tool. *J. Mol. Biol.*, Vol. 215, No. 3, pp. 403–410.

Azad, R. K. & Lawrence, J. G. (2005). Use of artificial genomes in assessing methods for atypical gene detection. *PLoS Comput. Biol.*, Vol. 1, No. 6, p. e56.

Baldi, P. & Baisnée, P.-F. (2000). Sequence analysis by additive scales: DNA structure for sequences and repeats of all lengths. *Bioinformatics*, Vol. 16, No. 10, pp. 865-889.

Becq, J., Gutierrez, M. C., Rosas-Magallanes, V., Rauzier, J., Gicquel, B., Neyrolles, O. & Deschavanne, P. (2007). Contribution of horizontally acquired genomic islands to the evolution of tubercle bacilli. *Mol. Biol. Evol.* 2007, Vol. 24, No. 8, pp. 1861-1871.

Beiko, R. G., Harlow, T. J. & Ragan, M. A. (2005). Highways of gene sharing in prokaryotes. *Proc. Natl. Acad. Sci. USA*, Vol. 102, No. 40, pp. 14332–14337.

Bohlin, J., Skjerve, E. & Ussery, D. (2008). Reliability and applications of statistical methods based on oligonucleotide frequencies in bacterial and archaeal genomes. *BMC Genomics*, Vol. 9, p. 104.

Chatterji, S., Yamazaki, I., Bai, Z. & Eisen J. A. (2008). CompostBin: a DNA composition based algorithm for binning environmental shortgun reads, In: *RECOMB*, Vingron, M. & Wong, L., pp. 17-28, LNBI 4955.

Coenye, T. & Vandamme, P. (2004). Use of the genomic signatures in bacterial classification and identification. *System. Appl. Microbiol.*, Vol. 27, No. 2, pp. 175-185.

Chapus, C., Dufraigne, C., Edwards, S., Giron, A., Fertil, B. & Deschavanne, P. (2005). Exploration of phylogenetic data using a global sequence analysis method. Vol. 9, No. 5, p. 63.

Choi, I.-G. & Kim, S.-H. (2007). Global extent of horizontal gene transfer. *Proc. Natl. Acad. Sci. USA* Vol. 104, No. 11, pp. 4489–4494.

Davenport, C. F., Wiehlmann, L., Reva, O. N. & Tümmler, B. (2009). Visualization of *Pseudomonas* genomic structure by abundant 8-14mer oligonucleotides. *Environ. Microbiol.*, Vol 11, No. 5, pp. 1092-1104.

Delwart, E. L. (2007). Viral metagenomics. *Rev. Med. Virol.* Vol. 17, No. 2, pp. 115-131.

Deschavanne, P. J., Giron, A., Vilain, J., Fagot, G. & Fertil, B. (1999). Genomic signature: characterization and classification of species assessed by chaos game representation of sequences. *Mol. Biol. Evol.*, Vol. 16, No. 10, pp. 1391-1399.

Dobrindt, U., Hochhut, B., Hentschel, U. & Hacker, J. (2004). Genomic islands in pathogenic and environmental microorganisms *Nat. Rev. Microbiol.* Vol. 2, No. 5, pp. 414–424.

Dufraigne, C., Fertil, B., Lespinats, S., Giron, A. & Deschavanne, P. (2005). Detection and characterization of horizontal transfers in prokaryotes using genomic signature. *Nucleic. Acids Res.*, Vol. 33, No. 1, p. e6.

Dutta, C. & Pan, A. (2002). Horizontal gene transfer and bacterial diversity. *J. Biosci.* Vol. 27, No. 1 Suppl. 1, pp. 27-33.

Hacker, J. & Carniel, E. (2001). Ecological fitness, genomic islands and bacterial pathogenicity. A Darwinian view of the evolution of microbes. *EMBO Rep.*, Vol. 2, No. 5, pp. 376–381.

Hsiao, W., Wan, I., Jones, S. J. & Brinkman, F. S. L. (2003). IslandPath: aiding detection of genomic islands in prokaryotes. *Bioinformatics*, Vol. 19, No. 3, pp. 418-420.

Jain, R., Rivera, M. C. & Lake, J. A. (1999). Horizontal gene transfer among genomes: the complexity hypothesis. *Proc. Natl. Acad. Sci. USA*, Vol. 96, No. 7, pp. 3801–3806.

Jernigan, R. W. & Baran, R. H. (2002). Pervasive properties of the genomic signature. *BMC Genomics*, Vol. 3, No. 1, p. 23.

Karlin, S. & Burge, C. (1995). Dinucleotide relative abundance extremes: a genomic signature, *Trends Genet.*, Vol. 11, No. 7, pp. 283-290.

Karlin, S. (1998). Global dinucleotide signatures and analysis of genomic heterogeneity. *Curr. Opin. Microbiol.*, Vol. 1, No. 5, pp. 598-610.

Karlin, S., Mrázek, J. & Campbell, A. (1997). Compositional biases of bacterial genomes and evolutionary implications. *J. Bacteriol.*, Vol. 179, No. 12, pp. 3899-3913.

Kirzhner, V., Bolshoy, A., Volkovich, Z., Korol A. & Nevo E. (2005). Large-scale genome clustering across life based on a linguistic approach. *BioSystems*, Vol. 81, No. 3, pp. 208-222.

Kislyuk, A., Bhatnagar, S., Dushoff, J. & Weitz J. S. (2009). Unsupervised statistical clustering of environmental shotgun sequences. *BMC Bioinformatics*, Vol. 10, p. 316.

Koski, L. B., Morton, R. A. & Golding, G. B. (2001). Codon bias and base composition are poor indicators of horizontally transferred genes. *Mol. Biol. Evol.*, Vol. 18, No. 3, pp. 404-412.

Lawrence, J. G. & Ochman H. (1997). Amelioration of bacterial genomes: rates of change and exchange. *J. Mol. Evol.*, Vol. 44, No. 4, pp. 383-397.

Lawrence, J. G. (1999). Gene transfer, speciation, and the evolution of bacterial genomes. *Curr. Opin. Microbiol.* Vol. 2, No. 5, pp. 519–523.

Mantri, Y. & Williams, K. P. (2004). Islander: a database of integrative islands in prokaryotic genomes, the associated integrases and their DNA site specificities. *Nucleic Acids Res.* Vol. 32, Database issue, pp. D55–D58.

Mrázek, J. & Karlin, S. (1999). Detecting alien genes in bacterial genomes. *Ann. NY Acad. Sci.*, Vol. 870, pp. 314-329.

Nakamura, Y., Itoh, T., Matsuda, H. & Gojobori, T. (2004). Biased biological functions of horizontally transferred genes in prokaryotic genomes. *Nat. Genet. Vol.* 36, No. 7, pp. 760-766.

Ochman, H., Lawrence, J. G. & Groisman, E. A. (2000). Lateral gene transfer and the nature of bacterial innovation. *Nature*, Vol. 405, No. 6784, pp. 299–304.

Pearson, W. R. (1995). Comparison of methods for searching protein sequence databases. *Protein Sci.*, Vol. 4, No. 6, pp. 1145-1160.

Pride, D. T. & Blaser, M. J. (2002). Identification of horizontally acquired elements in *Helicobacter pylori* and other prokaryotes using oligonucleotide difference analysis. *Genome Let.*, Vol. 1, No. 1, pp. 2–15.

Pride, D. T., Meinersmann, R. J., Wassenaar, T. M. & Blaser, M. J. (2003). Evolutionary implications of microbial genome tetanucleotide frequency biases. *Genome Res.*, Vol. 13, No. 2, pp. 145–155.

Reva, O. N. & Tümmler, B. (2004). Global features of sequences of bacterial chromosomes, plasmids and phages revealed by analysis of oligonucleotide usage patterns. *BMC Bioinformatics*, Vol. 5, p. 90.

Richter, D. C., Ott, F., Auch, A. F., Schmid, R. & Huson D. H. (2008). MetaSim: a sequencing simulator for genomics and metagenomics. *PLoS One*, Vol. 3, No. 10, p. e3373.

Saeed, I. & Halgamuge, K. (2009). The oligonucleotide frequency derived error gradient and its application to the binning of metagenome fragments. *BMC Genomics*, Vol. 10, Suppl. 3, p. S10.

Schbath, S. (2000). An overview on the distribution of word counts in Markov chains. *J. Comp. Biol.*, Vol. 7, No. ½, pp. 193-201.

Teeling, H., Meyerdierks, A., Bauer, M., Amann, R. & Glöckner, F. O. (2004). Application of tetranucleotide frequencies for the assignment of genomic fragments. *Environ. Microbiol.*, Vol. 6, No. 9, pp. 938-947.

Thomas, C. M. & Nielsen, K. M. (2005). Mechanisms of, and barriers to, horizontal gene transfer between bacteria. *Nat. Rev. Microbiol.*, Vol. 3, No. 9, pp. 711–721.

van Passel, M. W., Bart, A., Luyf, A. C., van Kampen, A. H. & van der Ende, A. (2006). The reach of the genome signature in prokaryotes. *BMC Evol. Biol.*, Vol. 6, p. 84.

Vlasblom, J. & Wodak, S. J. (2009). Markov clustering versus affinity propagation for the partitioning of protein interaction graphs. *BMC Bioinformatics*, Vol. 10, No. 99, p. 1.

Wang, B. (2001). Limitations of compositional approach to identify horizontally transferred genes. *J. Mol. Evol.*, Vol. 53, No. 3, pp. 244-250.

Weinel, C., Ussery, D. W., Ohlsson, H., Sicheritz-Ponten, T., Kiewitz, C. & Tümmler, B. (2002). Comparative genomics of *Pseudomonas aeruginosa* PAO1 and *Pseudomonas putida* KT2440: orthologs, codon usage, repetitive extragenic palindromic elements, and oligonucleotide motif signatures. *Genome Lett.*, Vol. 1, No. 4, pp. 175-187.

3

On the Structural Characteristics of the Protein Active Sites and Their Relation to Thermal Fluctuations

Shao-Wei Huang and Jenn-Kang Hwang
Institute of Bioinformatics and Systems Biology, National Chiao Tung University, HsinChu, Taiwan, R. O. C.

1. Introduction

Due to the advances in structural biology research, a large number of protein structures have been solved in the last decade. In the same time, we also witness a rapidly growing number of structures of *unknown function* being deposited in the PDB. As a result, the ability to predict protein function from its structure becomes increasingly important in computational biology.

The conventional comparative methods(for example, Laskowski, Watson et al. 2005; Watson, Sanderson et al. 2007) for identifying functional sites rely on evolutional information like homologous structures of known function or the known catalytic templates. However, these approaches are not applicable to these novel structures. One needs to develop ingenious approaches that do not rely on evolutionary information.

Recently, several groups(Amitai, Shemesh et al. 2004; Ben-Shimon & Eisenstein 2005; Sacquin-Mora, Laforet et al. 2007; Huang, Yu et al. 2011) developed novel approaches to predict the active sites of enzymes *from a single structure* without using any homologous structures or known catalytic templates. The basic idea of their approaches is simple: they first identify certain structural or dynamical features that are unique to the active sites; they then further refine this relationship such that it can be used to accurately predict the enzyme catalytic sites. For example, Pietrokovski and co-workers(Amitai, Shemesh et al. 2004) transformed the protein structure into residue interaction graphs, with each amino acid residue represented as a graph node and the interaction between them as a graph edge. They then computed the network closeness of each residue. They found that most catalytic residues are associated with the network centrality. Ben-Shimon and Eisenstein(Ben-Shimon & Eisenstein 2005), analyzing 175 enzymes, observed that most catalytic residues are near the enzyme centroid. Based on these results, they developed novel methods to predict catalytic sites from a single structure.

The aim of this review is to show that this peculiar relationship between catalytic sites and the structure centroid or its network centrality can be accounted for by the dynamical properties of the catalytic residues, which are in general more *rigid* than other residues. In addition, we will discuss the recent studies (Halle 2002; Shih, Huang et al. 2007; Huang, Shih et al. 2008; Lin, Huang et al. 2008; Lu, Huang et al. 2008) on the surprisingly close link between protein structure and its thermal fluctuations. These studies showed that the atomic thermal fluctuations and motional correlations can be extracted directly from protein structures without using any mechanical models.

The outline of this review is as follows: first, we will cover the recent works on extracting average dynamical properties directly from protein structures. This part occupies a significant portion of the text, since it provides not only theoretical foundations for but also physical insights on what we will introduce later - the prediction of the active sites from a single structure, which is a straightforward application of the results discussed in the first part. Finally, we will show that the rigidity of the active site residues can be inferred from the generally accepted theory of the mechanism of enzyme catalysis.

2. Atomic thermal fluctuations and motional correlations in protein

The motional correlation between atom i and j is given by

$$C_{ij} \sim \left\langle \delta \mathbf{r}_i \cdot \delta \mathbf{r}_j \right\rangle \tag{1}$$

where $\delta \mathbf{r}_i = \mathbf{r}_i - \langle \mathbf{r}_i \rangle$ is the displacement of the instantaneous position $\mathbf{r}_i$ of the atom i from its equilibrium position $\langle \mathbf{r}_i \rangle$. Knowledge of the motional correlations in protein provides valuable information about the relationship between protein dynamics and its function. For example, dynamical correlation networks may account for the long-range effects of faraway mutation sites on the functional site (Saen-Oon, Ghanem et al. 2008; Ishida 2010) or protein allostery(Fidelak, Ferrer et al. 2010; Amaro, Sethi et al. 2007; Tsai, del Sol et al. 2008). The diagonal terms or the auto-correlation terms in Eq. 1 describe atomic fluctuations. They can be obtained as B-factors from X-ray crystallography refinement or order parameters from NMR. The B-factor in its isotropic form is given formally as $B = (8\pi^2 / 3)\langle \delta \mathbf{r} \cdot \delta \mathbf{r} \rangle$.

2.1 Normal mode analysis

To compute the correlation matrix, one needs to evaluate the second derivatives of the potential energy of the protein structure. In molecular mechanics, the protein structure is usually modelled by analytical potential functions(Warshel 2002). In general, the bonding energy is approximated by a harmonic function, i.e., $\frac{1}{2}K_b(b-b_0)^2$, where K_b is the force constant, b and b_0 are the bond length and the equilibrium bond length, respectively. The bending interaction is approximated by $\frac{1}{2}K_\theta(\theta-\theta_0)^2$, where K_θ is the bending force constant, θ and θ_0 are the bending angle and the equilibrium bending angle, respectively. The torsional interaction is modelled by a sinusoidal functions: $K_\phi[1-\cos(n\phi+\delta)]^2$, where K_ϕ is the torsional force constant, ϕ and δ are the torsional angle and the reference torsional angles, respectively, and n is the torsional periodicity. The non-bonded van der Waals (VDW) interaction is approximated by the Lennard-Jones function (also referred to as a 6-12 function): $\varepsilon(r_0/r)^{12} - 2\varepsilon(r_0/r)^6$, where r is the distance between atoms, and ε and r_0 are the VDW parameters related to the potential depth and its minimum distance. The electrostatic interaction is described by the Coulomb equation: $332q_iq_j/r$, where q_i and q_j are the charges of the atom pair and r their separation. The complete potential function is usually referred to as a *force field*:

$$U = \sum_{\text{All Bonds}} \tfrac{1}{2}K_b(b-b_0)^2 + \sum_{\text{All Angles}} \tfrac{1}{2}K_\theta(\theta-\theta_0)^2 + \sum_{\text{All Torsional Angles}} K_\phi\left[1-\cos(n\phi+\delta)\right]$$
$$+ \sum_{\text{All nonbonded pairs}} \varepsilon\left[(r_0/r)^{12} - 2(r_0/r)^6\right] + \sum_{\text{All partial charges}} 332 q_i q_j / r \quad (2)$$

The Hessain matrix **H** is the square matrix of the second derivatives of the potential energy $\partial U / \partial x_i \partial x_j$, where x_i and x_j are the Cartesian coordinates of the atoms of the protein structure. Diagonalization of the Hessian matrix, i.e., $\mathbf{U}^{-1}\mathbf{H}\mathbf{U} = \mathbf{L}$, gives the normal mode frequencies **L** and normal mode vectors **U**, which describe harmonic vibrational motions of the structure. The overall molecular motion can be described as a linear combination of these normal modes. These frequencies and vectors define different modes of harmonic motions occurring in a protein structure. The above procedure is referred to as normal mode analysis (NMA) (Brooks & Karplus 1983; Levitt, Sander et al. 1985; Go 1990; Ma 2004). The correlations between atomic fluctuations given by Eq. 1 can be calculated from the normal modes. Mathematically, the correlation matrix is the pseudo-inverse of the Hessian matrix.

One limitation of NMA is its assumption that motions are harmonic. This may not be valid in the case of large-amplitude conformational dynamics in protein which are presumably highly anharmonic(Ma 2005). Furthermore, NMA needs to be carried out in an energy-minimized structure. This is to avoid the occurrence of the unphysical complex-valued normal mode frequencies. However, minimization may cause significant conformational deformations due to the inherent deficiencies of the force field.

2.2 Elastic network model

Elastic network model (ENM) (Tirion 1996), a simplified version of NMA, can be directly applied to the protein structures without energy minimization. In ENM, every atom is connected to any atoms (except itself, of course) as long as they are within a certain threshold distance (usually 12 Å) of the given atom. ENM does not distinguish between bonded interactions (such as bonding, bending or torsional interactions) and nonbonded interactions (such as VDW or electrostatic interactions). All interactions are represented by a harmonic function with a uniform force constant. Since all interactions in ENM are assumed to be of covalent nature, i.e., the atom pairs are connected by a harmonic force, this is equivalent to assuming a relatively rigid protein structure. We will comment more on that later.

A number of ENM variants has been developed: the Gaussian Network Model (GNM) (Bahar, Atilgan et al. 1997), Anisotropic Network Model (ANM) (Atilgan, Durell et al. 2001) and Quantized Elastic Deformational Model (Ming, Kong et al. 2002). Despite the simplicity of ENM, it predicts relatively accurate correlated motions in proteins. ENM has recently become a popular tool to analyze protein dynamics (Zheng, Brooks et al. 2007; Yang, Song et al. 2009; Zheng & Thirumalai 2009).

Recently, Hwang and co-workers (Shih, Huang et al. 2007; Huang, Shih et al. 2008; Lin, Huang et al. 2008; Lu, Huang et al. 2008) developed even simpler models to calculate the correlations between the atomic fluctuations in proteins directly from their structures without using mechanical models, or performing energy minimization or matrix diagonalization. They are the Protein Fixed-Point Model and the Weighted Contact Number Model, which will be discussed in the next sections.

2.3 The protein fixed-point model

In the Protein Fixed-Point (PFP) model, the protein structure is characterized by the PFP profile $\mathbf{R}$

$$\mathbf{R} = (\mathbf{r}_1 - \mathbf{r}_0, \mathbf{r}_2 - \mathbf{r}_0, \ldots, \mathbf{r}_N - \mathbf{r}_0) \tag{3}$$

where $\mathbf{r}_i$ is the coordinate of Cα atom of the ith residue, $\mathbf{r}_0$ is the fixed point and N is the number of residues. The fixed point $\mathbf{r}_0$ is identified with the centroid of the protein chain, $\mathbf{r}_0 = \sum_i \mathbf{r}_i / N$. Hwang and co-workers (Shih, Huang et al. 2007; Lu, Huang et al. 2008) showed that the correlation matrix $\mathbf{C}$ is well approximated by $\mathbf{R}^T\mathbf{R}$, where $\mathbf{R}^T$ is the transpose of $\mathbf{R}$, i.e.,

$$C_{ij} = \Delta\mathbf{r}_i \cdot \Delta\mathbf{r}_j \tag{4}$$

where $\Delta\mathbf{r}_i = \mathbf{r}_i - \mathbf{r}_0$ and $\Delta\mathbf{r}_j = \mathbf{r}_j - \mathbf{r}_0$. The motional correlation between atoms is the inner product of the vectors radiating from the centroid to the respective atoms. The B-factor, i.e., the diagonal element, is proportional to the square of the distance to the centroid(Kundu, Melton et al. 2002).

In the PFP model, the tricky issue is to determine the fixed points. In the case of a single domain protein, the fixed point is simply the centroid (or the center of mass) of the whole structure. However, in the case of multidomain protein or a protein complex, the fixed point is defined as the centroid of the *structure module*. There may be more than one fixed point since the protein structure may be composed of a number of modules. The structure modules are identified with either the structure domains or the biological units. The coordinates of biological units can be retrieved from either PDB or PQS(Henrick & Thornton 1998). Though the biological units are not uniquely defined, the PDB and PQS biological units agree on 82% of entries(Xu, Canutescu et al. 2006).

The general procedures to determine the structural modules of a protein structures go as follow: each protein chain is checked for its domains through the use of the Protein Domain Parser (PDP)(Alexandrov & Shindyalov 2003); if the PDP domain is not defined, the SCOP database(Murzin, Brenner et al. 1995) will be searched; if not found, the CATH database(Orengo, Michie et al. 1997) will be searched. If the chain is not a multidomain chain, it will be checked whether it is a part of a protein complex or a biological from PDB or PQS.

In Figure 1, we compare the computed PFP profiles with the experimental X-ray B-factors. Both the PFP and the B-factor profiles are expressed in terms of the Z-scores. In the case of the B-factor, the Z-score is defined as: $Z_B = (B - \overline{B}) / \sigma_B$, where $\overline{B}$ and σ_B are the mean and standard deviation of the B-factor. The Z-score of the PFP profile is defined similarly. For a dataset comprising 972 high-resolution X-ray structures with pairwise sequence identity ≤ 25%, the correlation coefficient between the computed and the X-ray B-factors is 0.59. There are 727 out of 972 of proteins (around 75%) having a correlation coefficient ≥ 0.5. In

comparison, GNM yields a correlation coefficient of 0.56 and the fraction of proteins with a correlation coefficient ≥ 0.5 is 69% for the same data set(Lu, Huang et al. 2008).
Figure 2 compares the PFP correlations with the NMA maps computed by GROMAC(Van Der Spoel, Lindahl et al. 2005). The agreements are excellent.

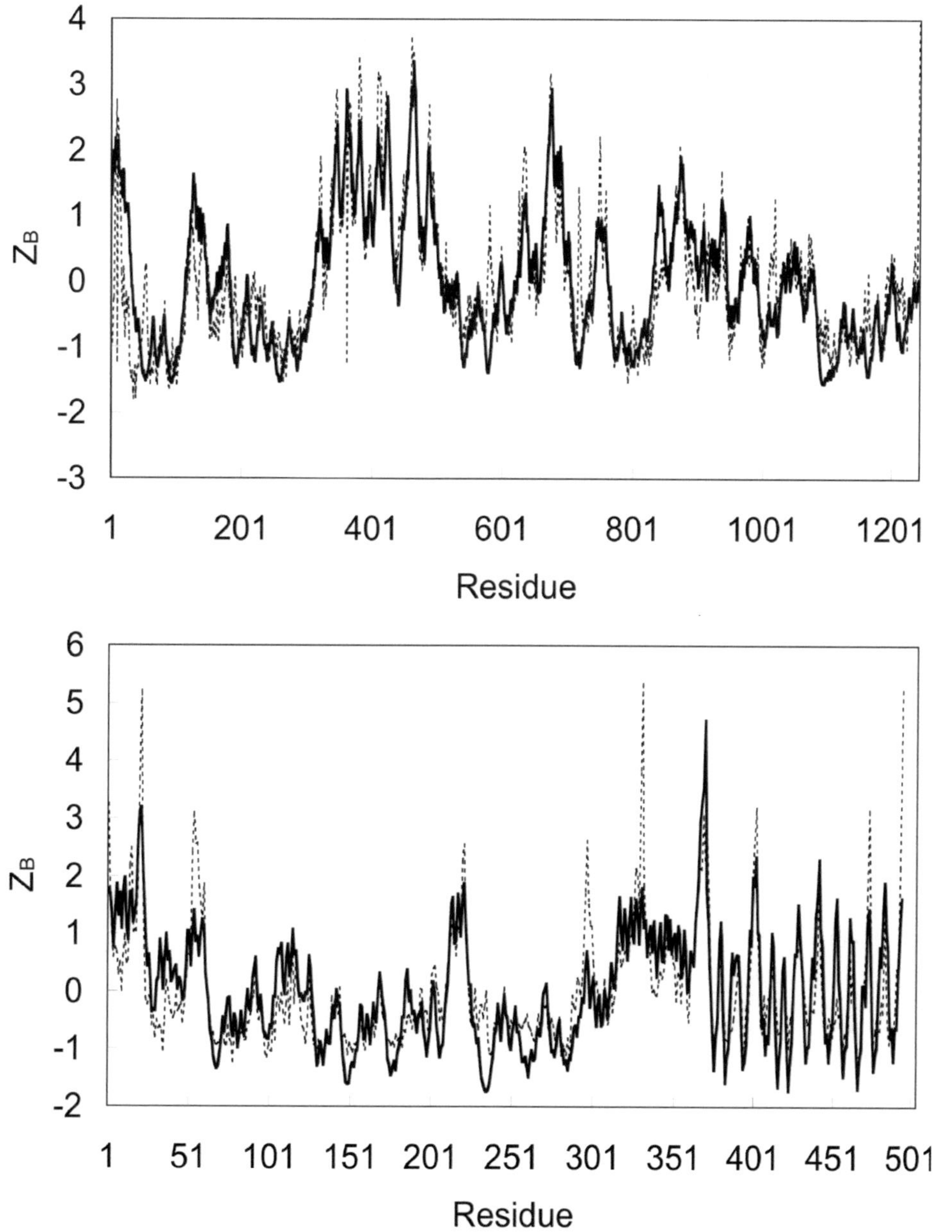

Fig. 1. Comparison of the computed PFP B-factor profile (solid line) and the X-ray B-factor profile (dotted line) of 1q16:A (top) and 2ffu (bottom). The vertical axis Z_b is the normalized B-factor.

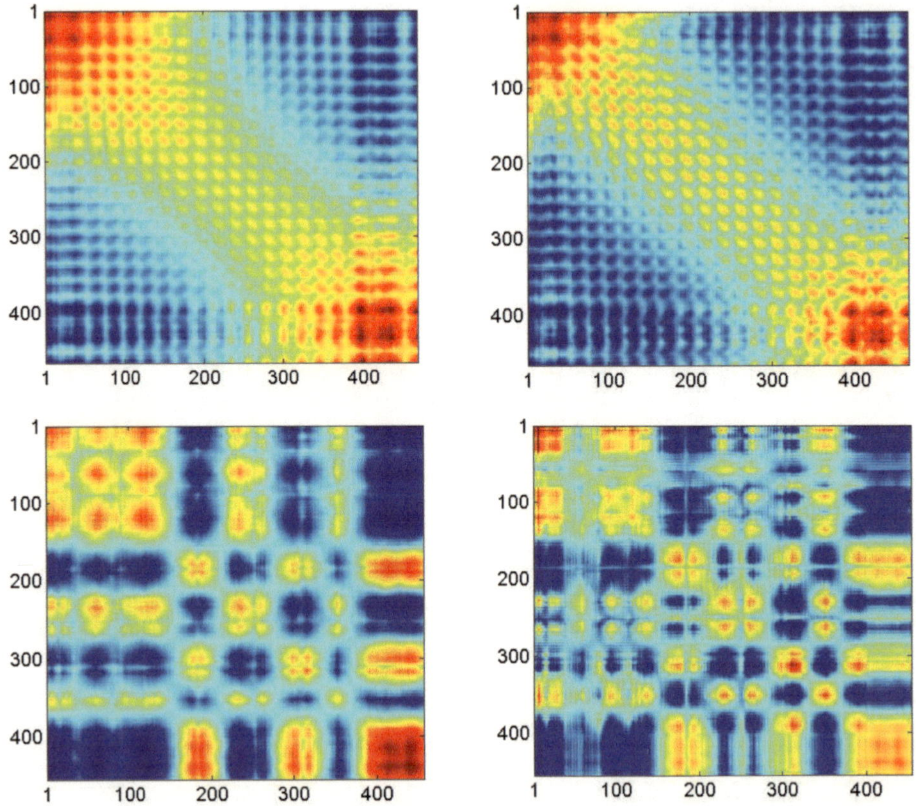

Fig. 2. Comparison of the PFP correlation map (left) and NMA correlation map computed by Gromacs (right) of 1o6v:a (top row) (this figure is adapted from Lu, Huang et al. 2008) and 1fup:a (bottom row). The colors are coded according to the rainbow spectrum. A more reddish color indicates a more negative correlation, while a more bluish color a more positive correlation.

2.4 The weighted contact number methods

The contact number (CN) of the residue i is defined as $n_i = \sum_{j \neq i}^{N} H(r_0 - r_{ij})$, where r_0 is the cutoff distance, r_{ij} is the distance between the Cα atoms of residue i and j, and $H(r)$ is the Heaviside step function defined as: $H(r) = 1$ if $r > 0$ and $H(r) = 0$ otherwise. Despite its popular uses in computational structural biology (Miyazawa & Jernigan 1984; Halle 2002), the CN has one major shortcoming: it treats the contribution *equally* of every contact atom, regardless of its distance to the center atom.

To take distance into consideration, Hwang and co-workers(Lin, Huang et al. 2008) define a weighted contact number (WCN) as

$$\nu_i = \sum_{j \neq i}^{N} \frac{1}{r_{ij}^2} \tag{5}$$

They showed that the reciprocal WCN profiles better reproduce the B-factor profiles and that the correlation between residue i and residue j is well approximated by

$$C_{ij} = \left(\sum_{k}^{N} \frac{1}{r_{ik} r_{jk}} \right)^{-1} \hat{\mathbf{x}}_i \cdot \hat{\mathbf{x}}_j \tag{6}$$

and $\hat{\mathbf{x}}_i$ is given by $(\mathbf{r}_i - \overline{\mathbf{r}})/|\mathbf{r}_i - \overline{\mathbf{r}}|$, where $\overline{\mathbf{r}}$ is $\sum_k^N \mathbf{r}_k / N$, and $\hat{\mathbf{x}}_j$ is defined similarly.

The WCN model may be a little more computationally expensive than the PFP model, but it can be complemented as an automatic procedure. This is in contrast to the PFP model, which requires manual intervention to define the centroids.

Figure 3 compares the computed WCN B-factor profiles with the experimental X-ray B-factors. The WCN model predicts better B-factors than the PFP model, the original CN model (Halle 2002) and the GNM. For the same dataset stated before, the WCN model yields a correlation coefficient of 0.61 with 79% of proteins having a correlation coefficient ≥ 0.5.

Figure 4 compares the WCN and the NMA correlation maps. The agreements are excellent.

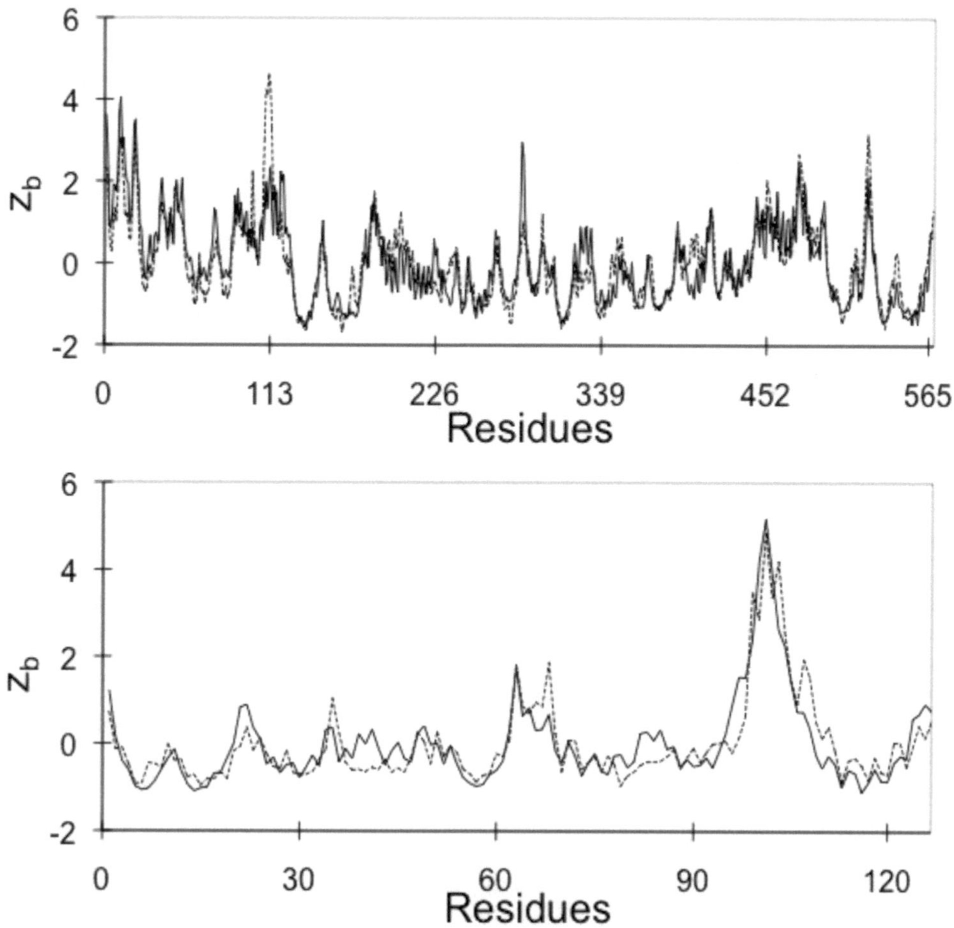

Fig. 3. Comparison of the computed WCN B-factor profile (solid line) and the X-ray B-factor profile (dotted line) of 1y0p:A (top) (this figure is adapted from Lin, Huang et al. 2008) and 1nu0 (bottom).

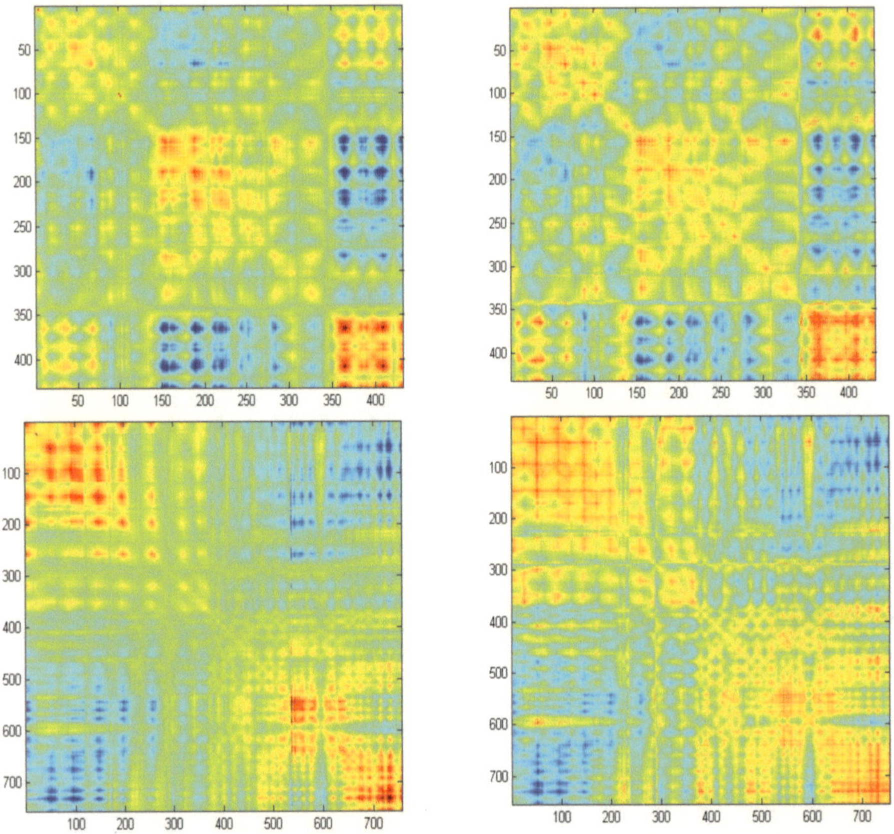

Fig. 4. Comparison of the WCN correlation map (left) and NMA correlation map computed by Gromacs (right) of 1cvr:a (top row) and 1rwh:a (bottom row) (this figure is adapted from Lin, Huang et al. 2008). The colors are coded according to the rainbow spectrum.

2.5 More on the structural and B-factor profiles

It is hardly surprising that an atom's thermal fluctuation as quantified by its B-factor should be somehow related in a qualitative way to its structural features, such as its packing environment or its position relative to the structure centroid. However, it is unexpected that the B-factor profiles and the structural profiles (i.e., the PFP and the WCN profiles) are so similar that, in some cases, one profile can be put on the top of the other with little discernible differences. It is even more surprising that this was done without information of the amino acid sequence.

Figure 5 compares the distributions of the correlation of B-factors with the original CN model, the WCN model and the PFP model for a dataset comprising 972 non-homologous structures. The B-factors distributions are computed using only Cα atoms.

The good agreement between the computed and the X-ray B-factors indicates that the B-factors can be derived solely from the protein backbone to a relatively high accuracy. A study(Lenin, Parthasarathy et al. 2000) showed that the variation in the atomic fluctuations, quantified by the B-factors, of a given protein segment only weakly depends on that of its amino acid. Note that the ENM (Tirion 1996; Bahar, Atilgan et al. 1997; Ming, Kong et al. 2002) also computes accurate B-factors without using the amino acid sequence.

While the amino acid sequence completely determines the 3-dimeniosnal structure of a protein, the thermal fluctuations and the motional correlations in a protein can be determined from its structure without its side-chain groups. The physical meaning of this is not clear and further study is definitely needed to clarify this issue.

3. The rigidity of enzyme catalytic sites

Enzymes accelerate chemical reactions by reducing the activation barrier. To achieve this, the enzyme structures are optimized through evolution to partially pre-organize their catalytic residues such that their charge distributions will stabilize the transition site complex, thus reducing the reorganization energy required for reaching the transition state(Warshel 1978; Warshel, Naray-Szabo et al. 1989; Warshel, Sharma et al. 2006). The reorganization energy is related to the activation free energy through the Marcus equation(Marcus 1956; Marcus 1956; Marcus 1957; Sumi & Marcus 1986):

$$\Delta G^{\neq} = \frac{\left(\lambda + \Delta G_0\right)^2}{4\lambda} \tag{7}$$

where $\Delta G^{\neq}$ is the activation free energy, ΔG_0 the free energy difference between the reactant and the product states and λ the reorganization energy. Eq. 7 states that the reduction of the reorganization energy λ will result in smaller activation free energy $\Delta G^{\neq}$.

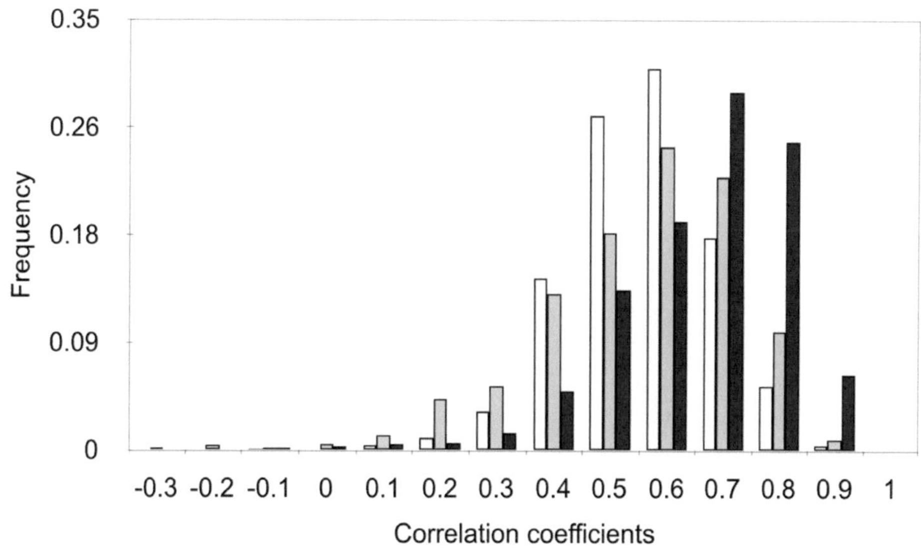

Fig. 5. Comparison of the correlation coefficients between experimental and the computed B-factor profiles based on the original CN model (white), the PFP (grey) and the WCN model (black) for the nonhomologous data set comprising 972 protein structures of length > 60. All of them are high-resolution X-ray structures with resolution ≤ 2.0 Å and R-factors ≤ 0.2. All chains are of pair-wise sequence identity ≤ 25%. This figure is adapted from Lin, Huang et al. 2008.

Hence, the active site residues are expected to be more rigid than the non-active site residues. And this has been verified by a number of studies: Yuan and co-workers (Yuan, Zhao et al. 2003) compared the B-factors of the active site sites with those of other non-active site residues of 69 apo-enzymes. They found that the active site residues indeed have lower B-factors. Analyzing a set of a set of 98 enzymes, Yang and Bahar (Yang & Bahar 2005) found that the catalytic sites usually occur in the global hinge centers and have low translational mobility. Recently, Lavery and co-workers(Sacquin-Mora, Laforet et al. 2007) showed that the force needed to displace a catalytic residue is usually larger than that to displace a non-catalytic residue.

3.1 Prediction of active site residue from a single structure

Since B-factors quantify structure flexibility, it may be tempting to use B-factors to distinguish between the active site residues and other residues. However, The B-factor values are affected by various experimental conditions such as temperature, crystallization and structural refinement.

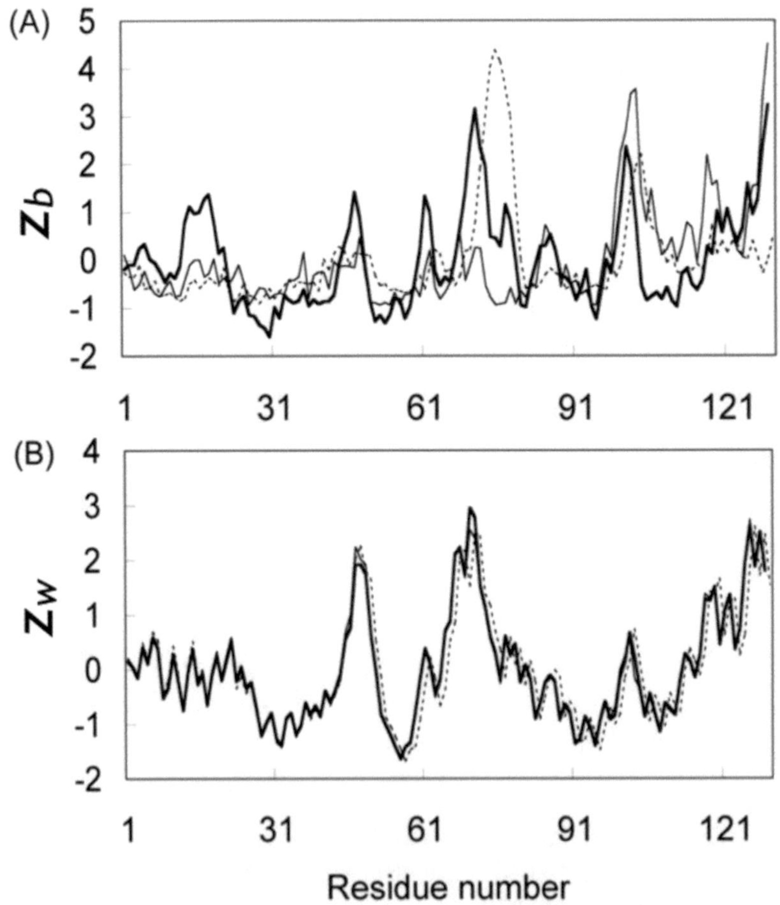

Fig. 6. (A) The Z_b profiles and (B) the Z_w profiles of 3 lysozymes: 6lyt (thick solid), 2bqo (dotted) and 2lzt (thin solid). Here Z_w is the normalized WCN. This figure is adapted from Huang, Yu et al. (2011).

Figure 6 compares the B-factor and the WCN profiles of several X-ray structures of lysozyme. Despite their almost identical structures -- their root-mean-square-deviations are within 0.6-0.8 Å, their B-factor profiles are very different.

On the other hand, their WCN profiles look almost identical. The B-factor is obviously not robust enough for predicting active site residues. The WCN profile, which correlates well with the B-factors but depends only on structure *per se*, is a much better discriminator for the catalytic residues.

Hwang and co-workers showed a straightforward application of the WCN profile to predicting the catalytic residues(Huang, Yu et al. 2011). We will illustrate this with an example: S-adenosylmethionine decarboxylase (AdoMetDC) (Ekstrom, Mathews et al. 1999), a critical regulatory enzyme of the polyamine synthetic pathway, has 5 catalytic residues. They are located in two chains: C82, S229 and H243 in chain A, and E11 and E67 in chain B. The active site of AdoMetDC is shown in Figure 7.

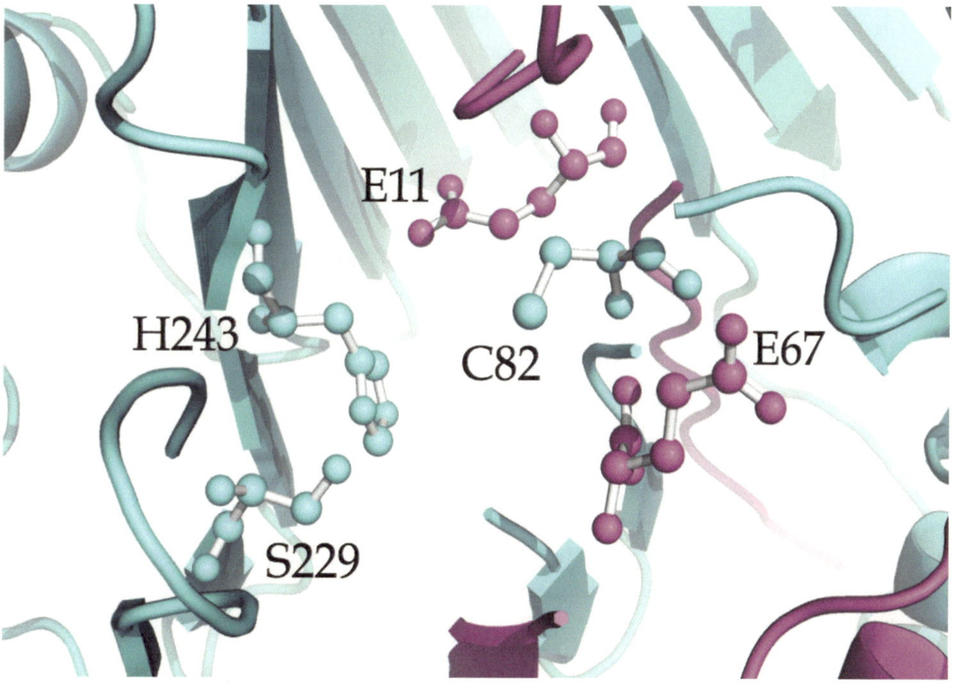

Fig. 7. The active site of AdoMetDC. Chain A is shown in cyan and chain B in magenta. The catalytic residues are labelled.

The WCN profile of AdoMetDC is shown in Figure 8. It should be noted that the WCN profile is computed for an incomplete structure -- many residues are missing in the X-ray structure of AdoMetDC. Despite this, most catalytic residues but one are located near the local minima of the WCN profile, as shown in Figure 8. The basic idea of the approach is simple: to obtain a threshold value for the WCN profile such that the residues whose Z-values are below the threshold are predicted to be catalytic residues. The threshold Z-value is determined by minimizing both the false positives and the false negatives of the predicted catalytic residues(Huang, Yu et al. 2011). The threshold for the WCN profile is -0.9.

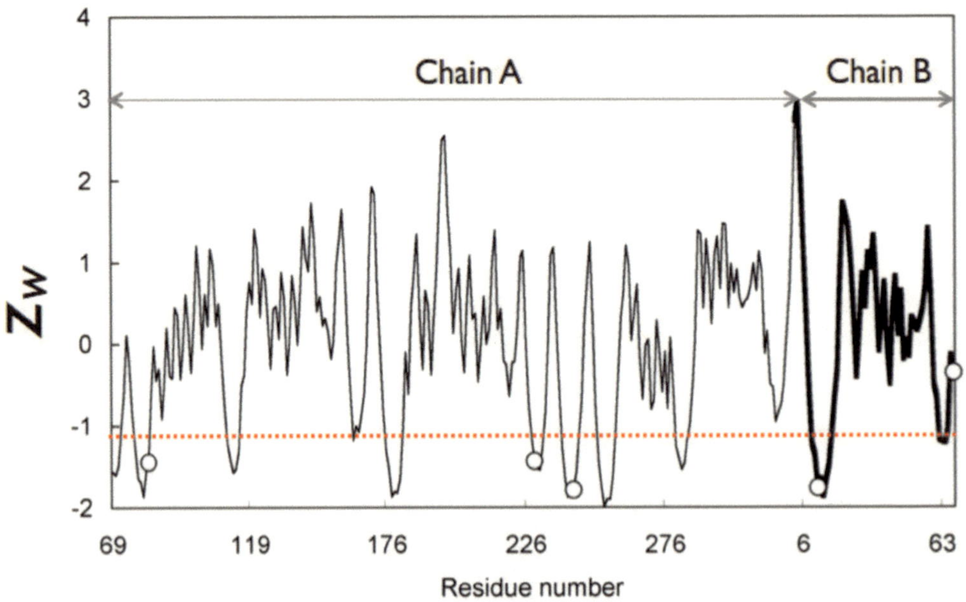

Fig. 8. The WCN profile of AdoMetDC (PDB ID: 1jen). The catalytic residues are marked in empty circles. The red dotted line indicates the threshold value.

3.2 Comparison of structural profiles

Figure 9 compares the receiver operating characteristics (ROC) curves of 4 structural profiles for their prediction of the active site residues for a data set comprising 760 X-ray nonhomologous enzyme structures(Huang, Yu et al. 2011). These profiles are the WCN profile, the PFP profile, the CN profile and the B-factor profile. The ROC curve is obtained by plotting the true positive rate (TPR) as a function of the false positive rate (FPR). TPR is defined as TPR = TP/P, where TP is true positives, i.e., number of correctly predicted catalytic residues, and P is the positive examples, i.e., the total number of catalytic residues. FPR is defined as FPR = FP/N, where FP is the false positives, i.e., the number of incorrectly predicted catalytic residues, and N is the negative examples, i.e., the number of non-catalytic residues. TPR is the *sensitivity*, while 1 – TPR is the *specificity*.

The WCN model performs the best among all the models. The PFP model comes in second. The relatively poor performance of the CN mode underlines the importance of attenuating the contributions of the neighboring atoms that depend on the distance between the interacting pair. The B-factor profile performs the worst. To have a feeling for the different performance between the WCN and the B-factor profiles, we can examine the case of highly specific predictions, i.e., at 95% specificity, the WCN model gives 52% sensitivity but the B-factor model gives only 11% sensitivity.

3.3 Other prediction methods

Here we will compare the WCN profile method with other methods in their prediction of the catalytic residues from a single structure.

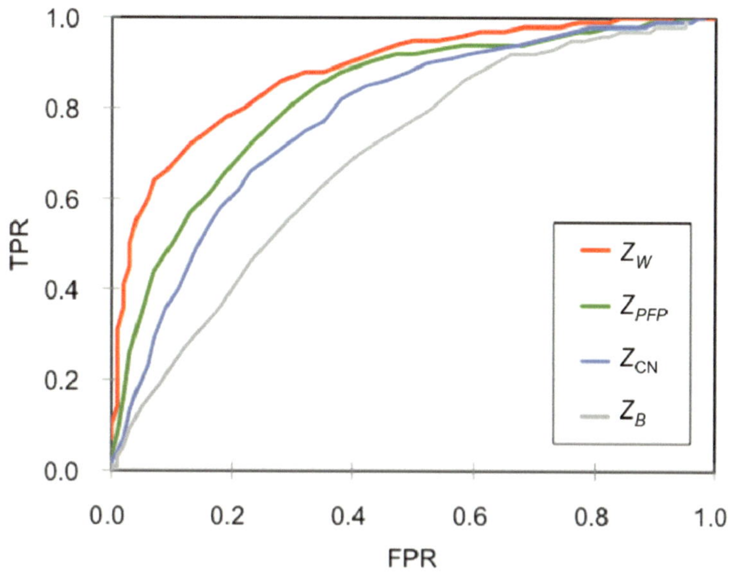

Fig. 9. The ROC curves of the WCN (Z_W), the PPF (Z_{PFP}), the CN (Z_{CN}) and the B-factor (Z_B) profiles for the prediction of catalytic residues.

Lavery and co-workers(Sacquin-Mora, Laforet et al. 2007), using Brownian dynamics simulation, computed the force needed to move any given amino acid residue with respect to other residues. They found that catalytic residues are invariably associated with high force constants. Their method gave 78% sensitivity and 74% specificity for predicting catalytic residues for a dataset of 98 non-homologous enzymes. In comparison, the WCN profile method yields 84% sensitivity and 82% specificity (Huang, Yu et al. 2011) for the same data set. This is quite remarkable, considering the simplicity of the WCN profile method as compared with the complexity of the Brownian simulation.

Analyzing the normal modes of a set of enzymes, Yang and Bahar (Yang & Bahar 2005) found that that the catalytic sites are usually of lower translational mobility than other residues, and that these catalytic sites generally coincide with the global hinge centers predicted by the GNM. Their method gave a sensitivity of 56% for a dataset of 24 enzymes. For the same data set, the WCN profile method gives 73% sensitivity and 82% specificity(Huang, Yu et al. 2011).

Ben-Shimon and Eisenstein(Ben-Shimon & Eisenstein 2005) developed a prediction algorithm called EnSite, based on the finding that the catalytic residues are often found among the 5% of residues closest to the enzyme centroid. The algorithm of EnSite is straightforward: it computes only the molecular surface that is close to the centroid, identifies continuous surface patches and finally ranks them by their area size. EnSite gives the result in terms of the rank of the correct prediction. This makes it difficult to compare the EnSite with the WCN profile method. The Pietrokovski's network centrality approach(Amitai, Shemesh et al. 2004) is similar to EnSite: both are based on the idea that the catalytic sites are near the *center* of the protein - be it represented by a structure or a network. Pietrokovski's approach gives 46.5% sensitivity and 9.4% specificity for a dataset of 178 structures. For a much larger dataset of 760 enzymes, the WCN profile gave 78% sensitivity and 80% specificity(Huang, Yu et al. 2011).

4. Conclusion

Catalytic residues are associated with a variety of structural or dynamic properties -- they are closer to the structure centroid, have higher packing density, or have smaller B-factors. Since the packing density and the centroid distances are closely related to the B-factors(Shih, Huang et al. 2007; Huang, Shih et al. 2008; Lin, Huang et al. 2008; Lu, Huang et al. 2008), all these seemingly diversified relationships can be reduced to one conclusion: the catalytic residues are more rigid. This is consistent with the present theory of enzyme catalysis(Warshel 1978; Warshel, Naray-Szabo et al. 1989; Warshel, Sharma et al. 2006; Sigala, Kraut et al. 2008): to accelerate the chemical reactions, the enzyme structures are optimized through evolution to partially pre-organize their catalytic residues to stabilize the transition state. As a result, catalytic residues tend to maintain similar conformations in both the reactant and the transition states. The catalytic residues will be more rigid than other non-catalytic residues.

5. Acknowledgements

This research was supported by the National Science Council and ATU from the Ministry of Education, Taiwan, R.O.C.

6. References

Alexandrov, N. and I. Shindyalov (2003). PDP: protein domain parser. *Bioinformatics,* Vol.19, No.3, pp. 429-30.

Amaro, R. E., A. Sethi, et al. (2007). A network of conserved interactions regulates the allosteric signal in a glutamine amidotransferase. *Biochemistry,* Vol.46, No.8, pp. 2156-73.

Amitai, G., A. Shemesh, et al. (2004). Network analysis of protein structures identifies functional residues. *Journal of Molecular Biology,* Vol.344, No.4, pp. 1135-1146.

Amitai, G., A. Shemesh, et al. (2004). Network analysis of protein structures identifies functional residues. *J Mol Biol,* Vol.344, No.4, pp. 1135-46.

Atilgan, A. R., S. R. Durell, et al. (2001). Anisotropy of fluctuation dynamics of proteins with an elastic network model. *Biophys J,* Vol.80, No.1, pp. 505-15.

Bahar, I., A. R. Atilgan, et al. (1997). Direct evaluation of thermal fluctuations in proteins using a single-parameter harmonic potential. *Fold Des,* Vol.2, No.3, pp. 173-81.

Ben-Shimon, A. and M. Eisenstein (2005). Looking at enzymes from the inside out: the proximity of catalytic residues to the molecular centroid can be used for detection of active sites and enzyme-ligand interfaces. *J Mol Biol,* Vol.351, No.2, pp. 309-26.

Brooks, B. and M. Karplus (1983). Harmonic dynamics of proteins: normal modes and fluctuations in bovine pancreatic trypsin inhibitor. *Proc Natl Acad Sci U S A,* Vol.80, No.21, pp. 6571-5.

Ekstrom, J. L., Mathews, II, et al. (1999). The crystal structure of human S-adenosylmethionine decarboxylase at 2.25 A resolution reveals a novel fold. *Structure,* Vol.7, No.5, pp. 583-95.

Fidelak, J., S. Ferrer, et al. (2010). Dynamic correlation networks in human peroxisome proliferator-activated receptor-gamma nuclear receptor protein. *Eur Biophys J,* Vol.39, No.11, pp. 1503-12.

Go, N. (1990). A theorem on amplitudes of thermal atomic fluctuations in large molecules assuming specific conformations calculated by normal mode analysis. *Biophys Chem,* Vol.35, pp. 105-112.

Halle, B. (2002). Flexibility and packing in proteins. *Proc Natl Acad Sci U S A,* Vol.99, No.3, pp. 1274-9.
Henrick, K. and J. M. Thornton (1998). PQS: a protein quaternary structure file server. *Trends Biochem Sci,* Vol.23, No.9, pp. 358-61.
Huang, S. W., C. H. Shih, et al. (2008). Prediction of NMR order parameters in proteins using weighted protein contact number model. *Theo. Chem. Acc.,* Vol.121, pp. 197-200.
Huang, S. W., S. H. Yu, et al. (2011). On the relationship between catalytic residues and their protein contact number. *Curr Protein Pept Sci,* Vol.In Press, pp.
Ishida, T. (2010). Effects of Point Mutation on Enzymatic Activity: Correlation between Protein Electronic Structure and Motion in Chorismate Mutase Reaction. *Journal of the American Chemical Society,* Vol.132, No.20, pp. 7104-7118.
Kundu, S., J. S. Melton, et al. (2002). Dynamics of proteins in crystals: comparison of experiment with simple models. *Biophys J,* Vol.83, No.2, pp. 723-32.
Laskowski, R. A., J. D. Watson, et al. (2005). Protein function prediction using local 3D templates. *J Mol Biol,* Vol.351, No.3, pp. 614-26.
Lenin, V. M. S., S. Parthasarathy, et al. (2000). Atomic displacement parameters of homologous proteins: Conservation of dynamics. *Current Science,* Vol.78, No.9, pp. 1098-1105.
Levitt, M., C. Sander, et al. (1985). Protein normal-mode dynamics: trypsin inhibitor, crambin, ribonuclease and lysozyme. *J Mol Biol,* Vol.181, No.3, pp. 423-47.
Lin, C. P., S. W. Huang, et al. (2008). Deriving protein dynamical properties from weighted protein contact number. *Proteins,* Vol.72, No.3, pp. 929-35.
Lu, C. H., S. W. Huang, et al. (2008). On the relationship between the protein structure and protein dynamics. *Proteins,* Vol.72, No.2, pp. 625-34.
Ma, J. (2004). New advances in normal mode analysis of supermolecular complexes and applications to structural refinement. *Curr Protein Pept Sci,* Vol.5, No.2, pp. 119-23.
Ma, J. (2005). Usefulness and limitations of normal mode analysis in modeling dynamics of biomolecular complexes. *Structure,* Vol.13, No.3, pp. 373-80.
Marcus, R. A. (1956). Electrostatic Free Energy and Other Properties of States Having Nonequilibrium Polarization. I. *J Chem Phys,* No.24, pp. 979.
Marcus, R. A. (1956). On the Theory of Oxidation-Reduction Reactions Involving Electron Transfer. I. *J Chem Phys,* Vol.24, No.966, pp.
Marcus, R. A. (1957). On the Theory of Oxidation-Reduction Reactions Involving Electron Transfer. II. Applications to Data on the Rates of Isotopic Exchange Reactions. *J Chem Phys,* Vol.26, pp. 867.
Ming, D., Y. Kong, et al. (2002). How to describe protein motion without amino acid sequence and atomic coordinates. *Proc Natl Acad Sci U S A,* Vol.99, No.13, pp. 8620-5.
Ming, D., Y. Kong, et al. (2002). Domain movements in human fatty acid synthase by quantized elastic deformational model. *Proc Natl Acad Sci U S A,* Vol.99, No.12, pp. 7895-9.
Miyazawa, S. and R. L. Jernigan (1984). Effective Inter-Residue Contact Energies from Protein Crystal-Structures. *Biophysical Journal,* Vol.45, No.2, pp. A130-A130.
Murzin, A. G., S. E. Brenner, et al. (1995). SCOP: a structural classification of proteins database for the investigation of sequences and structures. *J. Mol. Biol.,* Vol.247, pp. 536-540.
Orengo, C. A., A. D. Michie, et al. (1997). CATH- A Hierarchic Classification of Protein Domain Structures. *Structures,* Vol.5, No.8, pp. 1093-1108.
Sacquin-Mora, S., E. Laforet, et al. (2007). Locating the active sites of enzymes using mechanical properties. *Proteins,* Vol.67, No.2, pp. 350-9.

Sacquin-Mora, S., E. Laforet, et al. (2007). Locating the active sites of enzymes using mechanical properties. *Proteins-Structure Function and Bioinformatics,* Vol.67, No.2, pp. 350-359.

Saen-Oon, S., M. Ghanem, et al. (2008). Remote mutations and active site dynamics correlate with catalytic properties of purine nucleoside phosphorylase. *Biophys J,* Vol.94, No.10, pp. 4078-88.

Shih, C. H., S. W. Huang, et al. (2007). A simple way to compute protein dynamics without a mechanical model. *Proteins,* Vol.68, No.1, pp. 34-38.

Sigala, P. A., D. A. Kraut, et al. (2008). Testing geometrical discrimination within an enzyme active site: Constrained hydrogen bonding in the ketosterold isomerase oxyanion hole. *Journal of the American Chemical Society,* Vol.130, No.41, pp. 13696-13708.

Sumi, H. and R. A. Marcus (1986). Dyanmics effects in electron-transfer reactions. *J Chem Phys,* Vol.84, No.9, pp. 4894-4914.

Tirion, M. M. (1996). Large Amplitude Elastic Motions in Proteins from a Single-Parameter, Atomic Analysis. *Physical Review Letters,* Vol.77, No.9, pp. 1905-1908.

Tsai, C. J., A. del Sol, et al. (2008). Allostery: absence of a change in shape does not imply that allostery is not at play. *J Mol Biol,* Vol.378, No.1, pp. 1-11.

Van Der Spoel, D., E. Lindahl, et al. (2005). GROMACS: fast, flexible, and free. *J Comput Chem,* Vol.26, No.16, pp. 1701-18.

Warshel, A. (1978). Energetics of enzyme catalysis. *Proc Natl Acad Sci U S A,* Vol.75, No.11, pp. 5250-4.

Warshel, A. (2002). Molecular dynamics simulations of biological reactions. *Acc Chem Res,* Vol.35, No.6, pp. 385-95.

Warshel, A., G. Naray-Szabo, et al. (1989). How do serine proteases really work? *Biochemistry,* Vol.28, No.9, pp. 3629-37.

Warshel, A., P. K. Sharma, et al. (2006). Electrostatic basis for enzyme catalysis. *Chem Rev,* Vol.106, No.8, pp. 3210-35.

Watson, J. D., S. Sanderson, et al. (2007). Towards fully automated structure-based function prediction in structural genomics: a case study. *J Mol Biol,* Vol.367, No.5, pp. 1511-22.

Xu, Q., A. Canutescu, et al. (2006). ProtBuD: a database of biological unit structures of protein families and superfamilies. *Bioinformatics,* Vol.22, No.23, pp. 2876-82.

Yang, L., G. Song, et al. (2009). Protein elastic network models and the ranges of cooperativity. *Proceedings of the National Academy of Sciences of the United States of America,* Vol.106, No.30, pp. 12347-12352.

Yang, L. W. and I. Bahar (2005). Coupling between catalytic site and collective dynamics: a requirement for mechanochemical activity of enzymes. *Structure,* Vol.13, No.6, pp. 893-904.

Yuan, Z., J. Zhao, et al. (2003). Flexibility analysis of enzyme active sites by crystallographic temperature factors. *Protein Eng,* Vol.16, No.2, pp. 109-14.

Zheng, W., B. R. Brooks, et al. (2007). Allosteric transitions in the chaperonin GroEL are captured by a dominant normal mode that is most robust to sequence variations. *Biophys J,* Vol.93, No.7, pp. 2289-99.

Zheng, W. and D. Thirumalai (2009). Coupling between normal modes drives protein conformational dynamics: illustrations using allosteric transitions in myosin II. *Biophys J,* Vol.96, No.6, pp. 2128-37.

4

Decomposition of Intramolecular Interactions Between Amino-Acids in Globular Proteins - A Consequence for Structural Classes of Proteins and Methods of Their Classification

Boris Fackovec[1,2] and Jiri Vondrasek[1]
[1]*Institute of Organic Chemistry and Biochemistry, Czech Academy of Sciences, Prague,*
[2]*Faculty of Natural Sciences, Charles University in Prague, Prague,*
Czech Republic

1. Introduction

An amino-acid in proteins shows two different, yet mutually dependent faces connected through the polymer character of a protein in the final product. They are the amino-acid side-chain and its corresponding backbone part. On the level of the side-chains, we often refer to specific structural arrangements such as hydrophobic cluster motifs, salt-bridge motifs or hydrogen-bond motifs characterizing various parts of a protein and usually assigned to a certain function. The backbone on the other hand offers limited, yet general structural motifs - α, β and random coil patterns. All of these mentioned amino-acid features contribute to the synergy demonstrated observably by protein stability and protein function.

Thermal stability is one of the most important features of the structure of a fully folded protein. It is defined as the difference in the Gibbs free energy between its native and denaturated states and as such is a function of temperature and implicitly a function of protein composition and the effect of the environment. Nevertheless, it is necessary to say that for this function we do not know yet the precise and general form which could be applicable for a large set of proteins. There have been many attempts to propose an intuitive, yet productive decomposition of Gibbs free stabilization energy (GFSE) into simple terms. One of the scenarios utilized for such purposes is that the total free energy is the sum of the free energies of various atomic groups and the hydrophobic effect. However, as the free energy is not additive and the fractionation of free energy to independent terms is difficult, this attempt has been quite unsuccessful.

The utilization of molecular modeling methodology and tools has opened a more systematic and perhaps more promising approach - the evaluation of the enthalpy term in the equation for Gibbs free energy with reasonable accuracy (Lazaridis, Archontis, & Karplus, 1995). The remaining entropy term could be obtained by fitting the corresponding analytical form to the experimental data. There are basically three different enthalpy contributions that we can separate. The first comes from the intramolecular interactions between the atoms of proteins, producing the largest stabilizing enthalpy contribution. The second comes from the interactions between the molecules of a solvent, and finally the third contribution is the result of the interactions between the atoms of the solute (protein) and the solvent.

It is commonly believed that the dominant force of protein folding and therefore the main stabilizing force of the native structure is the hydrophobic effect (Dill, 1990). However, it has been insightfully pointed out (Makhatadze et al. 1995) that a water environment destabilizes folded protein structures and the decomposition of enthalpy shows that the solvation models introduce significant errors. In these studies, it has been assumed that the denatured state of a protein can be identified with the fully unfolded state (Makhatadze et al. 1989), where residues do not interact with each other. Even in light of this hypothesis, the intramolecular interactions between amino-acids in a protein are expected to contribute significantly to its overall stability. However, the hypothesis has never been proved and the importance of the intramolecular interactions would be much higher if the unfolding were considered as "core melting" rather than "oil-droplet dissolution". Regardless of the denatured form, the intramolecular organization of a protein is the result of a subtle balance between the rigidity/flexibility of the protein backbone and the noncovalent interactions between protein's side-chains. This result in conformational unique and stable protein structures as well as the ratio between the importance of the backbone/side-chain contributions can vary for different proteins.

The main problem of the enthalpy (or the potential energy) approach is that we are unable to evaluate the enthalpy-entropy compensation; therefore, the theoretically determined enthalpy contribution should be adjusted in some other way. A realistic method is to correlate the calculated values with the experimental data obtained by microcalorimetry, where both the enthalpy and the entropy terms can be determined. On the level of particular amino-acids, we face the problem of their "denatured-state" definition for the reasonable decomposition of the free energy on individual amino-acids.

The dissection of the enthalpy contribution which the intra-molecular noncovalent interaction energy (part of the potential energy) is a component of seems to be a reasonable approach for the study of the role of the composing amino-acids in protein stabilization. We can decompose this energy into individual pairwise amino-acid contributions and determine their importance for protein stability. The evaluation of the interaction energy (of noncovalent origin) between biomolecules or between their parts is a traditional field of the symbiosis between experiment and theory, and the methodology is well described and highly developed (Müller-Dethlefs & P Hobza, 2000). The crucial condition for the success of the theoretical methodology is the accuracy of the methods utilized. Recently, it has been quite common to evaluate the potential energy of a protein at the suitable *ab initio* methodology level, but we are still severely limited by the size of the protein. Therefore, the Density Functional Theory methods (DFT) are the most utilized for such purposes (Riley, 2010). Unfortunately, the DFT methods fail to describe the noncovalent interactions reasonably mostly because of the missing electron correlation term. Even the new functionals recently introduced (Kolář, 2010) have failed to describe properly the noncovalent potential curve mostly in the repulsion and asymptotic regions. Such inaccuracies can be tolerated at the energy minima, but only a limited number of the interactions between amino-acids in proteins meet such a requirement. Therefore, only high-level *ab initio* methods can be utilized – at least for benchmark studies. As was shown on a set of representative interactions between amino-acid side-chains in proteins in 2009, empirical force fields (namely OPLS and AMBER) are suitable for the description of their interaction (Berka, 2009). Kolar (Kolář, 2010) tested the performance of the energy calculations using MM on a representative set, S22, and found quite satisfactory agreement between the empirical force fields and high-level *ab initio* methods. It was later shown that we can use the empirical force field with satisfactory accuracy also for the description of the intramolecular interaction-energy distributions for pairs of amino-acid side-chains (Berka, 2010). Still, one has to be aware of the limitations of the force-field methods, namely for subtle cases of the interactions present in proteins. On the other hand, the utilization of empirical methods decreases the computational cost and provides an opportunity to investigate the trends presented in biomolecules if the highest accuracy is not the major issue.

The evaluation of the interaction energy between amino-acid residues resulted in the interaction energy matrix (IEM) concept being introduced in 2008 (Bendová-Biedermannová, 2008). The IEM approach was used to identify the key residues for protein stability in a model system - rubredoxin. The matrix carries information about the energy and the role of a residue in the protein structure, namely its interaction energy strength, which is more than the simple distance matrix concept. It also shows how much a certain residue is a hub within the context of the other interacting amino-acids. The IEM approach might also open new horizons for the investigations of proteins. The concept could be incorporated into the methods of protein-structure superpositions (similar to the DALI approach)(Holm & Sander, 1997) and can shed light on other protein-related issues - for example protein stability, folding kinetics, foldability and design.

The work presented in this study is based on the calculations of the amino-acid - amino-acid interaction energies (IEs) between all of the residues in approximately 1400 proteins to justify the roles of different amino-acids, their backbones and side-chains and their physical-chemical character for structural or stabilization preferences. We especially focused on the problem of how the interaction energy distributions are related to the secondary-structure content defined by the CATH (Orengo et al., 1997) and SCOP(Murzin, 1995) criteria.

2. Amino-acids in proteins and their distribution

2.1.1 Representative structure-set selection

All of the protein structures utilized in this study were obtained from the PDB database (download Jan 31, 2011). We selected only protein molecules with one chain, no ligands, resolved by the X-ray crystallography method at a minimum resolution of 2.0 A. We also omitted structures with a 70% sequence identity and higher. The database filter yielded 1531 structures. This number was slightly reduced by inconveniences with file processing to 1358. The characteristics of the set are illustrated in Figure 1 (size histogram, resolution histogram).

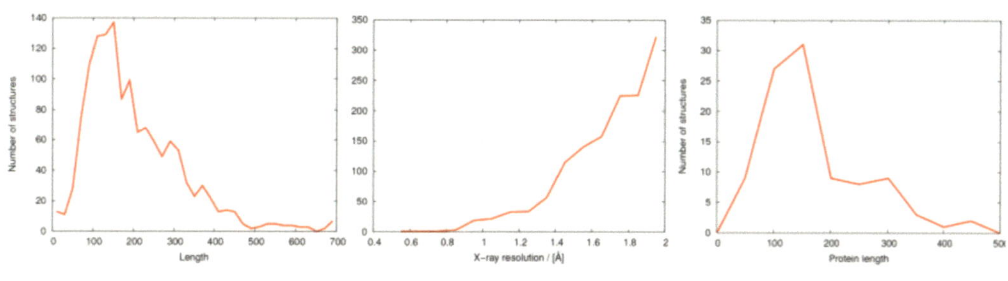

Fig. 1. a) Number of structures against protein length; binned by 20 AA; b) number of structures with a particular X-ray resolution; binned by 0.1 A; c) histogram of the sizes of the structures selected for secondary-structure studies.

Incomplete amino-acid side-chains (missing heavy atoms, disordered) were replaced by glycine in the cases where backbone atoms were available. Amino-acids with missing backbone atoms would have discredited the whole set and were therefore omitted. The missing hydrogen atoms were added by the Xleap program from the AMBER (Case et al. 2010) simulation package for pH 7 and the parameters were assigned according to the OPLS FF (Jorgensen & Rives 1988). The ambiguity of protonation, mainly in the case of histidine, is discussed later. The structures were optimized using the GROMACS (Hess et al. 2008) molecular simulation package with the steepest descent algorithm being employed. The hydrogen atoms were optimized first and then the full optimization of the whole protein in the gas phase was performed.

To address the question of the residue selectivity for secondary structure motifs, the structures were classified according to the CATH and SCOP categories and four representative sets were selected. To prevent the interference of the size and secondary structure effect, we assured that the structure sets possess the same size distribution.

Hence, the structures pertaining to particular secondary-structure sets were binned according to their chain length (bin size 50, see Figure 2) and were randomly removed from the bins until the number of structures in the corresponding bins was the same for all the sets. This procedure resulted in four sets, each containing 99 structures.

2.1.2 The fragmentation of proteins

To differentiate between the particular types of interactions which every amino-acid can maintain, we assigned every atom of a residue to one of four attributes according to their occurrence in the backbone or to their occurrence in certain types of amino-acid side-chains. The attributes were as follows – BB – backbone atoms, CH – side-chain atoms of charged residues (asp, glu, lys, arg, his), PO – side-chain atoms of polar residues (asn, gln, thr, ser) and NP – side-chain atoms of nonpolar and aromatic residues (gly, ala, leu, ile, val, pro, cys, met, phe, tyr, trp). Such classification provides the lowest number of groups necessary to discern between interactions characterized by different distance dependencies and orders of magnitude (different physical characters). On the other hand, breaking residues into more parts is restrained by the resulting charges of the fragments which would introduce significant but artificial electrostatic energies. The OPLS force field guarantees that the backbone (which includes C_α) and side-chain fragments are neutral. The physical character of the interaction energies of the aromatic residues is close to those of nonpolar residues. Hence, taking into account digestibility of presented data, we decided not to increase the number of attributes.

2.1.3 The Interaction Energy Matrix (IEM) calculation

After all of the structural optimizations, the pairwise interaction energies for all of the residues at the OPLS level were calculated excluding those between backbones of adjacent amino-acid in primary structure which were set to zero. The interactions were calculated separately for the backbones and side-chains as the sum of the interatomic Lennard-Jones and Coulombic contributions in the gas phase (ε_r=1) using an in-house developed Python program utilizing the standard libraries. The classification of the amino-acid atoms in four groups resulted in ten types of mutual interactions – BB-BB, BB-CH, BB-PO, BB-NP, CH-CH, CH-PO, CH-NP, PO-PO, PO-NP, NP-NP – reflecting the attributes of the atoms involved. For example, CH-CH represents salt bridges and all of the interactions between the side-chains of charged residues regardless of their relative distance and charge sign.

Each type of interaction for one protein was represented by one interaction energy matrix, namely a NxN (where N denotes the number of residues) matrix containing the interaction energy between the atoms of residues i and j with particular attributes assigned. It is guaranteed that no interaction energy is counted twice, so the sum of all of the matrices provides the interaction energy between the corresponding residues.
In order to compare the residual energy content, we have introduced a residue interaction energy (RIE) characteristic for each residue. The RIE of a certain type is defined as the sum of all of the interactions the residue can maintain - the sum of all the numbers in a particular row (or column) in the IEM of that type. At the end, we have ten (NxN dimension, where N is the number of amino-acids) IEMs of different types in one protein. Most of the IEs are of course almost zero; some are set as zero by definition.

2.1.4 Representation of data – cumulative distribution functions and histograms

There are two main data representation schemes in this work. Those are as follows:
The distributions of RIEs of a certain type in one protein. For one specific type and one specific protein set (for example CH-CH in SCOP β), the following procedure was performed to acquire an average distribution representing the whole set. The non-zero RIEs calculated from appropriate IEM were sorted independently for each protein and the distributions were obtained as a plot of the RIE against the residue rank in the sorted list normalized to one. To enable the averaging of the distributions, we represented each one by 1001 equally distant (on the rank coordinate) points between 0 and 1 (instead of for example N in the case of RIE BB). The RIE for each point was obtained by linear interpolation using the nearest two points of the calculated distribution. The averaged distribution was obtained by averaging the RIEs of the corresponding points of the curves of all of the proteins pertaining to the set. The inverse of the averaged distribution is a quite smooth cumulative distribution function representing the average for the set.
The distributions of the RIEs of a certain type for a particular amino-acid were sampled from all of the 1358 proteins. The RIEs of a particular type and AA were sampled from all the proteins and binned to yield quite smooth histograms.

2.2 Secondary-structure dependence

The RIE distribution of a particular type in a protein describes the distribution of the energetic importance of the residues. An average distribution also characterizes the particular type of interaction in the ensemble - the fraction of the key residues, their importance, and the fraction of the residues with repulsive interactions. The magnitude interval of a distribution is a very important parameter. It contains information about the interaction strength in the native states of the proteins. Unfortunately, this information does not denote the contribution of particular interactions to stability as it lacks information on the denatured state.
The shape of the distribution determines the pressure exerted on a residue and might help estimate the actual contribution of the corresponding interactions to protein stability. It is not surprising that the BB RIEs correlate with the secondary structures as the classifications indirectly use the BB RIEs. However, the differences are smaller than one might expect. It is also clear that none of the interactions other than BB is affected by the secondary-structure content.
From Figure 2, it can be concluded that the difference between the CATH and SCOP classifications is more significant mainly in the case of α proteins. Figures 3 and 4 show all of the types of distributions for a nonpolar (ALA, Figure 3) and a polar (THR, Figure 4)

amino-acid. It is obvious that the BB RIE cumulative distributions are the only distributions to have their shape affected by the secondary-structure content and the particular AA RIE distributions show more than one peak. The distinctive peaks might be assigned to special structural features and their identification remains a task for future studies.

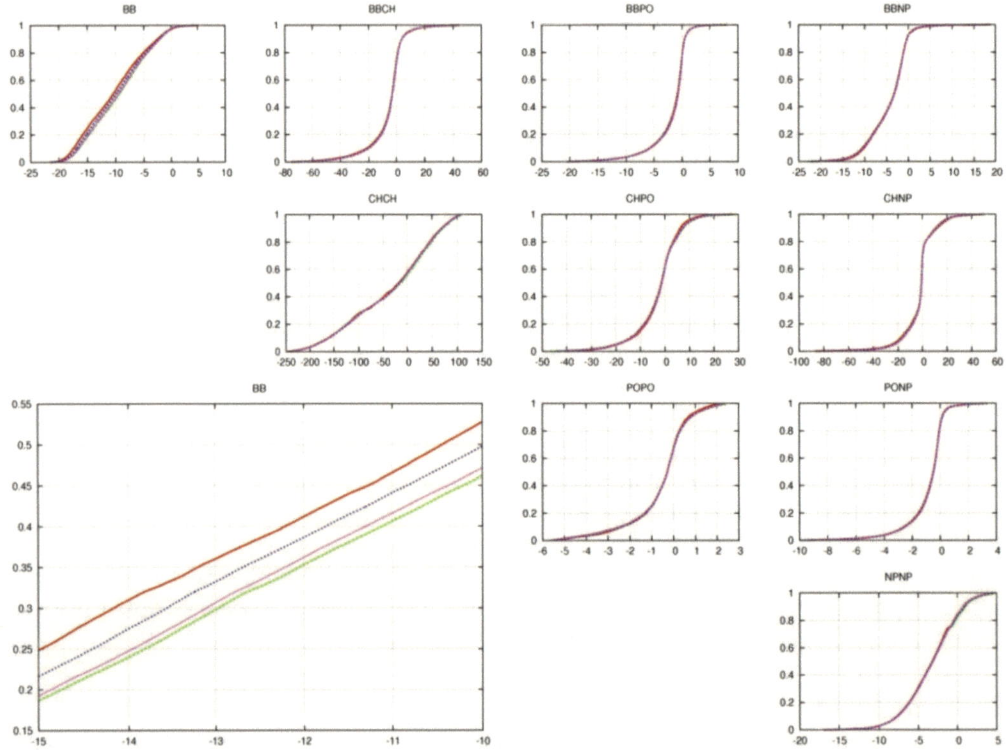

Fig. 2. The average RIE distributions of all ten types: a comparison of the secondary-structure classes. The colors of the lines correspond to the following structure sets: red – CATH α, blue – SCOP α, green – CATH β, magenta – SCOP β. The detail of the BB distribution in the bottom left corner is a zoom of the BB RIE distributions.

The fact that the CYS average NPNP RIE distribution is the only exception to the rule, because it has two peaks, can be explained by a different strength of the noncovalent interactions of the cystein SH group and cystine SS bridge.

The BB RIEs of particular AAs sampled through all of the structure sets are shown in Figure 5. There are remarkable differences between the shapes of the distributions corresponding to the α and β proteins as well as between the shapes of the distributions for particular AAs. Generally, the BB RIE distributions of the beta-structured proteins are shifted to a less attractive (less negative) noncovalent region.

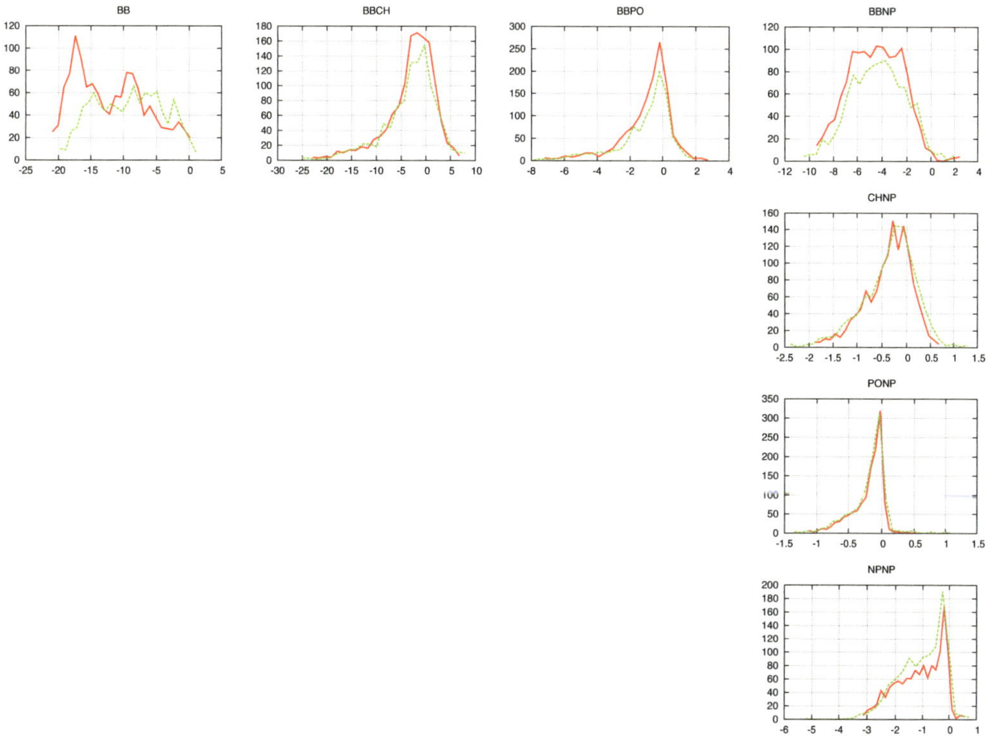

Fig. 3. All of the types of the RIE distributions of ALA. The red line corresponds to the CATH α set, the green line to the CATH β.

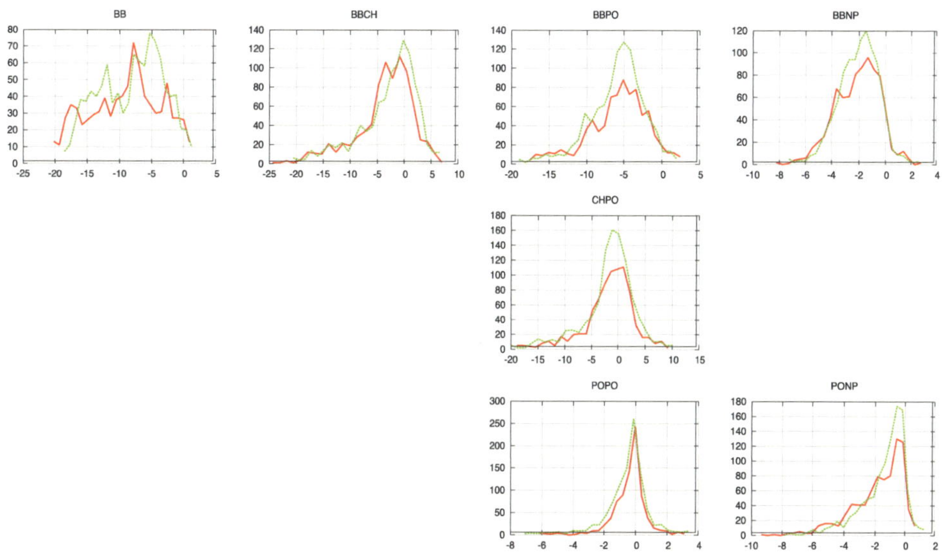

Fig. 4. All of the types of the RIE distributions of THR. The red line corresponds to the CATH α set, the green line to the CATH β.

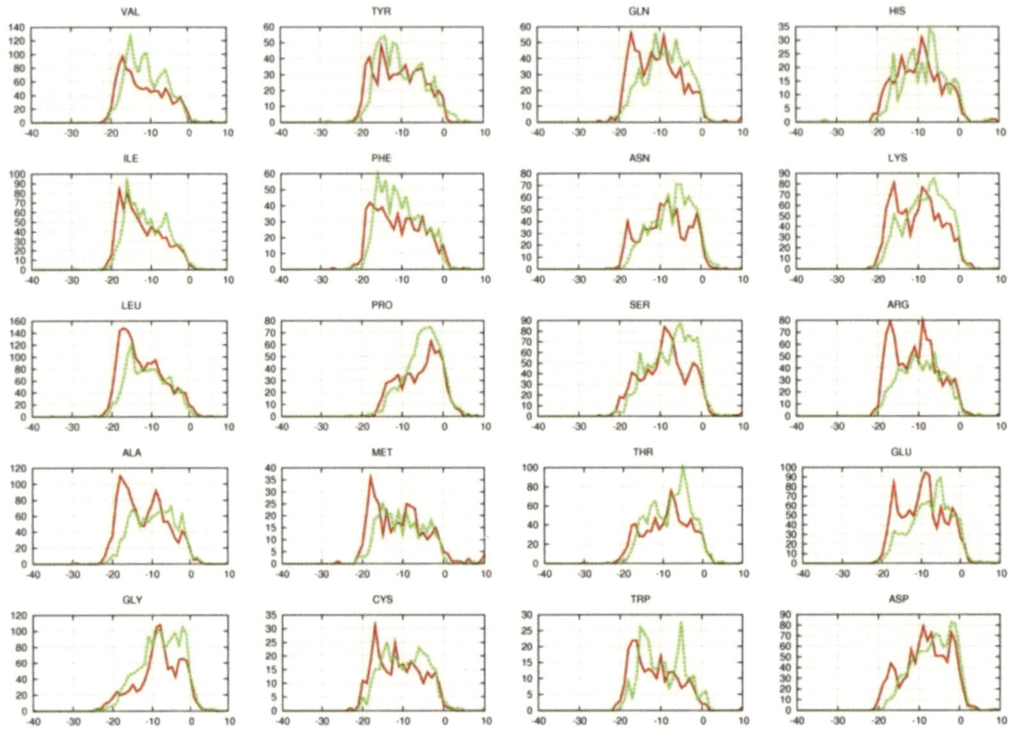

Fig. 5. The average BB RIE distributions for each AA. Sampled through proteins from the CATH α and CATH β sets. The red line corresponds to the CATH α set, the green line to the CATH β.

2.3 Size dependence

The proteins were selected based on their chain lengths up to fourteen groups regardless of their secondary-structure content. Their characteristics (chain-length range, average chain length, amino-acid type composition, number of proteins, number of residues of particular types, average surface area) are reviewed in Table 1.

min length	max length	average length	nonpolar resis	polar resis	charged resis	no. of structures	no. of residues
40	60	52.5	55.1	18.3	26.6	28	521
60	80	68.9	52.1	20.7	27.2	71	1342
80	100	90.7	53.0	18.5	28.5	105	1960
100	120	109.1	52.9	19.8	27.3	132	2492
120	140	129.1	53.9	18.8	27.3	125	2351
140	160	149.7	52.9	19.6	27.6	142	2655
160	180	169.6	53.8	19.5	26.6	85	1587
180	200	188.4	54.3	19.4	26.3	102	1909
200	220	209.9	53.1	19.9	26.9	65	1219
220	240	228.0	53.4	20.2	26.5	67	1253
240	260	249.3	54.3	18.9	26.8	57	1059
260	280	269.1	55.4	20.0	24.6	52	955
280	300	289.4	54.5	19.5	26.0	58	1061

Table 1. The characteristics of the structure sets used for the RIE-size dependence studies.

RIEs of a particular type were sampled from all of the proteins of a particular size group. The RIE averages were calculated separately for each interaction type of each size. The plots of the average RIEs against size are presented in four figures (Figures 6 to 9) in order to maintain the lucidity of the plots with lower magnitudes of average RIEs. The results reported in Figure 6 suggest that the RIE-size dependence varies significantly with the interaction type. On the one hand, the interaction of the polar residues with the backbone is almost independent of size. On the other hand, the interactions of the side-chains follow common rules, which are investigated later.

An interesting notion comes from a comparison of the magnitudes of the POPO and BBPO average RIEs. The lower RIE magnitudes in the case of POPO RIEs are probably caused by the lower probability of hydrogen-bond formation with polar side-chains in comparison with the backbone-polar side-chain because of the lower frequency of their occurrence.

A noticeable trend is the coupling of BBCH and CHPO interactions (see Figure 8). This binding may be ascribed to the same physical quality of these two types of interactions; they both represent charge–dipole interactions. The accuracy of the data can be estimated from the curve smoothness and is apparently lower in the case of charged residues. One possible reason for this trend is that the RIEs of charged residues are the products of a large compensation for the low amount of data.

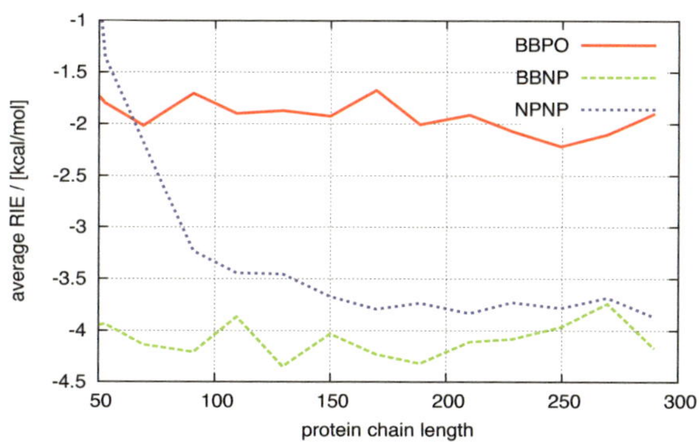

Fig. 6. The size dependence for BBPO, BBNP and NPNP interactions in the studied protein set. The NPNP differs significantly from the rest.

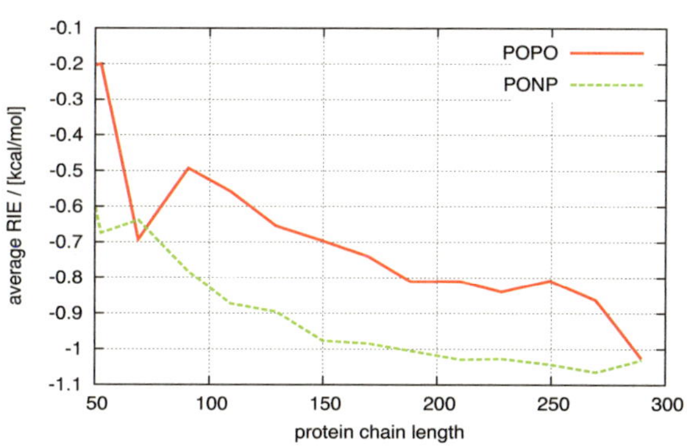

Fig. 7. The Size dependence for the POPO and PONP interactions in the studied protein set.

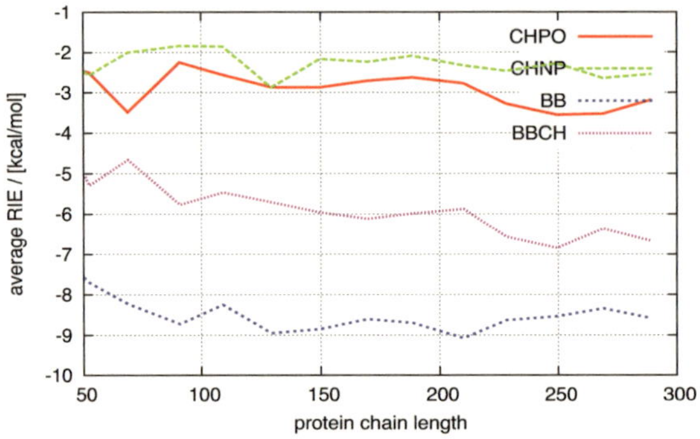

Fig. 8. The size dependence for the CHPO, CHNP, BB and BBCH interactions in the studied protein set.

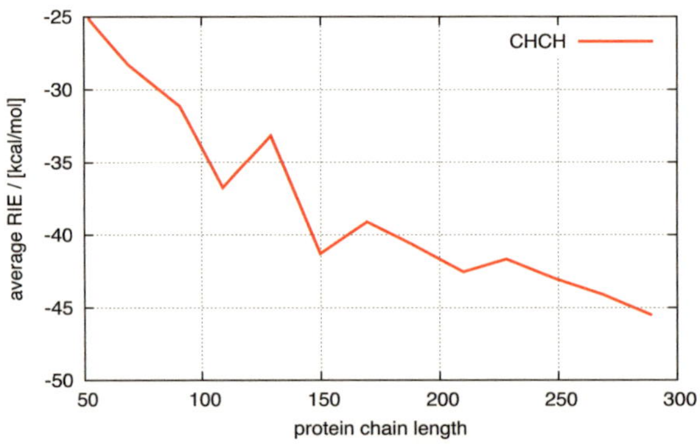

Fig. 9. The size dependence for the CHCH interactions in the studied protein set.

2.3.1 The model of size dependence for the interaction of nonpolar amino-acid side-chains

Two simple models were tested to explain the observed trends. In the first model, protein is assumed to be a sphere with nonpolar residues in its hydrophobic core and polar and charged residues forming its exterior shell. The size dependence of a NPNP RIE average is ascribed to the size dependence of the ratio between the core and surface residues. At infinite length, the NPNP RIE should reach its limit. The second model is more realistic in such a way that the core never behaves like a limitlessly increasing sphere and the volume occupied by the side-chains must reach its limit. This limits the NPNP RIE value which will not rise further with the increasing size of a protein and define a certain size of the most compact amino-acid arrangement.

2.3.1.1 The NPNP RIE Model 1

An average NP residue can be described by its characteristic length r, surface factor f_s (corresponding to its surface or interaction area $S_r = f_s r^2$), volume factor f_v (corresponding to its volume $V_r = f_v r^3$) and the NPNP RIE limit for the infinite bulk E_∞. As we are assuming that all of the nonpolar residues form the core which has a spherical shape, the core size is determined by the size of the protein and the ratio φ of nonpolar residues and all residues. A protein can be described by its porosity ε (determining the ratio of the gap volume to the volume of the whole protein) and at least its length N. Assuming that all of these quantities except for N are constants, the volume of each protein can be expressed as $V_p = NV_r/(1-\varepsilon) = Nf_v r^3/(1-\varepsilon)$ and the core volume as $V_c = V_p\varphi = N\varphi f_v r^3/(1-\varepsilon)$. The interaction surface of the core residues can be considered as $S_i = N\varphi S_r$ and the core surface is $S_c = 4\pi r_c^2$. E can be calculated as

$$E = E_\infty\left(1 - kN^{-1/3}\right), \tag{1}$$

where

$$k = \frac{1}{f_S}\sqrt[3]{\frac{1024\pi f_v^2}{9\varphi(1-\varepsilon)^2}} .$$

The k and E_1 parameters were fitted to the calculated data using Equation (1). As can be seen in Figure 10, the fitted curve does not represent the data very well.

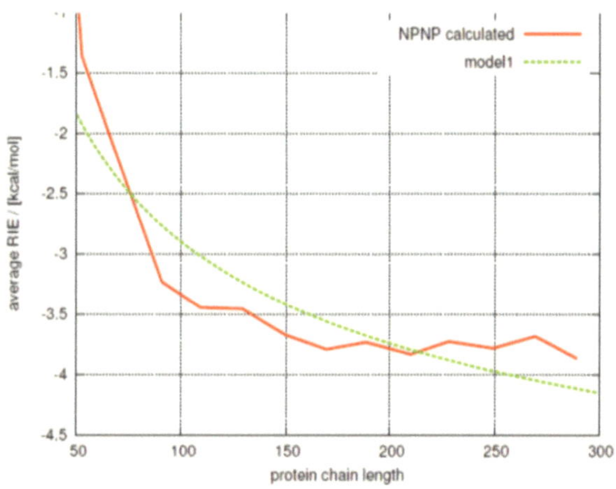

Fig. 10. The performance of Model 1

2.3.1.2 The NPNP RIE Model 2

The first model was extended by adding a new parameter, representing the domain size. The energy was represented by the following function:

$$E = \begin{cases} E = E_\infty\left(1 - kN^{-1/3}\right) & : N \le N_D \\ E_D & : N > N_D \end{cases}, \tag{2}$$

where N_D is the domain size and $E_D = E_\infty\,(1-kN_D^{-1/3})$ is NPNP RIE average at N_D. The parameters N_D, k and E_1 were fitted to the NPNP RIE averages. The agreement of the fitted curve with the data is satisfactory considering the simplicity of the model as one can see in Figure 11.

The coefficient k obtained by fitting the data is comparable to that obtained by a calculation using the estimated values of fv, fs, ε and the experimental value of φ. Other types of interactions seem to be unrelated to the domain size of a protein as there is no mechanism connected with size that we could follow.

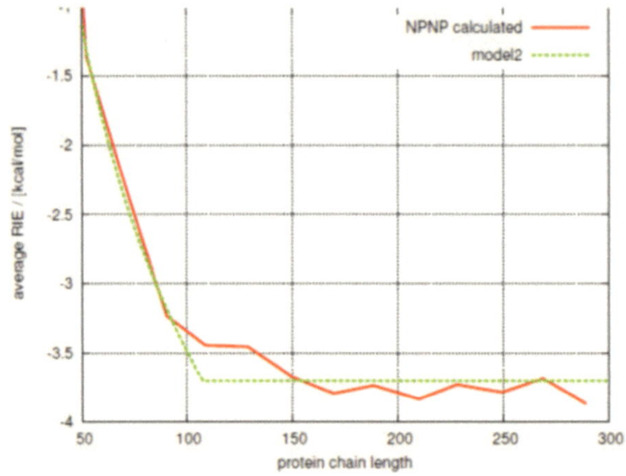

Fig. 11. The performance of Model 2

2.3.2 The reliability of the evaluated distributions

To adjust the reliability of our findings from computational point of view, we divided all of the proteins randomly into two groups. The distributions are indistinguishable, which proves that the distributions can be obtained by averaging even smaller sets of proteins. Additionally, we calculated the distributions using the OPLS force field in a Cα representation of the protein side-chains. Apparently (see Figure 12), the distributions for both FFs are the same. This not only proves that our results are robust against a FF parametrization error but also suggests that both FFs are within their limits equally good for RIE-distribution investigations.

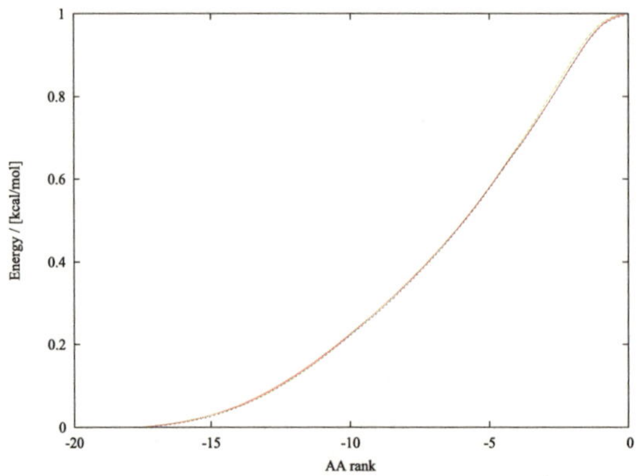

Fig. 12. A Comparison of the distributions obtained by averaging the distributions within the whole set using the OPLS Ca FF (dots) and Amber 03 C_a (full line) shows the robustness of the distributions against the FF used. The distributions obtained by averaging the distributions in two randomly chosen half-sets of structures calculated using the Amber C_a FF are indistinguishable, which proves that our set is sufficiently large.

3. Conclusion

RIE distributions in proteins, except for the BB RIE distributions, are not affected by secondary-structure content. The same applies for the distributions sampled for each amino-acid separately. Hence, we can claim that the strength and selectivity of the SC-SC and SC-BB interaction do not correlate with the secondary-structure content.

The size dependence of the RIEs can be satisfactorily described by the second model proposed. Its three parameters can be fitted to the results obtained by FF calculations of a high number of protein structures. One of the parameters obtained by fitting to the NPNP RIE averages represents the optimum definition of the domain size in globular proteins. Although the models proposed apply for all types of NP and PO SC-SC interactions, the models fail in the description of the BB and CH interactions. Many interesting facts about the size dependence of the RIE averages were revealed. First, the BBCH and CHPO interactions seem to be bound by some as-yet unknown rule. Second, the PO interactions exhibit "strange" behavior at a protein chain length of approximately seventy residues. These findings need to be investigated more deeply.

4. Acknowledgment

This work was supported by Grant No. P208/10/0725 from the Czech Science Foundation and Grants LH11020 and LC512 from the Ministry of Education, Youth and Sports of the Czech Republic. It was also a part of research projects No. Z40550506 and No. SM6198959216.

5. References

Bendová-Biedermannová, L., Hobza, Pavel, & Vondrásek, J. (2008). Identifying stabilizing key residues in proteins using interresidue interaction energy matrix. *Proteins*, *72*(1), 402-13. doi: 10.1002/prot.21938.

Berka, K., Laskowski, R. a, Hobza, Pavel, & Vondrášek, J. (2010). Energy Matrix of Structurally Important Side-chain/Side-chain Interactions in Proteins. *Journal of Chemical Theory and Computation*, *6*(7), 2191-2203. doi: 10.1021/ct100007y.

Berka, K., Laskowski, R., Riley, K., & Hobza, Pavel. (2009). Representative amino-acid side-chain interactions in proteins. a comparison of highly accurate correlated ab initio quantum chemical and empirical potential. *Journal of Chemical*, *5*(4), 982-992. doi: 10.1021/ct800508v.

D.A. Case, T.A. Darden, T.E. Cheatham, III, C.L. Simmerling, J. Wang, R.E. Duke, R. Luo, R.C. Walker, W. Zhang, K.M. Merz, B. Roberts, B. Wang, S. Hayik, A. Roitberg, G. Seabra, I. Kolossvai, K.F. Wong, F. Paesani, J. Vanicek, J. Liu, X. Wu, S.R. Brozell, T. Steinbrecher, H. Gohlke, Q. Cai, X. Ye, J. Wang, M.-J. Hsieh, G. Cui, D.R. Roe, D.H. Mathews, M.G. Seetin, C. Sagui, V. Babin, T. Luchko, S. Gusarov, A. Kovalenko, and P.A. Kollman (2010), *AMBER 11*, University of California, San Francisco.

Dill, K. A. (1990). Dominant forces in protein folding. *Biochemistry*, *29*(31), 7133–7155. ACS Publications.

Hess B, Kutzner C, van der Spoel D, Lindahl E. GROMACS 4: Algorithms for Highly Efficient, Load-Balanced, and Scalable Molecular Simulation. *Journal of Chemical Theory and Computation* 2008;4(3):435-47

Holm, L., & Sander, C. (1997). Dali/FSSP classification of three-dimensional protein folds. *Nucleic acids research*, *25*(1), 231-4. Retrieved from

Jorgensen WL, TiradoRives J. The Opls Potential Functions for Proteins - Energy Minimizations for Crystals of Cyclic-Peptides and Crambin. *Journal of the American Chemical Society* 1988;110(6):1657-66

Kolář, M., Berka, K., Jurečka, K., Hobza, P. (2010). Comparative Study of Selected Wave Function and Density Functional Methods for Noncovalent Interaction Energy

Calculations Using the Extended S22 Data Set. *Journal of Chemical Theory and Computation, 6*, 2365-2376.

Kolář, M., Berka, K., Jurecka, P., & Hobza, Pavel. (2010). On the Reliability of the AMBER Force Field and its Empirical Dispersion Contribution for the Description of Noncovalent Complexes. *Chemphyschem : a European journal of chemical physics and physical chemistry, 11*(11), 2399-408. doi: 10.1002/cphc.201000109.

Lazaridis, T., Archontis, G., & Karplus, M. (1995). Enthalpic contribution to protein stability: insights from atom-based calculations and statistical mechanics. *Advances in Protein Chemistry, 47*, 231-306.

Makhatadze, G. I., & Khechinashvili, N. N., with Venyaminov SYu & Griko YuV. (1989). Heat capacity and conformation of proteins in the denatured state. *Journal of molecular biology, 205*(4), 737-50..

Makhatadze, G., & Privalov, P. (1995). Energetics of protein structure. *Advances in Protein Chemistry, 47*, 307-425.

Müller-Dethlefs, K., & Hobza, P. (2000). Noncovalent interactions: a challenge for experiment and theory. *Chemical Reviews, 100*(1), 143-167.

Murzin, A.G., Brenner, S.E., Hubbard, T., Chothia, C. (1995). SCOP: a structural classification of proteins database for the investigation of sequences and structures. *Journal of Molecular Biology, 247*, 536-540..

Orengo, C. a, Michie, a D., Jones, S., Jones, D. T., Swindells, M. B., & Thornton, J. M. (1997). CATH--a hierarchic classification of protein domain structures. *Structure (London, England : 1993), 5*(8), 1093-108.

Riley, K. E., Pitoňák, M., Jurečka, P., & Hobza, Pavel. (2010). Stabilization and structure calculations for noncovalent interactions in extended molecular systems based on wave function and density functional theories. *Chemical Reviews, 110*(9), 5023-63. doi: 10.1021/cr1000173.

5

The Prediction and Analysis of Inter- and Intra-Species Protein-Protein Interaction

Theresa Tsun-Hui Tsao[1], Chen-Hsiung Chan[2],
Chi-Ying F. Huang[3] and Sheng-An Lee[4]
[1]*Department of Biomedical Electronics and Bioinformatics, National Taiwan University, Taipei*
[2]*Department of Medical Informatics, Tzu Chi University, Hualien*
[3]*Institute of Biomedical Informatics, National Yang Ming University, Taipei*
[4]*Department of Information Management, Kainan University, Taoyuan, Taiwan*
Taiwan

1. Introduction

Protein-protein interactions (PPIs) are essential to cellular processes. Recent developments of high-throughput technologies have uncovered vast numbers of PPIs. However, the experimental evidences are mostly for intra-species interactions of model organisms, especially human. Studies of non-human organisms and inter-species PPIs are few. For organisms such as *Arabidopsis thaliana*, the experimentally detected 5990 PPIs are estimated to be less than 3% of the entire *A. thaliana* interactome (M. Lin et al., 2011). The accuracy of high-throughput PPI experiments is also doubtful (Mrowka et al., 2001; Sprinzak et al., 2003; von Mering et al., 2002).
To resolve the above issues, several computational methods have been developed to evaluate and predict PPIs. This chapter focuses on direct PPIs which involve physical interactions of proteins, provides a brief overview of the reliabilities of high-throughput PPI detection technologies, and discusses the weakness and strength of important PPI computational prediction and evaluation methods. The major repositories which store, evaluate, and analyse both detected and predicted PPIs are also introduced.

2. Experimental detection of protein interactions

Not until the past decade, PPIs were identified by time consuming and labour intensive methods, such as low-throughput (small-scale) yeast 2-hybrid (Y2H). The development of high-throughput technologies brought studies of PPIs to an -omics level. Of all the technologies used for PPI detection, the high-throughput Y2H is most mature and commonly used. However, it is also one of the most inaccurate techniques, producing an estimated ~ 50% of false positives (Parrish et al., 2006; von Mering et al., 2002). The error rates of the other high- and medium-throughput technologies are summarized in Table 1 (Mrowka et al., 2001; Parrish et al., 2006; Sprinzak et al., 2003; von Mering et al., 2002).
In the past few years, BiFC has became one of the popular *in vivo* technologies as it has a medium throughput and reasonable cost, is technically straightforward, and provides

information on subcellular localizations of proteins. The drawback for BiFC is its occasional false postivies caused by non-specific interactions and background flurorescence. Split-luciferase system has an extremely low background, but does not disclose the subcellular localization of interactions. Protoplast Y2H is similar to Y2H, but technically more challenging as it has to be operated in a nuclei or protoplasts. SUS has high rates of false positives and background signals. The *in vitro* technologies are less favourable as the reactions do not occur in cellular environments and do not examine the cellular localization of proteins.

Technology	Throughput	Accuracy	References
In vivo			
High-throughput Y2H	High	×	(Causier, 2004; Causier & Davies, 2002)
Split-luciferase system	Medium-high	○	(Paulmurugan & Gambhir, 2005,2007)
Protoplast Y2H	Medium	×	(Fujikawa & Kato, 2007)
Split-ubiquitin system (SUS)	Medium	×	(Michnick, 2003; Obrdlik et al., 2004; Reinders et al., 2002; Schulze et al., 2003)
Bimolecular fluorescence complementation (BiFC)	Medium	△	(Citovsky et al., 2008; Hu et al., 2006; Ohad et al., 2007; Ohad & Yalovsky, 2010; Zhou et al., 2011)
In vitro			
Stable-isotope labeling of amino acids in cell culture (SILAC)	High	×	(Gruhler & Kratchmarova, 2008; Mann, 2006)
^{15}N-labeling	High	○	(Huttlin et al., 2007; Nelson et al., 2007)
Chemical crosslinking-MS using protein interaction reporter (PIRs)	High	×	(Anderson et al., 2007; Tang et al., 2005)
Protein microarrays	High	○	(Angenendt et al., 2006; Popescu et al., 2007; Ramachandran et al., 2004)
Single affinity purification-tagging	Medium-high	△	(Berggard et al., 2007)
Tandem affinity purification (TAP) tagging	Medium-high	△	(Rohila et al., 2006; Schoonheim et al., 2007)
Native chromatography or electrophoretic purification	Medium	×	(Liu et al., 2008)

Table 1. The technologies for high- and medium-throughput PPI detection (Morsy et al., 2008) Accuracy of each technology is summarised in this table. The symbol "×" indicates low accuracy, "○" indicates high accuracy, and "△" indicates sound accuracy.

3. Computational prediction and evaluation of protein interactions

As listed in Table 2, the methods for PPI prediction and evaluation can be classified into five categories based on the types of information required for analysis - (1) protein sequences, (2) Gene Ontology (GO), (3) gene expression profiles, (4) topology of the interaction network, and (5) experimental data.

Methods	References
Protein Sequences	
Interologs	(Matthews et al., 2001; Rhodes et al., 2005; von Mering et al., 2007)
Phylogenetic tree similarity	(Jothi et al., 2005)
Gene fusion	(Enright et al., 1999; Marcotte et al., 1999)
Gene neighbouring	(Dandekar et al., 1998; Overbeek et al., 1999; Tamames et al., 1997)
Domain-domain interactions	(Frishman, 2009; Ng et al., 2003)
Gene Expression Profile	
Co-expression correlation coefficience	(Ideker et al., 2002)
Shared Gene Ontology Annotation	
Protein functions	(De Bodt et al., 2009; Jain & Bader, 2010; Wu et al., 2006)
Protein localization	(De Bodt et al., 2009; Jain & Bader, 2010)
Topology Analysis	
Distance between proteins in a PPI network	(Dyer et al., 2007)
Experimental Data	
Cited literatures (text mining)	(Jaeger et al., 2008)
Detected PPI datasets	(von Mering et al., 2002)

Table 2. Computational methods for protein interaction prediction and evaluation

Methods for PPI prediction and evaluation is each developed based on an assumption which states certain criteria are more likely to occur between interacting proteins. These methods are often combined, usually in a Bayesian network (Huttenhower & Troyanskaya, 2006; Jansen et al., 2003; Lee et al., 2006; N. Lin et al., 2004; McDowall et al., 2009; Patil & Nakamura, 2005; Wang et al., 2009; Xu et al., 2011). Some criteria are more relevant to protein interactions than the others. When different methods are combined, the statistical confidence estimated by each method could be weighted according to the confidence level of corresponding assumption.

3.1 Interologs

Homologous proteins often conserve similar functions and PPIs across different organisms, especially in phylogenetically close-related species (Hirsh & Sharan, 2007). These conserved PPIs are designated as interologs.

In the interolog method, it is assumed that if a pair of proteins, A and B, interact and there are two other proteins, A′ and B′, of which A′ is homologous to A and B′ is homologous to B, then A′ and B′ are potentially an interolog to A and B. Interologs can occur among different species or in the same species (Mika & Rost, 2006). The conventional interolog method identifies homologous proteins by comparing the global sequences. For proteins of which only partial sequences are similar, sequence signatures may be compared instead of full sequences (Sprinzak & Margalit, 2001). Structurally similar proteins may also have similar protein interactions, but predicting PPIs by identifying proteins with similar structures is impeded by the limited structural information available (Aytuna et al., 2005; Ogmen et al., 2005). The interologous relationship between the two pairs of proteins, one pair predicted and one pair detected, could be evaluated by functions such as the *s* score. (He et al., 2008). Homologous genes of model organisms can be identified using BLAST or in HomoloGene database that automatically identifies and collects homologs from fully sequenced genomes (Sayers et al., 2011).

The interolog method has been used frequently. The first human PPI network, the Arabidopsis PPI network, and the rice blast fungus PPI network are a few examples constructed by predicted interologs (Geisler-Lee et al., 2007; He et al., 2008; Lehner & Fraser, 2004). Unfortunately, prediction of plant PPIs through a comparative interactome approach is challenged by the unique biology of plants which involves PPIs not commonly found in the other model organisms. Less than 50% of *A. thaliana* proteins have been found to have orthologs in the more extensively studied organisms such as yeast, *Caenorhabditis elegans*, fruit fly, or human (Gollery et al., 2006). Furthermore, the interolog method does not differentiate the functionally significant amino acid residues from the others; neglects the residue-specific requirements for interaction specificity and affinity (Uhrig & Hulskamp, 2006). For the highly homologous members of protein families, the interlog method could be prone to errors.

3.2 Phylogenetic relationship

Interacting proteins have been observed to have topologically similar phylogenetic trees for the corresponding protein families, presumably due to the co-evolution of cooperating proteins (Fryxell, 1996; Goh et al., 2000; Pages et al., 1997). Based on the above observation, the phylogenetic similarity method was proposed. To compare and construct the phylogenetic trees, firstly, the sequences of two potentially interacting protein families are aligned. Secondly, the evolutionary distance matrixes are calculated from the phylogenetic trees, one for each protein family. Finally, the Pearson's correlation coefficient between the two distance matrixes is calculated as an indication of the likelihood of interactions. Partial protein sequences could be used to construct the phylogenetic trees - for example, poorly conserved sequences have been removed to improve the performance of prediction (Kann et al., 2007).

A similar approach is the phylogenetic profile method. Phylogenetic profile is the profile which records the presence and absence of a protein across all species. Also due to the presumably co-evolution of proteins involved in the same biological process, proteins with similar phylogenetic profiles are more likely to have interactions. The profiles could be compared by Hamming distance (Pellegrini et al., 1999). Although this approach is powerful, it can be applied only to organisms which have been fully sequenced (Frishman, 2009). Additionally, there might be complications with essential proteins which are present in all organisms (Frishman, 2009). As the second generation sequencing (SGS) technologies exponentially accumulating full genome sequences of non-model organisms, this method is expected to become more favorable.

3.3 Gene fusion and neighboring

Genes which are previously separated in the genome of one organism can be fused into the same gene in another organism. Fused genes almost always encode functional related and physically interacting proteins (Enright et al., 1999; Marcotte et al., 1999). The fusion events might accelerate the formation of protein complexes by increasing the opportunity of correct physical contact between interacting sites.

Similarly, in bacteria, genes which are consistently located in the same operon across many species are likely to express functionally related, and often physically interacting, proteins (Dandekar et al., 1998; Overbeek et al., 1999; Tamames et al., 1997).

3.4 Domain-domain interactions

Just like protein interactions, domain interactions can be predicted by sequence homology among two pairs of interacting domains, by investigating the evolutionary traits of domains, or by identifying conserved neighboring relationship between domains (Frishman, 2009). Interacting proteins are also more likely to contain domains which have been detected or predicted to interact (Ng et al., 2003).

3.5 Co-expression

Interacting proteins are assumed to have similar expression patterns (Dyer et al., 2007). The co-expression correlation coefficients of seven model animals, including human, mouse, chicken, zebra fish, fruit fly, and *Coenorhabditis elegans*, and nematode, are recorded in COXPRESdb (Obayashi et al., 2008; Obayashi & Kinoshita, 2011). The co-expression correlation coefficients of *A. thaliana* and many other flowering plants are recorded in ATTED-II (Obayashi et al., 2011). High-throughput expression data are mostly available on Gene Expression Omnibus (GEO) or TAIR for *A. thaliana* experiments (Garcia-Hernandez et al., 2002; Sayers et al., 2011).

3.6 Gene Ontology (GO)

Interacting proteins are presumably to participate in related biological process and share similar cellular localization (Dyer et al., 2007; Shin et al., 2009). The GO project annotates the cellular components where a protein locates and the biological process in which a protein participates. The annotations are created by structured and controlled vocabularies. The semantic similarities between GO terms assigned to proteins are often used to evaluate the confidence levels of proposed PPIs (De Bodt et al., 2009; Jain & Bader, 2010).

3.7 Topology

As more and more PPIs are revealed, PPI networks can be constructed and analyzed by topology theories. It has been proposed that two proteins which interact with the same protein should have a shorter path between them on the PPI network (Dyer et al., 2007). It has also been proposed that interacting proteins might share more neighboring proteins on a PPI network (J. Chen et al., 2006; Chua et al., 2006).

3.8 Text mining

The protein interactions which have been reported repeatedly in more peer-reviewed literatures might be more trustworthy than the ones which have never or rarely been detected (Jaeger et al., 2008). However, it must be noted that proteins with more valuable functions, such as disease mechanisms, would have been studied more intensively and been documented more frequently.

PubMed and GeneRIF are common sources of text mining materials. The automated data gathering (e.g. text mining *via* natural language processing or biomedical language processing) is not as reliable as manually curated data. It must be noted that manual curation is neither 100% correct due to human errors and inconsistent standards for curation.

3.9 Experimental detections

PPIs detected by low-throughput technologies are generally considered as error free. For the medium- to high-throughput technologies, the reliability of the results varied as listed in Table 1. *In vivo* experiments are usually more accurate than the *in vitro* experiments, as *in vivo* experiments were conducted in cellular environments. Interactions supported by more than one method are generally believed to be more reliable (von Mering et al., 2002). PPI datasets which are more reliable are assumed to have more intersections with the other datasets and higher averaged numbers of documented protein interactions (Shin et al., 2009). Reliable PPI datasets should also contain greater proportion of interactions which have interacting domain pairs (He et al., 2008).

In silica protein docking is another approach which could be used for predicting protein interactions; however, it is impractical for high-throughput predictions due to the extremely large amount of required computation and the lack of detected or predicted structures for most proteins.

4. Protein interaction databases

More than 30 PPI databases have been published and are mostly available online (Fischer et al., 2005). Table 3 listed the frequently referenced databases. The contents of these databases are often overlapped and integrated to create larger non-redundant databases. These collections of PPIs can be used as the foundation for predicting and evaluating the reliability of PPIs.

Database	Validation	Organisms	Reference
MINT	Detected Provides confidence scores for PPIs.	Model organism	(Ceol et al., 2010)
DIP	Detected	Model organisms	(Salwinski et al., 2004)
BIND	Detected	~ 1500 organisms	(Gilbert, 2005; Isserlin et al., 2011; Willis & Hogue, 2006)
BioGRID	Detected	Model organisms	(Breitkreutz et al., 2008; Stark et al., 2011; Stark et al., 2006)

Database	Validation	Organisms	Reference
IntAct	Detected	Model organisms	(Aranda et al., 2010; Brandao et al., 2009)
HiPredict	Data from IntAct, BioGRID, and HPRD Provide confidence scores for PPIs.	Model organisms	(Patil et al., 2011)
MIPS	Detected	Model mammals	(Pagel et al., 2005)
HPRD	Detected	Human	(Goel et al., 2010)
STRING	Detected and predicted data from BIND, BioCarta, BioCyc, BioGRID,DIP, HPRD, IntAct, MINT, REACTOME, textmining, etc Provides confidence scores for PPIs.	~ 1000 organisms	(Szklarczyk et al., 2011; von Mering et al., 2007; von Mering et al., 2005)
HAPPI	Detected and predicted data from HPRD, BIND, MINT, STRING, OPHID, etc Provide confidence scores for PPIs.	Human	(J.Y. Chen et al., 2009)
AtPID	Predicted Provide confidence scores for PPIs.	*A. thaliana*	(Cui et al., 2008; Li et al., 2011)
PAIR	Detected (from IntAct, BioGRID and BIND) and predicted Provide confidence scores for PPIs.	*A. thaliana*	(M. Lin et al., 2011)
AtPIN	Experiemtnal (from BioGRID and IntAct) and predicted (from Geisler-Lee and AtPID) Provide confidence scores for PPIs.	*A. thaliana*	(Cui et al., 2008; Geisler-Lee et al., 2007; Li et al., 2011)
PIG	Data from BIND, IntAct, REACTOME, and MINT	Human-pathogen interacitons	(Driscoll et al., 2009)
HPIDB	Data from BIND, IntAct, REACTOME, MINT, GENERIF and PIG	Host-pathogen interactions, hosts are model organisms	(Kumar & Nanduri, 2010)

Table 3. Major PPI databases

MINT is one of the few repositories which provide confidence scores for experimentally detected PPIs. It uses the number and types of experiments in which a PPI is detected to estimate the confidence of data.

HiPredict is a repository which contains filtered high-confidence PPIs of nine model species from IntAct, BioGRID, and HPRD. While calculating the confidence of PPIs, HiPredict considers (1) the type of experiments which detect the PPIs, (2) the co-expression correlation coefficients of proteins, (3) shared GO terms of proteins, (4) presences of interologs in the same organisms, and (5) domain-domain interactions between proteins. These five criteria are combined in naïve Bayesian networks to give confidence scores.

STRING is one of the largest and most comprehensive PPI repositories. It evaluates PPIs using multiple criteria, including (1) the probability of finding the interacting proteins on the same KEGG pathway, (2) co-mentioning of gene/protein names in PubMed abstracts, (3) co-expression / co-regulation of proteins, (4) presence of interologs, and (5) presence of gene neighboring. Similar to HiPredict, the various criteria are also combined in naïve Bayesian networks.

HAPPI only collects human PPIs. For the PPIs which have been evaluated, such as data from STRING, the confidence scores are preserved. For the PPIs which have not evaluated, HAPPI calculates the confidence scores based on the type of experimental evidences and the source of data.

PAIR collects 5990 detected protein interactions and 145494 predicted interaction of *A. thaliana* (M. Lin et al., 2011). These PPIs were expected to cover 24% of the entire *A. thaliana* interactome, of which the size was estimated to be 200 000 PPIs (for 20 000 genes) based the size for yeast (18 000 PPIs for 6000 genes). An estimated 44% of the collected PPIs in PAIR are reliable. PAIR predicts PPIs using a machine learning approach with supports the vector machine (SVM) model. In the SVM model, indirect evidences of interactions (i.e. interologs, phylogenetic profile similarity, domain interactions, gene co-expression correlation, shared GO terms, and protein localizations) are combined. The model has been trained using Gold Standard Positives (GSPs), which are reliable PPIs from major repositories. The SVM scores also serve as the confidence scores for the predicted PPIs. The detected PPIs are collected from IntAct, BioGRID, and BIND.

AtPIN integrates (1) the predicted PPIs from Geisler-Lee and AtPID, (2) the curated PPIs from TAIR, and (3) the detected PPIs from BioGRID and IntAct (Brandao et al., 2009). Geisler-Lee (2007) predicts PPIs by identifying interologs. AtPIN calculates confidence scores of PPIs based on (1) the detected or predicted co-localization of interacting proteins and (2) the number of shared neighboring proteins of interacting proteins on the PPI network. It also provides the score calculated by AtPID. **AtPID** combines indirect evidences of interactions, including interologs, phylogenetic profiles, domain interactions, co-expression profile, shared protein functions, protein co-localisation, and gene fusion, in naïve Bayesian networks to predict and evaluate the PPIs of *A. thaliana* (Cui et al., 2008; Li et al., 2011). TAIR is a multi-tasking project which participates in a broad range of *A. thaliana* researches.

The data of protein interactions between hosts and pathogens are scarce. **PIG** integrates the manually curated human-pathogen PPIs from four databases, BIND, IntAct, REACTOME, and MINT, in one platform for searching, visualization, and analysis of PPI networks. The corresponding hyperlinks to UniProt database, Gene Ontology, InterProScan, and PubMed are filed under each protein entry in the user interface for convenient referencing.

Similar to PIG, **HPIDB** integrates several host-pathogen PPI databases, including BIND, IntAct, REACTOME, MINT, GENERIF, and PIG. However, unlike PIG, the PPIs in HPIDB are not limited to human host. Although the majority of data is for human (22386 PPIs), HPIDB also contains host-pathogen PPIs for mouse (147 PPIs), *A. thaliana* (99 PPIs), rat (53 PPIs), cattle (30 PPIs), and chicken (19 PPIs).

A few repositories collect genes which are involved in host-pathogen interactions, but do not contain data on physical protein interactions. **PHIDIAS** is a centralized respiratory for host-pathogen interactions. It collects information for 98 pathogens of two hosts, human, and mouse (Xiang et al., 2007). **PHI-Base** contains information for 405 fungal, oomycete, and bacterial genes which participate in pathogenicity, virulence, and induction of disease resistance (Baldwin et al., 2006; Winnenburg et al., 2006). 176 of these genes are from animal pathogens, 227 from plant pathogens, and 3 from pathogens of fungi. **PathoPlant** contains *A. thaliana* genes which are responsible in the defense against pathogens (Bulow et al., 2007).

5. Identification of drug targets within human-pathogen interactions network

The evolutionary history of human has never been parted with pathogens. Viruses, bacteria, fungi, and nematodes all play critical roles in shaping the human race. Recent advances in metagenomics and human microbiomes suggest that commensal microorganisms have significant influences to the metabolism, immune systems, general wellbeing, and even behaviour patterns of animal hosts.

Despite enormous efforts in preventing, diagnosing, and treating infectious diseases, pathogens still cause insurmountable burden and social-economical impacts to human. The developments of vaccines and drugs have helped to diminish several devastating diseases; however, emerging diseases caused by novel or previously unknown pathogens continuously lead to unexpected outbreaks. To account for current and future threats imposed by pathogens, it is necessary to understand human-pathogen interactions at the molecular level. Viruses require host factors for recognition, entrance, replication, and release. Their gene products form dense interaction networks with host proteins. Most bacteria, fungi, and nematodes, on the other hand, proliferate outside of human cells and interact with host cells with extracellular signals and receptors. The following sections of this chapter review previous works on high-throughput characterization of human-pathogen interactions interactions, mostly between human and viruses. Most works have focused on human-virus interactions.

5.1 Human-virus interactions

High-throughput characterization of intra-species interactions has been the focus of early day PPI studies. Inter-species interactions still constitute a minor part of most interactome databases. Beginning from 2007, several works on high-throughput human-virus interactions and host factor characterization have been published, including the ones for Epstein-Barr virus (EBV), hepatitis C virus (HCV), and influenza virus. Among these sparse inter-species interactions, those between human immunodeficiency virus 1 (HIV-1) and human are most abundant due to the research efforts devoted to this notorious virus. These datasets are summarized in Table 4.

Among these datasets, the HIV-1 human protein interaction database is so far the most comprehensive in terms of recorded interaction number and annotations. The number of human-virus PPIs can be estimated based on the number of human-HIV PPIs, which presumably have not been fully exposed, and the number of human viruses. A severe under-estimated number of human viruses is 200 ~ 1 000 species, which can be deduced to at least 1 ~ 5 million human-virus PPIs yet to be discovered. Despite the small number of human-virus PPIs being detected or predicted, this data is a start point to the research of the viral disease mechanisms and treatments.

Datasets	No. of Interactions	No. of Viral Strains	No. of Viral Proteins	No. of Human Proteins	Sources
Human Immunodeficiency Virus 1 (HIV-1)	5128	1	21	1433	(Pinney et al., 2009)
Epstein-Barr Virus	173	1	42	112	(Calderwood et al., 2007)
Hepatitis C Virus	481	1	11	414	(de Chassey et al., 2008)
Influenza Virus	400	1	10	246	(Konig et al., 2010)
NCBI Interactions	5370	39	86	1530	NCBI FTP Site[1]
IntAct	689	50	124	308	(Aranda et al., 2010)

Table 4. Summary of human-virus interaction datasets

5.2 HIV-1 interactions

The HIV-1, Human Protein Interaction Database (Pinney et al., 2009) was compiled by National Institute of Allergy and Infectious Diseases (NIAID), and hosted by NCBI. Interaction data in this database was collected from published literatures. Unlike other interaction data, entries in this dataset were associated with detailed annotations, including PubMed ID list for references, short phrases describing the interactions, and texts excerpted from the source literature. Interactions in this database are not just revealed by conventional Y2H or immune-co-precipitation, but 70 interactions were annotated with details. For example, the statement "HIV retropepsin cleaves human actin" is supported by four publications and attached with descriptions of the HIV retropepsin and human actin. Occassionally, the texts from the source literatures would provide additional information. In the case of "HIV retropepsin cleaves human alpha-2-macroglobulin precursor", the GeneRIF text states "the cleavage site of alpha 2-Macroglobulin by HIV-1 protease is the Phe684-Tyr685 bond", which depicts the interaction (cleavage) site. Interaction types include cleavage, binding, regulation/modulation, and post-translational modifications.

Analysis of this database found that there were 21 HIV gene products interacting with 1433 human proteins. The top 10 HIV and human proteins which participate in most HIV-human interactions are listed in Table 5.

By simply counting the numbers of PPIs in Table 5, critical host factors in HIV infections could be identified. The C-C chemokine receptor type 5 (CCR5) variants have been implicated in HIV-resistance and immunity (Blanpain et al., 2002). Stem cell-based gene therapy has successfully "cured" HIV with this genetic variant in early phase clinical trials (Symonds et al., 2010). Some host factors were also involved in various types of processes and diseases, such as tumour necrosis factor (TNF), which regulates cell proliferation, apoptosis, and has been implicated in cancer.

[1] ftp://ftp.ncbi.nlm.nih.gov/gene/GeneRIF/interactions.gz

Top 10 HIV Proteins	No. of Interactions	Top 10 Human Proteins	No. of Interactions
Tat	1685	T-cell surface glycoprotein CD4 isoform 1 precursor	177
gp120	1252	C-X-C chemokine receptor type 4 isoform b	54
Nef	496	Tumor necrosis factor	47
Vpr	342	Nuclear factor NF-kappa-B p105 subunit isoform 1	41
gp41	253	Nuclear factor NF-kappa-B p105 subunit isoform 1	41
gp160	232	C-C chemokine receptor type 5	40
Rev	139	Interferon gamma precursor	35
Matrix	132	Mitogen-activated protein kinase 1	30
Integrase	122	Major histocompatibility complex, class I, B precursor	30
Retropepsin	98	Cyclin-dependent protein kinase 9	30

Table 5. HIV and human proteins participate in largest numbers of human-HIV interactions.

Table 5 also suggests that Tat could be a potential drug target. The crystal structure of Tat which forms complex with cyclin-denpendent protein kinase 9 (CDK9) and cyclin T1 has been solved (Tahirov et al., 2010). The complex structure (PDB ID: 3MI9) reveals that the most part of Tat has physical contact with cyclin T1 and has only a small loop contacting CDK9. The structural information provides valuable insights to the design of Tat inhibitors.

5.3 Epstein-Barr virus interactions

Epstein-Barr virus (EBV) infects human epithelial cells, and is implicated in various types of cancer, such as Burkitt's lymphoma and nasopharyngeal carcinoma. The interactions within EBV proteins, and between EBV and human proteins, have been characterized using Y2H method (Calderwood et al., 2007). Overall, 43 EBV-EBV and 173 human-EBV interactions have been validated with experimental evidences.

Network analysis reveals that most EBV-EBV interactions take place among conserved "core" proteins, thus these interactions may be responsible for the general infection/replication of herpesviruses. On the other hand, most human proteins targeted by EBV are proposed as "hub" proteins, which participate in more human-human interactions and may have crucial roles in the underlying biological processes. The EBV protein targeting most human proteins is BFLF2, with 21 interaction partners. BFLF2 interacts with BFRF1 and changes cellular localization (Gonnella et al., 2005). Deletion of BFLF2 also impairs viral DNA packaging (Granato et al., 2008). The most targeted human proteins are HOMER3 and GRN. HOMER3, which binds to numerous receptors, is involved in diverse biological functions such as neuronal signalling and T-cell activation. Granulin (GRN) is a secreted glycosylated peptide which regulates cell growth and implicates in wound healing and tumorigenesis. BFLF2 interacts with both HOMER3 and GRN; however, the functional implications of interactions were not clear.

5.4 Hepatitis C virus interactions

Hepatitis C virus (HCV) is the pathogen which causes the chronic hepatitis infection. Infection with HCV may lead to cirrhosis and hepatocarcinoma if not properly treated with antiviral drugs or interferon. Unfortunately, current HCV treatments are expensive and can have severe adverse effects. The human-HCV interaction map would allow us to understand the mechanisms of HCV infection and its chronic nature.

The human-HCV interaction network is constituted by 481 HCV-human interactions (de Chassey et al., 2008). Among these interactions, 314 were determined with Y2H experiments, and others were identified from literature reviews. The most connected HCV proteins include NS3, NS5A, and CORE. Human proteins targeted by most HCV proteins include nuclear receptor subfamily 4, group A, member 1 (NR4A1), homeobox D8 (HOXD8), and SET domain containing 2 (SETD2). NR4A1 is a nuclear transcription factor, which is highly expressed in adrenal cortex, lung, and prostate; however its expression level in liver is low. HOXD8 is important to development; its deletion leads to limb deformation. Expression level of HOXD8 is highest in kidney. SETD2 is a histone methyltransferase and also contains transcription activation domain. Recently, SETD2 has been found as a tumour suppressor gene (Duns et al., 2010). However, the roles of HCV-SETD2 interactions in tumorigenesis remain elusive.

The analyses of EBV-human and HCV-human interaction networks found that viral proteins tend to interact with "hubs" in human protein-protein interaction networks. In human proteins targeted by HCV, three KEGG pathways were significantly enriched, including insulin signalling pathway, TGFβ signalling pathway, and Jak-STAT signalling pathway (de Chassey et al., 2008). Also, "focal adhesion" pathway has been identified as a novel pathway targeted by HCV. In our own analysis using bootstrap to estimate the statistical significance of HCV targeted gene numbers, we have also identified that "focal adhesion" and "ECM-receptor interaction" pathways may be perturbed by HCV infection (Table 6).

Path ID	Title	No. of genes in pathway	# of HCV targets	Random (mean)	Random (SD)	Z-stat	p-value
5160	**Hepatitis C**	134	23	1.82	1.35	15.70	$< 2.2 \times 10^{-16}$
5200	**Pathways in cancer**	328	31	4.60	2.20	12.02	$< 2.2 \times 10^{-16}$
5212	**Pancreatic cancer**	71	12	0.99	0.95	11.54	$< 2.2 \times 10^{-16}$
4510	**Focal adhesion**	202	22	2.80	1.68	11.42	$< 2.2 \times 10^{-16}$
5222	**Small cell lung cancer**	84	13	1.19	1.08	10.98	$< 2.2 \times 10^{-16}$
4520	**Adherents junction**	75	12	1.07	1.03	10.59	$< 2.2 \times 10^{-16}$
4722	**Neurotrophin signaling pathway**	127	15	1.78	1.32	10.04	$< 2.2 \times 10^{-16}$
5215	**Prostate cancer**	89	12	1.32	1.12	9.56	$< 2.2 \times 10^{-16}$
4512	**ECM-receptor interaction**	84	11	1.16	1.05	9.35	$< 2.2 \times 10^{-16}$

Table 6. Top 10 KEGG pathways potentially perturbed by HCV infection.

5.5 Influenza virus host factors

Influenza A virus causes epidemics every now and then. The high transmission rate of influenza virus makes it one of the greatest threats to public health, especially when long diminished strains or emerging strains turned to the surface. The rapidly evolving virus makes it difficult to predict and prepare seasonal vaccines. Drug-resistant strains also challenge our ability to treat and control the disease.

The identification of host factors required by influenza virus may contribute to the prevention and treatment of the virus. Host factors involved in early stage influenza virus replication have been characterized with genome-wide RNA interference (RNAi) screening (Konig et al., 2010). Unlike Y2H experiments, host factors identified with RNAi do not necessary interact with viral proteins directly. Nevertheless, these findings imply that viral diseases may be treated by regulating some of these host factors. One example is the inhibitor for the host factors, CAMK2B, which impedes viral growth and may be developed to new antiviral drugs.

5.6 Human-bacteria interactions

Bacteria cells can reproduce without the cellular machinery of hosts. Studies on human-bacteria interactions thus have been focused on cellular-level interactions. So far, only limited efforts have been devoted to the identification of human-bacteria interactions at the molecular-level. The interactions between three pathogenic bacteria, *Bacillus anthracis*, *Francisella tularensis*, and *Yersinia pestis*, have been characterized using high-throughput Y2H experiments (Dyer et al., 2010). The reported dataset includes 3,073, 1,383, and 4,059 interactions between human or *B. anthracis*, *F. tularensis* or *Y. pestis*, respectively.

The topology of these human-bacteria and human protein-protein interaction networks revealed that many bacteria proteins target "hubs" in human PPI networks. Specifically, several host defence pathways have been identified, including innate immunity and inflammation. Comparative analysis of the three human-pathogen interaction networks also confirmed these findings. Several methods have been used to identify the conserved protein interaction modules (CPIM), and found that these bacteria may have interfered host innate immune responses, including antigen binding and processing, and several immune response pathways.

Analysis of these interactions faces some obstacles. Large proportions of the bacteria proteins (285/943, 66/349, 630/1,218 protiens for *B. anthracis*, *F. tularensis*, and *Y. pestis*, respectively) which interact with human proteins are putative, hypothetical, or uncharacterized. Without sufficient functional annotations, interpretations of these interactions can be superficial. Furthermore, bacterial proteins and human proteins were confined within the membrane of respective cells, and only certain types of proteins can be exported or internalized by host cells. Thus, the annotations or predictions of protein subcellular localization are important tasks for the interpretation and refinements of human-bacteria interaction networks.

5.7 Systematic analysis of host-pathogen interactions

Human-pathogen interactions have been collected and analyzed for their network properties (Dyer et al., 2008). Pooling together human-pathogen interactions should enable identification of common targets and biological processes perturbed by pathogens. A total of

10477 interactions between human and the 190 pathogen strains have been collected from several public databases. Networks for human-bacteria and human-virus interactions have been constructed separately. Most of the interactions were human-virus interactions, notably human-HIV interactions.

Special attention has been paid to human proteins which interact with multiple pathogen groups. Such proteins are believed to be the common targets of these pathogens, and may be the highlights of critical events during pathogen invasion.

The analysis of human proteins targeted by multiple viral pathogens have revealed that viruses perturb host cells mainly through controlling cell cycle, regulating apoptosis, and transporting viral particles across membrane (Dyer et al., 2008). Human-bacteria interactions, on the other hand, perturb Gene Ontology processes like "immune system process" and "immune response". It is notable that much of these perturbed pathways were linked to inflammatory and cancer, suggesting multiple roles of pathogens in various diseases.

Another analysis on human-viral interaction network also highlighted the mechanisms of non-infectious diseases (Navratil et al., 2011). Totally 2,099 manually curated interactions of 416 viral proteins from 110 species have been collected. This human-virus interaction network has been integrated with human PPI network. Disease gene annotations from OMIM have been evaluated for their associations with viral proteins. Links between virus and auto-immune diseases have been found, including type 1 diabetes. A comparison between human-virus interaction network and human type I interferon network also revealed that viruses attack host at multiple levels, from receptors to transcription factors (Navratil et al., 2010).

We have also performed similar analysis with human-virus interactions collected from NCBI interactions[2], IntAct (Aranda et al., 2009), and other sources. The association of KEGG (Kanehisa et al., 2010) disease pathways and human-virus interactions have been analyzed. Several KEGG pathways have been identified with high significance, including "systemic lupus erythematosus", "pathways in cancer", "chemokine signalling pathway", "focal adhesion", and "T cell receptor signalling pathway". These findings are *in par* with studies described in previous sections; all pointing to pathogens gain their foothold in host cells through modulating host defence mechanisms. In the meantime, inflammation, autoimmune diseases and cancers may arise as results of these modulations.

6. Conclusion

At the present, the number of confident PPI data is scarce, especially for non-human organisms and inter-species interactions. The prediction of PPIs, as well as the evaluation of accuracy of detected and predicted PPIs, are important topics which require further advances in methodology, tools and data generation. It is believed that, in recent years, as the second generation sequencing (SGS) rapidly discloses full genome sequences and exponentially accumulates high-throughput expression data, more and more inter- and intra-species networks PPI will be constructed for, not only model organisms, but also crops, biofuel producing algae and bacteria, between host-pathogens, and between symbiotic organisms.

[2] ftp://ftp.ncbi.nlm.nih.gov/gene/GeneRIF/interactions.gz

7. References

Anderson, G. A.; Tolic, N.; Tang, X.; Zheng, C. & Bruce, J. E. (2007). Informatics strategies for large-scale novel cross-linking analysis. *J Proteome Res*, Vol. 6, No. 9, pp. 3412-21

Angenendt, P.; Kreutzberger, J.; Glokler, J. & Hoheisel, J. D. (2006). Generation of high density protein microarrays by cell-free in situ expression of unpurified pcr products. *Mol Cell Proteomics*, Vol. 5, No. 9, pp. 1658-66

Aranda, B.Achuthan, P.Alam-Faruque, Y.Armean, I.Bridge, A.Derow, C.Feuermann, M.Ghanbarian, A. T.Kerrien, S.Khadake, J., et al. (2009). The intact molecular interaction database in 2010. *Nucleic Acids Res*, Vol. 38, No. Database issue, pp. D525-31

Aranda, B.Achuthan, P.Alam-Faruque, Y.Armean, I.Bridge, A.Derow, C.Feuermann, M.Ghanbarian, A. T.Kerrien, S.Khadake, J., et al. (2010). The intact molecular interaction database in 2010. *Nucleic Acids Res*, Vol. 38, No. Database issue, pp. D525-31

Aytuna, A. S.; Gursoy, A. & Keskin, O. (2005). Prediction of protein-protein interactions by combining structure and sequence conservation in protein interfaces. *Bioinformatics*, Vol. 21, No. 12, pp. 2850-5

Baldwin, T. K.; Winnenburg, R.; Urban, M.; Rawlings, C.; Koehler, J. & Hammond-Kosack, K. E. (2006). The pathogen-host interactions database (phi-base) provides insights into generic and novel themes of pathogenicity. *Mol Plant Microbe Interact*, Vol. 19, No. 12, pp. 1451-62

Berggard, T.; Linse, S. & James, P. (2007). Methods for the detection and analysis of protein-protein interactions. *Proteomics*, Vol. 7, No. 16, pp. 2833-42

Blanpain, C.; Libert, F.; Vassart, G. & Parmentier, M. (2002). Ccr5 and hiv infection. *Receptors Channels*, Vol. 8, No. 1, pp. 19-31

Brandao, M. M.; Dantas, L. L. & Silva-Filho, M. C. (2009). Atpin: Arabidopsis thaliana protein interaction network. *BMC Bioinformatics*, Vol. 10, p 454

Breitkreutz, B. J.Stark, C.Reguly, T.Boucher, L.Breitkreutz, A.Livstone, M.Oughtred, R.Lackner, D. H.Bahler, J.Wood, V., et al. (2008). The biogrid interaction database: 2008 update. *Nucleic Acids Res*, Vol. 36, No. Database issue, pp. D637-40

Bulow, L.; Schindler, M. & Hehl, R. (2007). Pathoplant: A platform for microarray expression data to analyze co-regulated genes involved in plant defense responses. *Nucleic Acids Res*, Vol. 35, No. Database issue, pp. D841-5

Calderwood, M. A.Venkatesan, K.Xing, L.Chase, M. R.Vazquez, A.Holthaus, A. M.Ewence, A. E.Li, N.Hirozane-Kishikawa, T.Hill, D. E., et al. (2007). Epstein-barr virus and virus human protein interaction maps. *Proc Natl Acad Sci U S A*, Vol. 104, No. 18, pp. 7606-11

Causier, B. & Davies, B. (2002). Analysing protein-protein interactions with the yeast two-hybrid system. *Plant Mol Biol*, Vol. 50, No. 6, pp. 855-70

Causier, B. (2004). Studying the interactome with the yeast two-hybrid system and mass spectrometry. *Mass Spectrom Rev*, Vol. 23, No. 5, pp. 350-67

Ceol, A.; Chatr Aryamontri, A.; Licata, L.; Peluso, D.; Briganti, L.; Perfetto, L.; Castagnoli, L. & Cesareni, G. (2010). Mint, the molecular interaction database: 2009 update. *Nucleic Acids Res*, Vol. 38, No. Database issue, pp. D532-9

Chen, J.; Hsu, W.; Lee, M. L. & Ng, S. K. (2006). Increasing confidence of protein interactomes using network topological metrics. *Bioinformatics*, Vol. 22, No. 16, pp. 1998-2004

Chen, J. Y.; Mamidipalli, S. & Huan, T. (2009). Happi: An online database of comprehensive human annotated and predicted protein interactions. *BMC Genomics*, Vol. 10 Suppl 1, p S16

Chua, H. N.; Sung, W. K. & Wong, L. (2006). Exploiting indirect neighbours and topological weight to predict protein function from protein-protein interactions. *Bioinformatics*, Vol. 22, No. 13, pp. 1623-30

Citovsky, V.; Gafni, Y. & Tzfira, T. (2008). Localizing protein-protein interactions by bimolecular fluorescence complementation in planta. *Methods*, Vol. 45, No. 3, pp. 196-206

Cui, J.; Li, P.; Li, G.; Xu, F.; Zhao, C.; Li, Y.; Yang, Z.; Wang, G.; Yu, Q. & Shi, T. (2008). Atpid: Arabidopsis thaliana protein interactome database--an integrative platform for plant systems biology. *Nucleic Acids Res*, Vol. 36, No. Database issue, pp. D999-1008

Dandekar, T.; Snel, B.; Huynen, M. & Bork, P. (1998). Conservation of gene order: A fingerprint of proteins that physically interact. *Trends Biochem Sci*, Vol. 23, No. 9, pp. 324-8

De Bodt, S.; Proost, S.; Vandepoele, K.; Rouze, P. & Van de Peer, Y. (2009). Predicting protein-protein interactions in arabidopsis thaliana through integration of orthology, gene ontology and co-expression. *BMC Genomics*, Vol. 10, p 288

de Chassey, B.Navratil, V.Tafforeau, L.Hiet, M. S.Aublin-Gex, A.Agaugue, S.Meiffren, G.Pradezynski, F.Faria, B. F.Chantier, T., et al. (2008). Hepatitis c virus infection protein network. *Mol Syst Biol*, Vol. 4, p 230

Driscoll, T.; Dyer, M. D.; Murali, T. M. & Sobral, B. W. (2009). Pig--the pathogen interaction gateway. *Nucleic Acids Res*, Vol. 37, No. Database issue, pp. D647-50

Duns, G.; van den Berg, E.; van Duivenbode, I.; Osinga, J.; Hollema, H.; Hofstra, R. M. & Kok, K. (2010). Histone methyltransferase gene setd2 is a novel tumor suppressor gene in clear cell renal cell carcinoma. *Cancer Res*, Vol. 70, No. 11, pp. 4287-91

Dyer, M. D.; Murali, T. M. & Sobral, B. W. (2007). Computational prediction of host-pathogen protein-protein interactions. *Bioinformatics*, Vol. 23, No. 13, pp. i159-66

Dyer, M. D.; Murali, T. M. & Sobral, B. W. (2008). The landscape of human proteins interacting with viruses and other pathogens. *PLoS Pathog*, Vol. 4, No. 2, p e32

Dyer, M. D.; Neff, C.; Dufford, M.; Rivera, C. G.; Shattuck, D.; Bassaganya-Riera, J.; Murali, T. M. & Sobral, B. W. (2010). The human-bacterial pathogen protein interaction networks of bacillus anthracis, francisella tularensis, and yersinia pestis. *PLoS One*, Vol. 5, No. 8, p e12089

Enright, A. J.; Iliopoulos, I.; Kyrpides, N. C. & Ouzounis, C. A. (1999). Protein interaction maps for complete genomes based on gene fusion events. *Nature*, Vol. 402, No. 6757, pp. 86-90

Fischer, T. B.; Paczkowski, M.; Zettel, M. F. & Tsai, J. (2005). A guide to protein interaction databases, In: *The proteomics protocols handbook,* ed. J. M. Walker, pp (753-799), Humana Press, 978-1-59259-890-8,

Frishman, D., Albrecht, M., Blankenburg, H., Bork, P., Harrington, E. D., Hermjakob, H., Jensen, L. J., Juan, D. A., Lengauer, T., Pagel, P., Schachter, V. and Valencia, A.

(2009). Protein-protein interactions: Analysis and prediction, In: *Modern genome annotation: The biosapiens network.* , ed. D. a. V. Frishman, A., pp (353-410), Springer, Wien, Austria

Fryxell, K. J. (1996). The coevolution of gene family trees. *Trends Genet*, Vol. 12, No. 9, pp. 364-9

Fujikawa, Y. & Kato, N. (2007). Split luciferase complementation assay to study protein-protein interactions in arabidopsis protoplasts. *Plant J*, Vol. 52, No. 1, pp. 185-95

Garcia-Hernandez, M.Berardini, T. Z.Chen, G.Crist, D.Doyle, A.Huala, E.Knee, E.Lambrecht, M.Miller, N.Mueller, L. A., et al. (2002). Tair: A resource for integrated arabidopsis data. *Funct Integr Genomics*, Vol. 2, No. 6, pp. 239-53

Geisler-Lee, J.; O'Toole, N.; Ammar, R.; Provart, N. J.; Millar, A. H. & Geisler, M. (2007). A predicted interactome for arabidopsis. *Plant Physiol*, Vol. 145, No. 2, pp. 317-29

Gilbert, D. (2005). Biomolecular interaction network database. *Brief Bioinform*, Vol. 6, No. 2, pp. 194-8

Goel, R.; Muthusamy, B.; Pandey, A. & Prasad, T. S. (2010). Human protein reference database and human proteinpedia as discovery resources for molecular biotechnology. *Mol Biotechnol*,

Goh, C. S.; Bogan, A. A.; Joachimiak, M.; Walther, D. & Cohen, F. E. (2000). Co-evolution of proteins with their interaction partners. *J Mol Biol*, Vol. 299, No. 2, pp. 283-93

Gollery, M.; Harper, J.; Cushman, J.; Mittler, T.; Girke, T.; Zhu, J. K.; Bailey-Serres, J. & Mittler, R. (2006). What makes species unique? The contribution of proteins with obscure features. *Genome Biol*, Vol. 7, No. 7, p R57

Gonnella, R.Farina, A.Santarelli, R.Raffa, S.Feederle, R.Bei, R.Granato, M.Modesti, A.Frati, L.Delecluse, H. J., et al. (2005). Characterization and intracellular localization of the epstein-barr virus protein bflf2: Interactions with bfrf1 and with the nuclear lamina. *J Virol*, Vol. 79, No. 6, pp. 3713-27

Granato, M.; Feederle, R.; Farina, A.; Gonnella, R.; Santarelli, R.; Hub, B.; Faggioni, A. & Delecluse, H. J. (2008). Deletion of epstein-barr virus bflf2 leads to impaired viral DNA packaging and primary egress as well as to the production of defective viral particles. *J Virol*, Vol. 82, No. 8, pp. 4042-51

Gruhler, S. & Kratchmarova, I. (2008). Stable isotope labeling by amino acids in cell culture (silac). *Methods Mol Biol*, Vol. 424, pp. 101-11

He, F.; Zhang, Y.; Chen, H.; Zhang, Z. & Peng, Y. L. (2008). The prediction of protein-protein interaction networks in rice blast fungus. *BMC Genomics*, Vol. 9, p 519

Hirsh, E. & Sharan, R. (2007). Identification of conserved protein complexes based on a model of protein network evolution. *Bioinformatics*, Vol. 23, No. 2, pp. e170-6

Hu, C. D.; Grinberg, A. V. & Kerppola, T. K. (2006). Visualization of protein interactions in living cells using bimolecular fluorescence complementation (bifc) analysis. *Curr Protoc Cell Biol*, Vol. Chapter 21, p Unit 21 3

Huttenhower, C. & Troyanskaya, O. G. (2006). Bayesian data integration: A functional perspective. *Comput Syst Bioinformatics Conf*, pp. 341-51

Huttlin, E. L.; Hegeman, A. D.; Harms, A. C. & Sussman, M. R. (2007). Comparison of full versus partial metabolic labeling for quantitative proteomics analysis in arabidopsis thaliana. *Mol Cell Proteomics*, Vol. 6, No. 5, pp. 860-81

Ideker, T.; Ozier, O.; Schwikowski, B. & Siegel, A. F. (2002). Discovering regulatory and signalling circuits in molecular interaction networks. *Bioinformatics*, Vol. 18 Suppl 1, pp. S233-40

Isserlin, R.; El-Badrawi, R. A. & Bader, G. D. (2011). The biomolecular interaction network database in psi-mi 2.5. *Database (Oxford)*, Vol. 2011, p baq037

Jaeger, S.; Gaudan, S.; Leser, U. & Rebholz-Schuhmann, D. (2008). Integrating protein-protein interactions and text mining for protein function prediction. *BMC Bioinformatics*, Vol. 9 Suppl 8, p S2

Jain, S. & Bader, G. D. (2010). An improved method for scoring protein-protein interactions using semantic similarity within the gene ontology. *BMC Bioinformatics*, Vol. 11, p 562

Jansen, R.; Yu, H.; Greenbaum, D.; Kluger, Y.; Krogan, N. J.; Chung, S.; Emili, A.; Snyder, M.; Greenblatt, J. F. & Gerstein, M. (2003). A bayesian networks approach for predicting protein-protein interactions from genomic data. *Science*, Vol. 302, No. 5644, pp. 449-53

Jothi, R.; Kann, M. G. & Przytycka, T. M. (2005). Predicting protein-protein interaction by searching evolutionary tree automorphism space. *Bioinformatics*, Vol. 21 Suppl 1, pp. i241-50

Kanehisa, M.; Goto, S.; Furumichi, M.; Tanabe, M. & Hirakawa, M. (2010). Kegg for representation and analysis of molecular networks involving diseases and drugs. *Nucleic Acids Res*, Vol. 38, No. Database issue, pp. D355-60

Kann, M. G.; Jothi, R.; Cherukuri, P. F. & Przytycka, T. M. (2007). Predicting protein domain interactions from coevolution of conserved regions. *Proteins*, Vol. 67, No. 4, pp. 811-20

Konig, R.Stertz, S.Zhou, Y.Inoue, A.Hoffmann, H. H.Bhattacharyya, S.Alamares, J. G.Tscherne, D. M.Ortigoza, M. B.Liang, Y., et al. (2010). Human host factors required for influenza virus replication. *Nature*, Vol. 463, No. 7282, pp. 813-7

Kumar, R. & Nanduri, B. (2010). Hpidb--a unified resource for host-pathogen interactions. *BMC Bioinformatics*, Vol. 11 Suppl 6, p S16

Lee, H.; Deng, M.; Sun, F. & Chen, T. (2006). An integrated approach to the prediction of domain-domain interactions. *BMC Bioinformatics*, Vol. 7, p 269

Lehner, B. & Fraser, A. G. (2004). A first-draft human protein-interaction map. *Genome Biol*, Vol. 5, No. 9, p R63

Li, P.; Zang, W.; Li, Y.; Xu, F.; Wang, J. & Shi, T. (2011). Atpid: The overall hierarchical functional protein interaction network interface and analytic platform for arabidopsis. *Nucleic Acids Res*, Vol. 39, No. Database issue, pp. D1130-3

Lin, M.; Shen, X. & Chen, X. (2011). Pair: The predicted arabidopsis interactome resource. *Nucleic Acids Res*, Vol. 39, No. Database issue, pp. D1134-40

Lin, N.; Wu, B.; Jansen, R.; Gerstein, M. & Zhao, H. (2004). Information assessment on predicting protein-protein interactions. *BMC Bioinformatics*, Vol. 5, p 154

Liu, X.; Yang, W. C.; Gao, Q. & Regnier, F. (2008). Toward chromatographic analysis of interacting protein networks. *J Chromatogr A*, Vol. 1178, No. 1-2, pp. 24-32

Mann, M. (2006). Functional and quantitative proteomics using silac. *Nat Rev Mol Cell Biol*, Vol. 7, No. 12, pp. 952-8

Marcotte, E. M.; Pellegrini, M.; Thompson, M. J.; Yeates, T. O. & Eisenberg, D. (1999). A combined algorithm for genome-wide prediction of protein function. *Nature*, Vol. 402, No. 6757, pp. 83-6

Matthews, L. R.; Vaglio, P.; Reboul, J.; Ge, H.; Davis, B. P.; Garrels, J.; Vincent, S. & Vidal, M. (2001). Identification of potential interaction networks using sequence-based searches for conserved protein-protein interactions or "Interologs". *Genome Res*, Vol. 11, No. 12, pp. 2120-6

McDowall, M. D.; Scott, M. S. & Barton, G. J. (2009). Pips: Human protein-protein interaction prediction database. *Nucleic Acids Res*, Vol. 37, No. Database issue, pp. D651-6

Michnick, S. W. (2003). Protein fragment complementation strategies for biochemical network mapping. *Curr Opin Biotechnol*, Vol. 14, No. 6, pp. 610-7

Mika, S. & Rost, B. (2006). Protein-protein interactions more conserved within species than across species. *PLoS Comput Biol*, Vol. 2, No. 7, p e79

Morsy, M.; Gouthu, S.; Orchard, S.; Thorneycroft, D.; Harper, J. F.; Mittler, R. & Cushman, J. C. (2008). Charting plant interactomes: Possibilities and challenges. *Trends Plant Sci*, Vol. 13, No. 4, pp. 183-91

Mrowka, R.; Patzak, A. & Herzel, H. (2001). Is there a bias in proteome research? *Genome Res*, Vol. 11, No. 12, pp. 1971-3

Navratil, V.; de Chassey, B.; Meyniel, L.; Pradezynski, F.; Andre, P.; Rabourdin-Combe, C. & Lotteau, V. (2010). System-level comparison of protein-protein interactions between viruses and the human type i interferon system network. *J Proteome Res*, Vol. 9, No. 7, pp. 3527-36

Navratil, V.; de Chassey, B.; Combe, C. R. & Lotteau, V. (2011). When the human viral infectome and diseasome networks collide: Towards a systems biology platform for the aetiology of human diseases. *BMC Syst Biol*, Vol. 5, p 13

Nelson, C. J.; Huttlin, E. L.; Hegeman, A. D.; Harms, A. C. & Sussman, M. R. (2007). Implications of 15n-metabolic labeling for automated peptide identification in arabidopsis thaliana. *Proteomics*, Vol. 7, No. 8, pp. 1279-92

Ng, S. K.; Zhang, Z. & Tan, S. H. (2003). Integrative approach for computationally inferring protein domain interactions. *Bioinformatics*, Vol. 19, No. 8, pp. 923-9

Obayashi, T.; Hayashi, S.; Shibaoka, M.; Saeki, M.; Ohta, H. & Kinoshita, K. (2008). Coxpresdb: A database of coexpressed gene networks in mammals. *Nucleic Acids Res*, Vol. 36, No. Database issue, pp. D77-82

Obayashi, T. & Kinoshita, K. (2011). Coxpresdb: A database to compare gene coexpression in seven model animals. *Nucleic Acids Res*, Vol. 39, No. Database issue, pp. D1016-22

Obayashi, T.; Nishida, K.; Kasahara, K. & Kinoshita, K. (2011). Atted-ii updates: Condition-specific gene coexpression to extend coexpression analyses and applications to a broad range of flowering plants. *Plant Cell Physiol*, Vol. 52, No. 2, pp. 213-9

Obrdlik, P.El-Bakkoury, M.Hamacher, T.Cappellaro, C.Vilarino, C.Fleischer, C.Ellerbrok, H.Kamuzinzi, R.Ledent, V.Blaudez, D., et al. (2004). K+ channel interactions detected by a genetic system optimized for systematic studies of membrane protein interactions. *Proc Natl Acad Sci U S A*, Vol. 101, No. 33, pp. 12242-7

Ogmen, U.; Keskin, O.; Aytuna, A. S.; Nussinov, R. & Gursoy, A. (2005). Prism: Protein interactions by structural matching. *Nucleic Acids Res*, Vol. 33, No. Web Server issue, pp. W331-6

Ohad, N.; Shichrur, K. & Yalovsky, S. (2007). The analysis of protein-protein interactions in plants by bimolecular fluorescence complementation. *Plant Physiol*, Vol. 145, No. 4, pp. 1090-9

Ohad, N. & Yalovsky, S. (2010). Utilizing bimolecular fluorescence complementation (bifc) to assay protein-protein interaction in plants. *Methods Mol Biol*, Vol. 655, pp. 347-58

Overbeek, R.; Fonstein, M.; D'Souza, M.; Pusch, G. D. & Maltsev, N. (1999). Use of contiguity on the chromosome to predict functional coupling. *In Silico Biol*, Vol. 1, No. 2, pp. 93-108

Pagel, P.Kovac, S.Oesterheld, M.Brauner, B.Dunger-Kaltenbach, I.Frishman, G.Montrone, C.Mark, P.Stumpflen, V.Mewes, H. W., et al. (2005). The mips mammalian protein-protein interaction database. *Bioinformatics*, Vol. 21, No. 6, pp. 832-4

Pages, S.; Belaich, A.; Belaich, J. P.; Morag, E.; Lamed, R.; Shoham, Y. & Bayer, E. A. (1997). Species-specificity of the cohesin-dockerin interaction between clostridium thermocellum and clostridium cellulolyticum: Prediction of specificity determinants of the dockerin domain. *Proteins*, Vol. 29, No. 4, pp. 517-27

Parrish, J. R.; Gulyas, K. D. & Finley, R. L., Jr. (2006). Yeast two-hybrid contributions to interactome mapping. *Curr Opin Biotechnol*, Vol. 17, No. 4, pp. 387-93

Patil, A. & Nakamura, H. (2005). Filtering high-throughput protein-protein interaction data using a combination of genomic features. *BMC Bioinformatics*, Vol. 6, p 100

Patil, A.; Nakai, K. & Nakamura, H. (2011). Hitpredict: A database of quality assessed protein-protein interactions in nine species. *Nucleic Acids Res*, Vol. 39, No. Database issue, pp. D744-9

Paulmurugan, R. & Gambhir, S. S. (2005). Firefly luciferase enzyme fragment complementation for imaging in cells and living animals. *Anal Chem*, Vol. 77, No. 5, pp. 1295-302

Paulmurugan, R. & Gambhir, S. S. (2007). Combinatorial library screening for developing an improved split-firefly luciferase fragment-assisted complementation system for studying protein-protein interactions. *Anal Chem*, Vol. 79, No. 6, pp. 2346-53

Pellegrini, M.; Marcotte, E. M.; Thompson, M. J.; Eisenberg, D. & Yeates, T. O. (1999). Assigning protein functions by comparative genome analysis: Protein phylogenetic profiles. *Proc Natl Acad Sci U S A*, Vol. 96, No. 8, pp. 4285-8

Pinney, J. W.; Dickerson, J. E.; Fu, W.; Sanders-Beer, B. E.; Ptak, R. G. & Robertson, D. L. (2009). Hiv-host interactions: A map of viral perturbation of the host system. *AIDS*, Vol. 23, No. 5, pp. 549-54

Popescu, S. C.; Popescu, G. V.; Bachan, S.; Zhang, Z.; Seay, M.; Gerstein, M.; Snyder, M. & Dinesh-Kumar, S. P. (2007). Differential binding of calmodulin-related proteins to their targets revealed through high-density arabidopsis protein microarrays. *Proc Natl Acad Sci U S A*, Vol. 104, No. 11, pp. 4730-5

Ramachandran, N.; Hainsworth, E.; Bhullar, B.; Eisenstein, S.; Rosen, B.; Lau, A. Y.; Walter, J. C. & LaBaer, J. (2004). Self-assembling protein microarrays. *Science*, Vol. 305, No. 5680, pp. 86-90

Reinders, A.; Schulze, W.; Kuhn, C.; Barker, L.; Schulz, A.; Ward, J. M. & Frommer, W. B. (2002). Protein-protein interactions between sucrose transporters of different affinities colocalized in the same enucleate sieve element. *Plant Cell*, Vol. 14, No. 7, pp. 1567-77

Rhodes, D. R.; Tomlins, S. A.; Varambally, S.; Mahavisno, V.; Barrette, T.; Kalyana-Sundaram, S.; Ghosh, D.; Pandey, A. & Chinnaiyan, A. M. (2005). Probabilistic model of the human protein-protein interaction network. *Nat Biotechnol*, Vol. 23, No. 8, pp. 951-9

Rohila, J. S.Chen, M.Chen, S.Chen, J.Cerny, R.Dardick, C.Canlas, P.Xu, X.Gribskov, M.Kanrar, S., et al. (2006). Protein-protein interactions of tandem affinity purification-tagged protein kinases in rice. *Plant J*, Vol. 46, No. 1, pp. 1-13

Salwinski, L.; Miller, C. S.; Smith, A. J.; Pettit, F. K.; Bowie, J. U. & Eisenberg, D. (2004). The database of interacting proteins: 2004 update. *Nucleic Acids Res*, Vol. 32, No. Database issue, pp. D449-51

Sayers, E. W.Barrett, T.Benson, D. A.Bolton, E.Bryant, S. H.Canese, K.Chetvernin, V.Church, D. M.DiCuccio, M.Federhen, S., et al. (2011). Database resources of the national center for biotechnology information. *Nucleic Acids Res*, Vol. 39, No. Database issue, pp. D38-51

Schoonheim, P. J.; Veiga, H.; Pereira Dda, C.; Friso, G.; van Wijk, K. J. & de Boer, A. H. (2007). A comprehensive analysis of the 14-3-3 interactome in barley leaves using a complementary proteomics and two-hybrid approach. *Plant Physiol*, Vol. 143, No. 2, pp. 670-83

Schulze, W. X.; Reinders, A.; Ward, J.; Lalonde, S. & Frommer, W. B. (2003). Interactions between co-expressed arabidopsis sucrose transporters in the split-ubiquitin system. *BMC Biochem*, Vol. 4, p 3

Shin, C. J.; Wong, S.; Davis, M. J. & Ragan, M. A. (2009). Protein-protein interaction as a predictor of subcellular location. *BMC Syst Biol*, Vol. 3, p 28

Sprinzak, E. & Margalit, H. (2001). Correlated sequence-signatures as markers of protein-protein interaction. *J Mol Biol*, Vol. 311, No. 4, pp. 681-92

Sprinzak, E.; Sattath, S. & Margalit, H. (2003). How reliable are experimental protein-protein interaction data? *J Mol Biol*, Vol. 327, No. 5, pp. 919-23

Stark, C.; Breitkreutz, B. J.; Reguly, T.; Boucher, L.; Breitkreutz, A. & Tyers, M. (2006). Biogrid: A general repository for interaction datasets. *Nucleic Acids Res*, Vol. 34, No. Database issue, pp. D535-9

Stark, C.Breitkreutz, B. J.Chatr-Aryamontri, A.Boucher, L.Oughtred, R.Livstone, M. S.Nixon, J.Van Auken, K.Wang, X.Shi, X., et al. (2011). The biogrid interaction database: 2011 update. *Nucleic Acids Res*, Vol. 39, No. Database issue, pp. D698-704

Symonds, G. P.; Johnstone, H. A.; Millington, M. L.; Boyd, M. P.; Burke, B. P. & Breton, L. R. (2010). The use of cell-delivered gene therapy for the treatment of hiv/aids. *Immunol Res*, Vol. 48, No. 1-3, pp. 84-98

Szklarczyk, D.Franceschini, A.Kuhn, M.Simonovic, M.Roth, A.Minguez, P.Doerks, T.Stark, M.Muller, J.Bork, P., et al. (2011). The string database in 2011: Functional interaction networks of proteins, globally integrated and scored. *Nucleic Acids Res*, Vol. 39, No. Database issue, pp. D561-8

Tahirov, T. H.; Babayeva, N. D.; Varzavand, K.; Cooper, J. J.; Sedore, S. C. & Price, D. H. (2010). Crystal structure of hiv-1 tat complexed with human p-tefb. *Nature*, Vol. 465, No. 7299, pp. 747-51

Tamames, J.; Casari, G.; Ouzounis, C. & Valencia, A. (1997). Conserved clusters of functionally related genes in two bacterial genomes. *J Mol Evol*, Vol. 44, No. 1, pp. 66-73

Tang, X.; Munske, G. R.; Siems, W. F. & Bruce, J. E. (2005). Mass spectrometry identifiable cross-linking strategy for studying protein-protein interactions. *Anal Chem*, Vol. 77, No. 1, pp. 311-8

Uhrig, J. F. & Hulskamp, M. (2006). Plant gtpases: Regulation of morphogenesis by rops and ros. *Curr Biol*, Vol. 16, No. 6, pp. R211-3

von Mering, C.; Krause, R.; Snel, B.; Cornell, M.; Oliver, S. G.; Fields, S. & Bork, P. (2002). Comparative assessment of large-scale data sets of protein-protein interactions. *Nature*, Vol. 417, No. 6887, pp. 399-403

von Mering, C.; Jensen, L. J.; Snel, B.; Hooper, S. D.; Krupp, M.; Foglierini, M.; Jouffre, N.; Huynen, M. A. & Bork, P. (2005). String: Known and predicted protein-protein associations, integrated and transferred across organisms. *Nucleic Acids Res*, Vol. 33, No. Database issue, pp. D433-7

von Mering, C.; Jensen, L. J.; Kuhn, M.; Chaffron, S.; Doerks, T.; Kruger, B.; Snel, B. & Bork, P. (2007). String 7--recent developments in the integration and prediction of protein interactions. *Nucleic Acids Res*, Vol. 35, No. Database issue, pp. D358-62

Wang, C.; Cheng, J. & Su, S. (2009). Prediction of interacting protein pairs from sequence using a bayesian method. *Protein J*, Vol. 28, No. 2, pp. 111-5

Willis, R. C. & Hogue, C. W. (2006). Searching, viewing, and visualizing data in the biomolecular interaction network database (bind). *Curr Protoc Bioinformatics*, Vol. Chapter 8, p Unit 8 9

Winnenburg, R.; Baldwin, T. K.; Urban, M.; Rawlings, C.; Kohler, J. & Hammond-Kosack, K. E. (2006). Phi-base: A new database for pathogen host interactions. *Nucleic Acids Res*, Vol. 34, No. Database issue, pp. D459-64

Wu, X.; Zhu, L.; Guo, J.; Zhang, D. Y. & Lin, K. (2006). Prediction of yeast protein-protein interaction network: Insights from the gene ontology and annotations. *Nucleic Acids Res*, Vol. 34, No. 7, pp. 2137-50

Xiang, Z.; Tian, Y. & He, Y. (2007). Phidias: A pathogen-host interaction data integration and analysis system. *Genome Biol*, Vol. 8, No. 7, p R150

Xu, Y.; Hu, W.; Chang, Z.; Duanmu, H.; Zhang, S.; Li, Z.; Yu, L. & Li, X. (2011). Prediction of human protein-protein interaction by a mixed bayesian model and its application to exploring underlying cancer-related pathway crosstalk. *J R Soc Interface*, Vol. 8, No. 57, pp. 555-67

Zhou, J.; Lin, J.; Zhou, C.; Deng, X. & Xia, B. (2011). An improved bimolecular fluorescence complementation tool based on superfolder green fluorescent protein. *Acta Biochim Biophys Sin (Shanghai)*, Vol. 43, No. 3, pp. 239-44

6

Computational Prediction of Post-Translational Modification Sites in Proteins

Yu Xue[1], Zexian Liu[2], Jun Cao[2] and Jian Ren[3]
[1]*Huazhong University of Science & Technology, Wuhan, Hubei,*
[2]*University of Science & Technology of China, Hefei, Anhui,*
[3]*Sun Yat-sen University, Guangzhou, Guangdong,*
China

1. Introduction

The last several decades have witnessed rapid progresses in identification and functional analysis of post-translational modifications (PTMs) in proteins. Through temporal and spatial modification of proteins by covalent attachment of additional chemical groups and other small proteins, proteolytic cleavage or intein splicing, PTMs greatly expand the proteome diversity and play important roles in regulating the stability and functions of the proteins (Mann and Jensen, 2003; Walsh, 2005; Walsh and Jefferis, 2006). To date, more than 350 types of distinct PTMs were experimentally discovered *in vivo*, while subsequently functional assays have detected a number of exciting observations. In 1992, the Nobel Prize in Physiology or Medicine was awarded to Edmond H. Fischer and Edwin G. Krebs for their seminal discovery that reversible protein phosphorylation is a biological regulatory mechanism (Kresge et al., 2011), while Leland H. Hartwell, Tim Hunt, and Paul M. Nurse shared the Nobel Prize in Physiology or Medicine 2001 for their profound contributions in identification of key regulators including cyclin-dependent kinases (CDKs) and cyclins that precisely orchestrate the cell cycle process through phosphorylation (Balter and Vogel, 2001). Moreover, Aaron Ciechanover, Avram Hershko and Irwin Rose became laureates of the Nobel Prize in Chemistry 2004 for their discovery of ubiquitin-mediated protein degradation (Vogel, 2004).

Although virtually all PTMs play their major roles as regulating the biological processes, different ones have their aspects with emphasis. For example, phosphorylation is preferentially implicated in signal-transduction cascades, while ubiquitination regulates the lifetime of proteins by targeting specific substrates for degradation. Recently, protein lysine acetylation was observed to play a predominant role in regulation of cellular metabolism (Wang et al., 2010; Zhao et al., 2010). Other types of PTMs such as sumoylation, glycosylation and palmitoylation are also critical for exactly orchestrating distinct cellular processes (Fukata and Fukata, 2010; Linder and Deschenes, 2007). Furthermore, the crosstalk among different PTMs is ubiquitous, especially on histones, which is regarded as the "histone code" (Jenuwein and Allis, 2001). The aberrances of PTMs are highly associated in diseases and cancers, while a variety of regulatory enzymes involved in PTMs have been drug targets (Lahiry et al., 2010; Norvell and McMahon, 2010). In this regard, elucidation of PTMs regulatory roles is fundamental for understanding molecular mechanisms of diseases and cancers, and further biomedical design.

Recently, with the developments of "state-of-the-art" techniques especially the high-throughput mass spectrometry (HTP-MS), large-scale identification of PTMs substrates with their sites has become a popular and near-routine assay (Choudhary and Mann, 2010). For example, combined with efficient isolation and enrichment methods such as antibodies which specifically recognize modified peptides and subsequently HTP-MS profiling, thousands of PTMs sites, eg., phosphorylation (Olsen et al., 2006; Villen et al., 2007), acetylation (Choudhary et al., 2009), or glycosylation (Zielinska et al., 2010) sites can be accurately determined in a single experiment. These high-throughout approaches can provide systematic insights into the biological roles of PTMs, especially a global view. However, due to the technical limitations such as low-sensitive detection of modifications in low expressed proteins (Ackermann and Berna, 2007; Boschetti and Righetti, 2009; Yates et al., 2009), and error-prone determination of multiple modified proteins (Hunter, 2007; Young et al., 2010), it is still a great challenge for fully charactering the whole PTM events *in vivo*.

In contrast with conventional experimental methods, computational analysis of PTMs has also been an alternative and attractive approach for its accuracy, fast-speed and convenience. The computational predictors can narrow down the number of potentially candidates and rapidly generate useful information for further experimental investigations. In one of our recent reviews, we specifically summarized more than 50 computational resources including public databases and prediction tools for phosphorylation (Xue et al., 2010a). Currently, although there have been ~170 databases and computational tools developed for PTM analysis (http://www.biocuckoo.org/link.php), accurate prediction of PTM sites in given proteins is still not a simple job. Again, although protein 3D structure information can be helpful for prediction of PTMs sites (Kumar and Mohanty, 2010), mainstream computational approaches were designed mainly based on protein primary sequence features (Xue et al., 2010a). A variety of algorithms have been introduced into this field, such as position-specific scoring matrix (PSSM) (Obenauer et al., 2003), support vector machines (SVMs) (Kim et al., 2004), artificial neural network (ANN) (Blom et al., 2004), Hidden Markov Model (HMM) (Huang et al., 2005), Bayesian decision theory (Xue et al., 2006a), and Conditional Random Field (CRF) (Dang et al., 2008). These methods were largely introduced from the fields of informatics or statistics and originally designed for general propose.

Previously, we developed a series of GPS algorithms (Initially defined as Group-based Phosphorylation Scoring and later renamed as Group-based Prediction System), which have been exclusively and successfully used for the prediction of kinds of PTM sites, such as phosphorylation (Xue et al., 2005; Xue et al., 2008; Xue et al., 2011; Zhou et al., 2004), sumoylation (Ren et al., 2009; Xue et al., 2006b), palmitoylation (Ren et al., 2008; Zhou et al., 2006a), *S*-Nitrosylation (Xue et al., 2010b) and nitration (Liu et al., 2011). The prediction performance of GPS 1.x (1.0 and 1.1) could be comparative to other analogous approaches, while GPS 2.x versions (2.0 and 2.1) are much better than other strategies (Xue et al., 2010a). Recently, we greatly improved our previous method and released the GPS 3.0 algorithm, while has been successfully adopted for predicting *S*-nitrosylation (Xue et al., 2010b) and nitration sites (Liu et al., 2011), with further enhanced performances. During the past several years, a considerable number of bioinformatists or experimentalists have communicated with us on GPS algorithm details, which were never thoroughly described in our previously research articles due to page limitation. In this regard, the aim of this chapter is to provide a comprehensive description of GPS series algorithms. First, we gave a historical introduction of GPS 1.x, GPS 2.x and the latest GPS 3.0 algorithms. Also, we used palmitoylation as an application to design a site-specific predictor of CSS-Palm 3.0 with GPS 3.0. The procedures of benchmark data preparation, scoring strategies, performance evaluation and comparison, and software package implementation were clearly described. The online service and local packages of CSS-Palm 3.0 are freely available at: http://csspalm.biocuckoo.org/. For convenience, the computational tools developed in GPS series algorithms were summarized in Table 1.

Name	PTM type	Website Link	Reference
GPS 1.0 & 1.1	Phosphorylation	http://gps.biocuckoo.org/1.1/	Xue *et al.*, 2005; Zhou *et al.*, 2004
CSS-Palm 1.0	Palmitoylation	http://csspalm.biocuckoo.org/1.0/	Zhou *et al.*, 2006a
SUMOsp 1.0	Sumoylation	http://sumosp.biocuckoo.org/1.0/	Xue *et al.*, 2006b
GPS 2.0 & 2.1	Phosphorylation	http://gps.biocuckoo.org/	Xue *et al.*, 2008; Xue *et al.*, 2011
CSS-Palm 2.0	Palmitoylation	http://csspalm.biocuckoo.org/	Ren *et al.*, 2008
SUMOsp 2.0	Sumoylation	http://sumosp.biocuckoo.org/	Ren *et al.*, 2009
GPS-SNO 1.0	*S*-Nitrosylation	http://sno.biocuckoo.org/	Xue *et al.*, 2010b
GPS-YNO2 1.0	nitration	http://yno2.biocuckoo.org/	Liu *et al.*, 2011

Table 1. The computational tools constructed with GPS series algorithms.

2. GPS series algorithms

In this section, we described the theoretical basis and developmental history of GPS series algorithms. The chief hypothesis of the algorithm is established on consensus experimental observations that if two short peptides share high sequence homology, they may exhibit similar 3D structures and biochemical properties. This hypothesis is widely adopted by conventional experimentalists, who usually compare a given protein to homologous modified proteins by sequence alignment. If a conserved peptide is observed around aligned modified residue, they may obtain confidence that the peptide in the given protein can also be modified. We borrowed this hypothesis and implemented it into an automatic algorithm. First, we used the amino acid substitution matrix BLOSUM62 to evaluate the similarity between two *phosphorylation site peptides* PSP(m, n) with m residues upstream and n residues downstream flanking the phosphorylated site, while m and n were arbitrarily determined as 3 for phosphorylation site peptide (Xue et al., 2005; Zhou et al., 2004). In GPS (Group-based Phosphorylation Scoring) 1.0 and 1.1, we clustered the phosphorylated peptides with Markov cluster algorithm (MCL for short) with an additional hypothesis of that one protein kinase (PK) can recognize more than one motif in substrates (Xue et al., 2005; Zhou et al., 2004). Later, based on the observation of different matrix generating different performance, we developed the matrix mutation (MaM) approach for the performance improvement in GPS 2.0, which was refined as Group-based Prediction System (Xue et al., 2008). In GPS 2.0, the MCL clustering was discarded for its low efficiency while the informative peptide was selected as PSP(7, 7) (Xue et al., 2008). For the prediction of sumoylation (Ren et al., 2009) and palmitoylation sites (Ren et al., 2008), we testingly classified modification sites based on known linear motifs together with GPS 2.0 algorithm, and achieved increased performances. Later, we improved the algorithm to version 2.1 for the prediction of phosphorylation sites with a additional motif length selection method (MLS) (Xue et al., 2011). Recently, with two additional approaches of k-means clustering and weight training (WT), we further designed GPS 3.0 algorithm for the prediction of S-nitrosylation (Xue et al., 2010b) and nitration sites (Liu et al., 2011). The details of GPS series algorithms were described below.

2.1 GPS 1.x algorithms

The GPS 1.x algorithms include two versions of GPS 1.0 and GPS 1.1. In 2004, we initially designed the GPS 1.0 algorithm for the prediction of kinase-specific phosphorylation sites, while the full name of GPS was "Group-based Phosphorylation Scoring" (Zhou et al., 2004). From public databases and literature curation, we collected 2,001 experimentally identified phosphorylation sites with their cognate PKs (Zhou et al., 2004). Since similar PKs can modify similar sequences in the substrates, we clustered the available PKs into 52 PK groups according to the BLAST results and Swiss-Prot/TrEMBL annotations. Then the known phosphorylation sites were classified into one of multiple of the 52 PK groups based on their regulatory PK information (Zhou et al., 2004). Based on the hypothesis of similar short peptides bearing similar biological properties, we designed a simple scoring strategy. Given a *phosphorylation site peptide* PSP(m, n) as a serine (S), threonine (T) or tyrosine (Y) amino acid flanked by m residues upstream and n residues downstream, we employed the amino acid substitution matrix BLOSUM62 to evaluate the similarity of peptides. In GPS 1.0 (Zhou et al., 2004), the m and n were arbitrarily chosen as 3. For two PSP(m, n), we scored the similarity of two PSP(m, n) as:

$$S(A,B) = \sum_{-m \le i \le n} Score(A[i], B[i]) \tag{1}$$

Score($A[i]$, $B[i]$) represents the substitution score of the two amino acid of $A[i]$ and $B[i]$ in BLOSUM62. If S(A, B)<0, we simply redefined it as S(A, B)=0. With an additional hypothesis that one PK can recognize multiple motifs, we automatically clustered the phosphorylation site peptides into more than one groups with Markov cluster algorithm (MCL for short). Thus, given any peptide for the prediction of kinase-specific phosphorylation sites, we calculated the average score between the peptide and the experimentally identified phosphorylated site peptides in each cluster, while the maximum score among the clusters was decided as the final score. The prediction performance can be comparative with other tools such as Scansite (Obenauer et al., 2003). Later, we slightly refined the algorithm and released GPS 1.1 version, which can predict phosphorylation sites for 216 unique kinases in 71 kinase groups (Xue et al., 2005). Again, the m and n in GPS 1.1 were still selected as 3. The only improvement in GPS 1.1 is that the classification of PKs is better than GPS 1.0 (Xue et al., 2005).

Furthermore, we applied the GPS 1.x algorithm to predict a variety of PTMs such as sumoylation and palmitoylation, with additional refinement if necessary (Xue et al., 2006b; Zhou et al., 2006a). For the prediction of sumoylation sites in SUMOsp 1.0 (Xue et al., 2006b), both GPS 1.x and Motif-X algorithms (Schwartz and Gygi, 2005) were employed because a large proportion of sumoylation sites follow a consensus motif ψ-K-X-E (ψ is a hydrophobic amino acid) or ψ-K-X-E/D (Johnson, 2004). Thus, all known sumoylation sites were classified into two groups with consensus and non-consensus. A given peptide will be compared to known sumoylation sites of both two groups by calculating the average similarity scores, respectively. The maximum score was decided as the final score. And if the score is higher than a pre-determined threshold, the peptide will be predicted as potential sumoylation site (Xue et al., 2006b). In SUMOsp 1.0, the sumoylation site peptide for the prediction was arbitrarily selected as SSP(7, 7). For the prediction of palmitoylation sites in

CSS-Palm 1.0 (Zhou et al., 2006a), the palmitoylation site peptide was casually chosen as PSP(7, 7). Because no common canonical consensus sequence/motif for palmitoylation was reported, we developed a BLOSUM62-based Clustering method (BBC) based on the graph theory, and classified all known palmitoylation sites into three clusters (Zhou et al., 2006a).

2.2 GPS 2.x algorithms

The GPS 1.x algorithms were too preliminary, while a variety of issues were not addressed. By personal communications, several researchers asked us a number of questions. For example, why we arbitrarily chose BLOSUM62 rather than other amino acid substitution matrices? Why we classified the PKs based on BLAST searching rather than using pre-established classification information? Why we selected PSP(3, 3) or PSP(7, 7)? The aim of GPS 2.x algorithms was to resolve these problems.

To evaluate the prediction performance and robustness of a predictor, we usually preformed the self-consistency validation, the leave-one-out validation (LOO) and *n*-fold cross-validations. The self-consistency validation used the training positive data (+) and negative data (-) directly to evaluate the prediction performance, and represented the computational power of the prediction system. However, the prediction system might be overtrained and only perfect for the training data set, with low prediction ability for new data. In this regard, the LOO validation and *n*-fold cross-validations should be performed to evaluate the robustness and the stability on an independent data set. In the LOO validation, each site in the data set was picked out in turn as an independent test sample, and all the remaining sites were regarded as training data. This process was repeated until each site was used as test data one time. In *n*-fold cross-validations, all the (+) sites and (-) sites were combined and then divided equally into *n* parts, keeping the same distribution of (+) and (-) sites in each part. Then *n*-1 parts were merged into a training data set while the remnant part was taken as a testing data set. This process was repeated 20 times and the average performance of *n*-fold cross-validations was used to estimate the performance. In our previous study, when the training data set is large enough (number of positive sites ≥ 30), the results of *n*-fold cross-validations are similar with the LOO result. In this regard, we merely used the LOO validation to evaluate the robustness and stability.

In GPS 1.x algorithms, the amino acid substitution matrix was arbitrarily chosen as BLOSUM62, while performances of other matrices were not evaluated. For the sake of better performance, we tested other matrices such as BLOSUM30, 45, 62, 90 and PAM10, 90, 250, 500, and found different matrices could generate various performances (Xue et al., 2008). For example, PAM10-based scoring can easily generate a perfect self-consistency result with sensitivity (*Sn*) of 100% and specificity (*Sp*) of 100%, while the LOO result is very poor that denotes the prediction model is highly over-fitting and unstable. In this regard, a key challenge is that whether we can obtain an optimal or near-optimal matrix with the highest LOO values. To address this issue, we developed GPS 2.0 algorithm with an additional approach of matrix mutation (MaM) (Xue et al., 2008). First BLOSUM62 matrix was chosen as the initial matrix. The performance (*Sn* and *Sp*) of LOO validation was calculated. For the prediction of kinase-specific phosphorylation sites, we fixed *Sp* at 90% to improve *Sn* by randomly picking out an element of the matrix for +1 or -1. The procedure was terminated when the *Sn* value was not increased any further. Although matrix mutation in other types was also valid, the MaM strategy can improve the LOO result significantly, whereas the self-consistency was only influenced moderately. Thus, such a procedure made the predictor more robust and stable. By comparison, the GPS 2.0 exhibited superior performance against other analogous tools (Xue et al., 2008).

In GPS 1.x algorithms, numerous PKs were casually classified into several groups simply based on sequence comparison by BLAST, based on the hypothesis that PKs in a same group/subfamily will recognize similar sequence patterns of substrates for modification (Xue et al., 2005; Zhou et al., 2004). Because the kinomes of several eukaryotic organisms have been comprehensively identified, phylogenetically analyzed, and classified into a hierarchical structure, including group, family, subfamily, and single PK (Caenepeel et al., 2004; Manning et al., 2002), and because most of the phosphorylation sites in the public database have been experimentally verified in mammals (~97.6%), we adopted the well established rule for human PK classification (Xue et al., 2008) in GPS 2.0 to cluster various PKs with their verified sites into a hierarchical structure with four levels, including group, family, subfamily, and single PK. The PK groups with less than three sites were singled out (Xue et al., 2008). The training data could be reused several times and included in different PK clusters based on their known PK information. GPS 2.0 can predict kinase-specific phosphorylation sites for 408 human PKs in hierarchy.

In GPS 2.0, the PSP(7, 7) was arbitrarily determined (Xue et al., 2008). Later, we carefully studied how different combinations of PSP(m, n) influenced prediction performance and robustness (Xue et al., 2011). The self-consistency validation and LOO validation were thoroughly carried out for each PK group. We observed that the self-consistency results will be always increased with longer PSP(m, n). However, when the phosphorylated peptide was elongated, the LOO results will first reach a peak value then decrease. In this regard, we developed GPS 2.1 algorithm with an additional approach of motif length selection (MLS) (Xue et al., 2011), which could automatically detect the optimal length of PSP(m, n) with the highest LOO performance. We exhaustively tested all combinations of PSP(m, n) ($m = 1,\ldots, 15$; $n = 1, \ldots, 15$). The Sn values were calculated under the Sp of 85, 90 and 95%. Then the average Sn was calculated as the final Sn value. By comparing to GPS 2.0 software (Xue et al., 2008), the average Sn of the LOO was significantly increased by 15.62%, whereas the average Sn value of the self-consistency was slightly reduced by 2.28%. The Sp score was fixed at 90% for comparison. In this regard, the MLS method could efficiently narrow down the difference between the LOO validation and self-consistency validation to improve the robustness of prediction system (Xue et al., 2011).

The GPS 2.1 algorithm was also adopted for the prediction of sumoylation and palmitoylation sites, with additional improvements (Ren et al., 2009; Ren et al., 2008). The experimentally identified sumoylation sites were classified into two types including Type I (consensus) and Type II (non-consensus) sites in SUMOsp 2.0 (Ren et al., 2009). Type I sites followed the ψKXE (ψ is A, I, L, M, P, F, or V and X is any amino acid residue) motif, while Type II sites contained other non-canonical sites. Also, we clustered known palmitoylation sites in CSS-Palm 2.0 into three groups, including Type I (sites follow a -CC– motif, C is a cysteine residue), Type II (sites follow a -CXXC– motif, C is a cysteine residue and X is a random residue) and Type III (other sites) group (Ren et al., 2008). In SUMOsp 1.0 and CSS-Palm 1.0, the threshold is the same for different clusters. However, for SUMOsp 2.0 and CSS-Palm 2.0, we set different threshold for each group, separately. The prediction performance of GPS 2.x is much better than GPS 1.x (Ren et al., 2009; Ren et al., 2008; Xue et al., 2008).

2.3 GPS 3.0 algorithm

Although GPS 1.x and 2.x algorithms were successfully applied in the prediction of phosphorylation, sumoylation and palmitoylation sites, they exhibited poor performance for other PTMs, such as *S*-nitrosylation (Xue et al., 2010b) and nitration sites (Liu et al., 2011). Thus, the GPS algorithm still need further improvements.

We hypothesized that one type of PTM can recognize multiple sequence patterns/motifs. If this hypothesis is correct, the prediction performance can be enhanced by clustering known PTM sites into multiple groups. In GPS 1.x, the MCL algorithm was adopted to automatically classify the known phosphorylation site peptides into multiple clusters if available. However, only eight PK groups obtained more than one cluster (Xue et al., 2005; Zhou et al., 2004). In CSS-Palm 1.0, we adopted a graph-based BBC method for clustering known palmitoylation sites, but it can not significantly improve performance for other types of PTMs (unpublished). In this regard, the clustering strategy was dropped for its low efficiency in GPS 2.0 for the prediction of phosphorylation sites (Xue et al., 2008). For the prediction of sumoylation (SUMOsp 2.0) and palmitoylation (CSS-Palm 2.0) sites, we clustered known sumoylation and palmitoylation into two and three groups based on reported motifs, although the palmitoylation motifs are much weak (Ren et al., 2009; Ren et al., 2008). However, for *S*-nitrosylation and nitration sites, even very weak motifs are not available (Liu et al., 2011; Xue et al., 2010b). In this regard, an interesting question is how to classify PTM sites without any obvious motifs? To address this problem, we developed GPS 3.0 algorithm with an additional *k*-mean clustering method (Liu et al., 2011; Xue et al., 2010b), which was extensively used in a variety of aspects (Herwig et al., 1999; Yoon et al., 2007). With the algorithm, we successfully classified 504 experimentally identified *S*-nitrosylation into 3 groups in GPS-SNO 1.0 (Xue et al., 2010b), while the 1,066 known nitration sites were clustered into 5 groups in GPS-YNO2 1.0 (Liu et al., 2011).

Again, in GPS 1.x and GPS 2.x, the contribution of each residue for substrate recognition by enzymes was regarded as equal. However, there were various amino acid preferences in the residues around the phosphorylation sites for different PKs (Schwartz and Gygi, 2005). For example, the substrates of CDKs follow a pS-P-X-K motif (pS is the phosphorylated serine), which indicates that the adjacent proline is critical for the CDK-specific phosphorylation (Schwartz and Gygi, 2005). Furthermore, the glutamine residue adjacent to the serine/threonine (S/T-Q) was found to be important for ATM (ataxia telangiectasia mutated)/ATR (ATM and Rad3-related) recognition (Matsuoka et al., 2007). In this regard, the different contributions of distinct positions around the PTM sites should be considered and included in the computational model. In this regard, an additional approach of weight training (WT) was added in GPS 3.0 algorithm. We optimized the weight of each position in the *PTM site peptide* PSP(m, n) for every cluster according to the leave-one-out performance (Liu et al., 2011; Xue et al., 2010b).

Together with MaM and MLS approaches, we determined the order of training processes to be: *k*-means clustering, MLS, WT and MaM. By exhaustively testing, it was found that this training order cannot be changed (Liu et al., 2011; Xue et al., 2010b). The prediction performance of GPS 3.0 is much better than GPS 1.x and GPS 2.x algorithms. The GPS 3.0 was firstly introduced and described in the construction of GPS-SNO 1.0 and GPS-YNO2 1.0 (Liu et al., 2011; Xue et al., 2010b). Below, we used palmitoylation as an example to depict the implementation process in detail.

3. An application: Prediction of palmitoylation sites with GPS 3.0 algorithm

In order to describe the GPS series algorithms thoroughly, here we employed the GPS 3.0 algorithm to predict palmitoylation sites as an example. Palmitoylation is the only type of reversible lipid modification, and dynamically regulates protein trafficking and functions through addition of saturated 16-carbon palmitic acids to specific cysteine residues by DHHC palmitoyltransferases (Fukata and Fukata, 2010; Linder and Deschenes, 2007). First, we manually collected the experimentally identified palmitoylation sites from scientific literatures in PubMed. Redundant homologous sites were cleared, while the positive and negative data sets were prepared. The procedures of performance improvement with an order of *k*-means clustering, MLS, WT and MaM were described in detail. Finally, the CSS-Palm 3.0 software packages were implemented in JAVA. The full process of CSS-Palm 3.0 construction is shown in Fig. 1.

3.1 Data preparation

Previously, we manually collected the experimental identified palmitoylation sites from scientific literature which was published before October 8th, 2007 (Ren et al., 2008). Since a large number of experimental studies were reported after CSS-Palm 2.0 was developed, here we further searched the literature in PubMed with the keywords of "palmitoylation" and "palmitoylated" to obtain additional verified palmitoylation sites (before February 14th, 2010). The protein sequences were retrieved from the UniProt database (UniProt, 2010).

In general, if the training data set is highly redundant with too many homologous sites, the prediction accuracy will be overestimated. To avoid such overestimation, we clustered the protein sequences with a threshold of 40% identity by CD-HIT (Li and Godzik, 2006). If two proteins were similar with ≥40% identity, we re-aligned the proteins with BL2SEQ, a program in the BLAST package (Johnson et al., 2008), and checked the results by hand. If two palmitoylation sites from two homologous proteins were at the same position after sequence alignment, only one item was preserved, the other was discarded. Finally, the non-redundant benchmark data set for training contained 439 positive sites from 194 unique substrates.

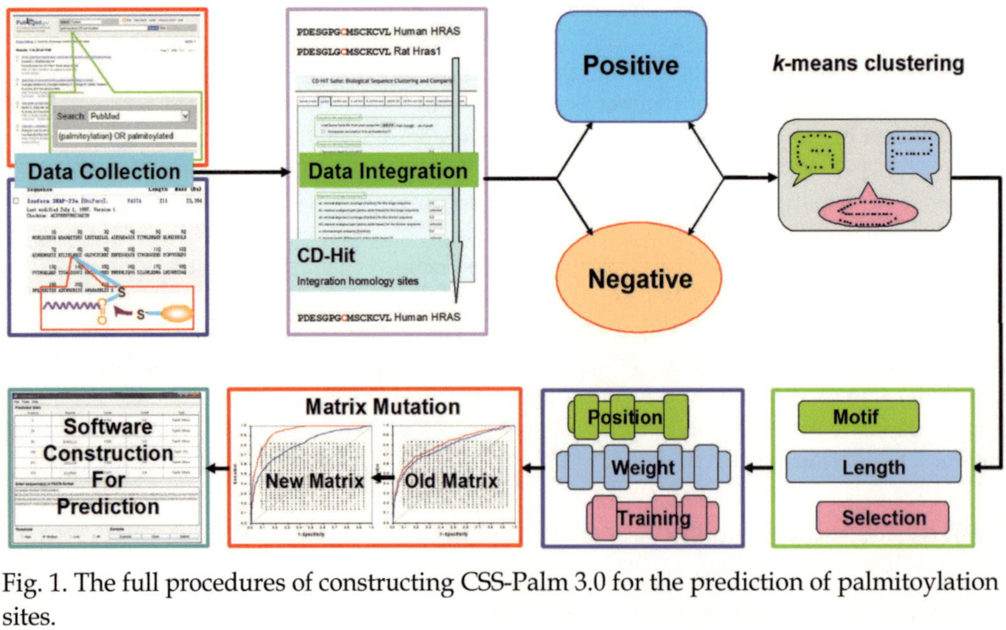

Fig. 1. The full procedures of constructing CSS-Palm 3.0 for the prediction of palmitoylation sites.

We defined a *palmitoylation site peptide* PSP(m, n) as a palmitoylated cysteine flanked by m residues upstream and n residues downstream. As previously described (Ren et al., 2008; Zhou et al., 2006a), we regarded all experimentally verified palmitoylation sites as positive data (+), while all other non-palmitoylated cysteine sites in the same substrates were taken as negative data (-). If a palmitoylated cysteine locates at the N- or C-terminus of the protein and the length of the peptide is smaller than $m+n+1$, we added one or multiple "*" characters as pseudo amino acids to complement the PSP(m, n). Finally, we got 439 positive sites and 2,171 negative sites.

3.2 The GPS 3.0 algorithm

The GPS 3.0 algorithm contains two major components of scoring strategy and performance improvement.

Given the hypothesis of similar short peptides bearing similar biochemical properties, the similarity between two PSP(m, n) of A and B can be calculated with equation (1). Again, if S (A, B) <0, we simply redefined it as S (A, B)=0. A putative PSP(m, n) is compared with each of the experimentally verified palmitoylated peptides in a pairwise manner to calculate the similarity score. The average value of the substitution scores is regarded as the final score. The schematic description of the scoring strategy with examples was shown in Fig. 2. The performance improvement processes with of four sequential steps of k-means clustering, MLS, WT and MaM were presented below.

The Scoring Strategy

- **e.g. 1, Given two Peptides:**
  - **AQECIL (Palmitoylated)**
  - **IQECLI (Unknown)**
  - Similarity score: -1+5+5+9+2+2=22
- **e.g. 2, Given two Peptides:**
  - **AQESILR (Palmitoylated)**
  - ****ESLIR (Unknown)**
  - Similarity score: 0+0+5+9+2+2=18
- **Note:**
  - (1) The given peptides will be compared with each known palmitoylated peptide to calculate similarity score
  - (2) Setting the score as zero, if <0
  - (3) Final score: average

BLOSUM62 (modified, partial)

	A	R	N	D	C	Q	E	G	H	I	L	*
A	4	-1	-2	-2	0	-1	-1	0	-2	-1	-1	0
R	-1	5	0	-2	-3	1	0	-2	0	-3	-2	0
N	-2	0	6	1	-3	0	0	0	1	-3	-3	0
D	-2	-2	1	6	-3	0	2	-1	-1	-3	-4	0
C	0	-3	-3	-3	9	-3	-4	-3	-3	-1	-1	0
Q	-1	1	0	0	-3	5	2	-2	0	-3	-2	0
E	-1	0	0	2	-4	2	5	-2	0	-3	-3	0
G	0	-2	0	-1	-3	-2	-2	6	-2	-4	-4	0
H	-2	0	1	-1	-3	0	0	-2	8	-3	-3	0
I	-1	-3	-3	-3	-1	-3	-3	-4	-3	4	2	0
L	-1	-2	-3	-4	-1	-2	-3	-4	-3	2	4	0
*	0	0	0	0	0	0	0	0	0	0	0	1

The substitution scores between pseudo amino acid "*" and any other residues were redefined as zero

Fig. 2. Schematic description of the scoring strategy in GPS algorithm.

3.2.1 *k*-means clustering

In CSS-Palm 2.0, we clustered experimental identified palmitoylation sites into three groups based on known motifs (Ren et al., 2008). Because the palmitoylation motifs are very weak, the prediction performance was only considerably improved (Ren et al., 2008). For grouping known palmitoylation sites, here we used the *k*-mean clustering approach, which was widely adopted and got successful performance in our previous studies (Liu et al., 2011; Xue et al., 2010b). The clustering process was described as below:

Given two PSP(m, n) peptides A and B, the similarity was defined and measured as: $s(A, B) = N_s/N$. The N is the number of all substitutions, whereas the N_s is the number of conserved substitutions with $Score(a, b) > 0$ in the BLOSUM62 matrix. The $s(A, B)$ ranges from 0 to 1. Thus, the distance between them can be defined as: $D(A, B) = 1/s(A, B)$. If $s(A, B) = 0$, $D(A, B) = \infty$. By exhaustive testing, the k was roughly set to 3, while PSP(7, 7) was adopted. First, three palmitoylation sites from the positive data (+) were randomly chosen as the centroids. Second, the other positive sites were compared in a pairwise manner with the three centroids and clustered into groups with the highest similarity values. Third, the centroid of each cluster was updated with the highest average similarity (HAS). The second and third steps were iteratively repeated until the clusters did not change any longer. After the three clusters for the positive sites had been determined, we put each negative site into the cluster with the HAS.

Given a potential PSP(7, 7) for prediction, we firstly determined which cluster it belongs to, by calculating the average similarity score of the PSP(7, 7) against each cluster (Fig. 3). For example, the PSP(7, 7) P_1 will be regarded as Cluster 1 type site, while the P_2 and P_3 will be determined to be Cluster 2 and 3 type sites, respectively (Fig. 3).

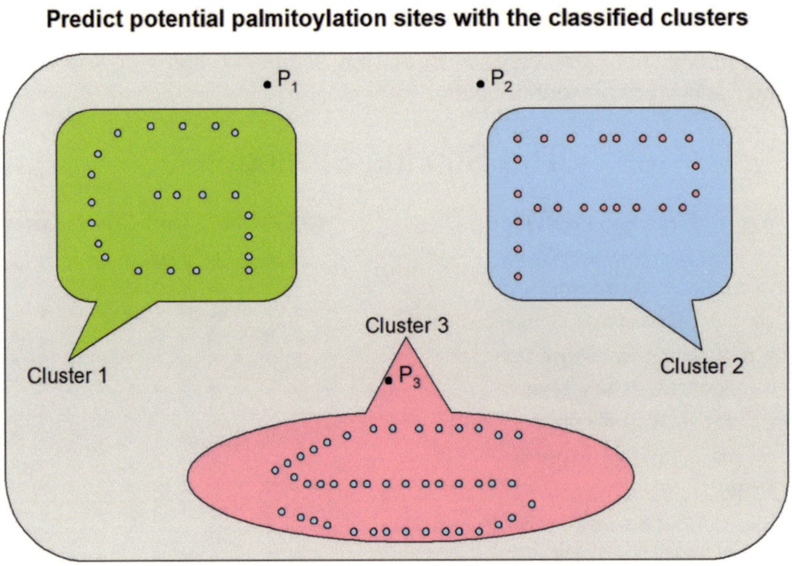

Fig. 3. Prediction of potential palmitoylation sites with the classified clusters.

3.2.2 Motif length selection (MLS)

Previously, the m and n in PSP(m, n) was arbitrarily determined (Xue et al., 2008; Zhou et al., 2004; Zhou et al., 2006a). In this step, we determined the optimized combination of PSP(m, n) for optimal performance. The combinations of PSP(m, n) (m = 1, ..., 30; n = 1, ..., 30) were extensively tested. The optimal PSP(m, n) for each cluster was separately selected, with the highest LOO performance. From our previous experience, a higher Sp value is more important than a higher Sn to avoid too many false positive hits (Ren et al., 2009; Ren et al., 2008; Xue et al., 2005). Thus, to improve the prediction performance and robustness in the region of high Sp is more important than other regions. In this study, we fixed the Sp at 90% to compare Sn values.

3.2.3 Weight training (WT)

We updated the substitution score between two PSP(m, n) peptides A and B as below:

$$S'(A,B) = \sum_{-m \le i \le n} w_i Score(A[i], B[i]) \tag{2}$$

The w_i is the weight of position i. Again, if $S'(A, B)<0$, we redefined it as $S'(A, B)=0$. Initially, the w was chosen as 1 for each position. We randomly picked out the weight of any position for +1 or -1, and adopted the manipulation if the *Sn* score of the re-calculated LOO result with the *Sp* fixed at 90% was increased. The process was repeated until convergence was reached.

3.2.4 Matrix mutation (MaM)

Previously, we chose the BLOSUM62 matrix to evaluate the similarity between PSP(m, n). Later, we observed that different matrices generate various performances (Xue et al., 2008). For palmitoylation, we also tested a variety of matrices such as BLOSUM30, 45, 62, 90, and PAM 10, 90, 250 and 500. The self-consistency (red) and LOO (blue) validations were preformed (Fig. 4). To balance the prediction performance and robustness of the prediction system, the BLOSUM62 matrix was adopted as the initial matrix in CSS-Palm 3.0.

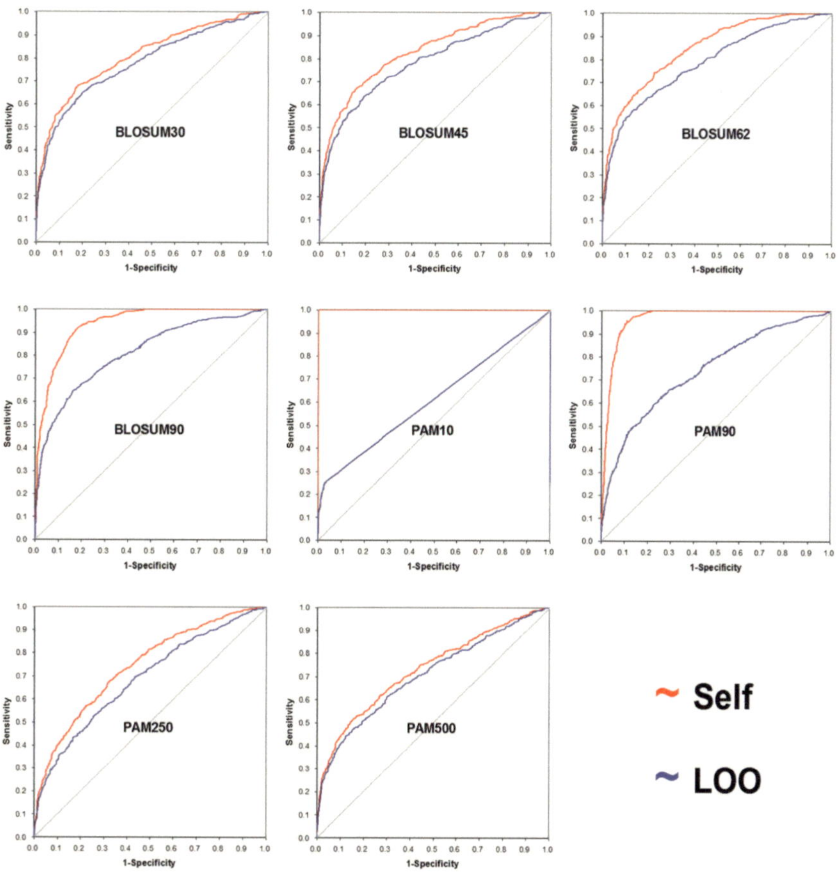

Fig. 4. Different matrices generate various performances. The ROC curves of self-consistency and LOO validations were drawn in red and blue, respectively.

In the MaM process, the LOO result with BLOSUM62 was first calculated. In BLOSUM62, the substitution score between "*" and other residues is -4 but redefined as 0. Then we fixed the *Sp* at 90% to improve *Sn* by randomly picking out one value from the BLOSUM62 matrix for mutation (+1 or -1). If the *Sn* value increased, the mutation was adopted. This process was terminated when the *Sn* value was not increased any further. Interestingly, we observed that when the training time of MaM is long enough, the mutated matrix generated from other matrices, e.g., BLOSUM45, is exactly identical to the one from BLOSUM62 (Data not shown). In this regard, the final mutated matrix is not dependent on the initial matrix.

3.3 Software construction

Previously, we only developed the online services for the predictions with PHP/PERL. We also discussed the general user interface for the predictors of PTMs sites (Zhou et al., 2006b). When the number of users becomes large, the server will bear high burden with very low speed. In this regard, we recently constructed computational tools in JAVA, whereas the online service and local packages were both provided (Liu et al., 2011; Ren et al., 2008; Ren et al., 2009; Xue et al., 2008; Xue et al., 2010b; Xue et al., 2011). The online service and local packages of CSS-Palm 3.0 were implemented in JAVA. For the online service, we tested the CSS-Palm 3.0 on a variety of internet browsers, including Internet Explorer 6.0, Netscape Browser 8.1.3 and Firefox 2 under the Windows XP Operating System (OS), Mozilla Firefox 1.5 of Fedora Core 6 OS (Linux), and Safari 3.0 of Apple Mac OS X 10.4 (Tiger) and 10.5 (Leopard). For the Windows and Linux systems, the latest version of Java Runtime Environment (JRE) package (JAVA 1.4.2 or later versions) of Sun Microsystems should be pre-installed. However, for Mac OS, CSS-Palm 3.0 can be directly used without any additional packages. For convenience, we also developed local packages of CSS-Palm 3.0, which worked with the three major Operating Systems, Windows, Linux and Mac. The software and the online sever are freely available at: http://csspalm.biocuckoo.org/.

3.4 Performance evaluation and comparison

As previously described (Ren et al., 2008; Zhou et al., 2006a), We adopted four standard measurements, including accuracy (*Ac*), sensitivity (*Sn*), specificity (*Sp*) and Mathew correlation coefficient (*MCC*). *Ac* illustrates the correct ratio between both positive (+) and negative (-) data sets, while *Sn* and *Sp* represent the correct prediction ratios of positive (+) and negative data (-) sets respectively. However, when the number of positive data and negative data differ too much from each other, *MCC* should be included to evaluate the prediction performance. The value of *MCC* ranges from -1 to 1, and a larger *MCC* value stands for better prediction performance.

Among the data with positive hits by CSS-Palm 3.0, the real positives are defined as true positives (*TP*), while the others are defined as false positives (*FP*). Among the data with negative predictions by CSS-Palm 3.0, the real positives are defined as false negatives (*FN*), while the others are defined as true negatives (*TN*). The performance measurements of *Ac*, *Sn*, *Sp* and *MCC* are defined as below:

$$Sn = \frac{TP}{TP + FN} \tag{3}$$

$$Sp = \frac{TN}{TN + FP} \tag{4}$$

$$Ac = \frac{TP + TN}{TP + FP + TN + FN} \tag{5}$$

$$MCC = \frac{(TP \times TN) - (FN \times FP)}{\sqrt{(TP + FN) \times (TN + FP) \times (TP + FP) \times (TN + FN)}} \tag{6}$$

To avoid overestimation, the LOO validation and 4-, 6-, 8-, 10-fold cross-validations were performed to evaluate the prediction robustness and performance of CSS-Palm 3.0. Receiver Operating Characteristic (ROC) curves are presented in Fig. 5A, while the AROCs (area under ROCs) were calculated as 0.889 (leave-one-out), 0.877 (4-fold), 0.879 (6-fold), 0.887 (8-fold) and 0.906 (10-fold), respectively (Fig. 5A). Since the 4-, 6-, 8-, 10-fold cross-validations were close to the leave-one-out validation, it was demonstrated that CSS-Palm 3.0 is a robust predictor of palmitoylation sites with promising performance.

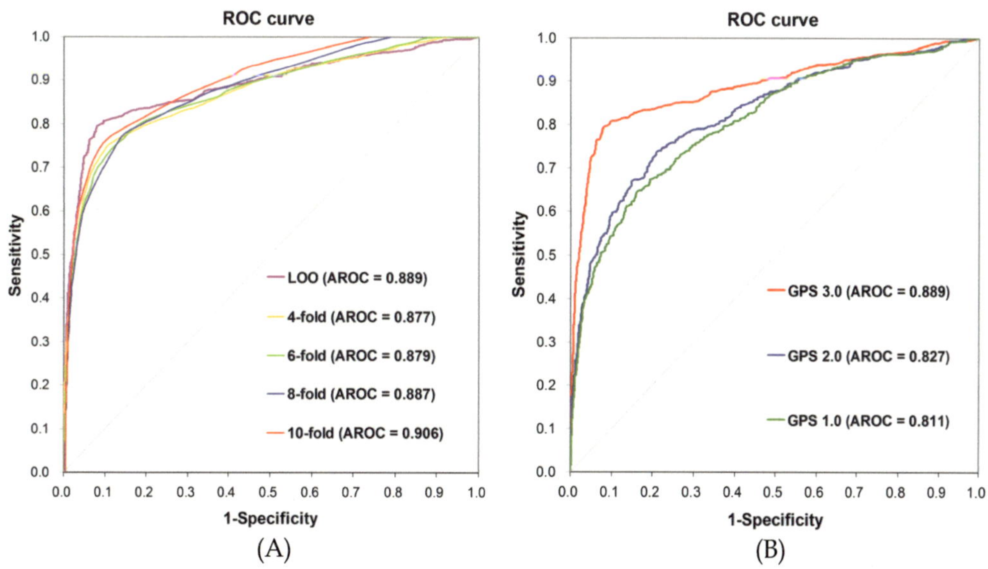

Fig. 5. Performance evaluation and comparison. (A) The LOO validation and 4-, 6-, 8-, 10-fold cross-validations were performed to evaluate the prediction robustness and performance of CSS-Palm 3.0. (B) Comparison of LOO results of GPS 1.0, GPS 1.0 and GPS 3.0 algorithms.

For comparison, we also calculated the performance of GPS 1.0 and 2.0 algorithms. To avoid any bias, the same training data set used in GPS 3.0 was also employed in the two methods. The LOO validations were carried out for GPS 1.0 and 2.0 algorithms, while the ROC curves were drawn (Fig. 5B). The AROC values were calculated as 0.811 (GPS 1.0), 0.827 (GPS 2.0) and 0.889 (GPS 3.0), respectively. In addition, we fixed the *Sp* values and compared *Sn* scores (Table 2). When the *Sp* value was ~85%, the *Sn* values of GPS 3.0, GPS 2.0 and GPS 1.0 were 72.44%, 48.29% and 43.05%, respectively (Table 2). Also, when the *Sp* value was ~90%, the *Sn* values of GPS 3.0, GPS 2.0 and GPS 1.0 were 80.64%, 59.68% and 54.21%, separately (Table 2). In addition, when the *Sp* value was ~95%, the *Sn* of GPS 3.0 (82.23%) was still much better than GPS 2.0 (67.20%) and GPS 1.0 (62.41%) (Table 2). Taken together, our results exhibited that the performance of GPS 3.0 is better than GPS 1.0 and 2.0. Finally, the

CSS-Palm 3.0 software was constructed with three thresholds of High, Medium and Low, with the *Sp* values of ~95%, ~90% and ~85%, respectively.

Method	Threshold	*Ac*	*Sn*	*Sp*	*MCC*
GPS 3.0	High	91.30%	72.44%	95.12%	0.6850
	Medium	88.74%	80.64%	90.37%	0.6458
	Low	84.75%	82.23%	85.26%	0.5749
GPS 2.0		87.24%	48.29%	95.12%	49.64%
		84.94%	59.68%	90.05%	48.09%
		82.03%	67.20%	85.03%	45.90%
GPS 1.0		86.28%	43.05%	95.03%	0.4485
		84.21%	54.21%	90.28%	0.4410
		81.23%	62.41%	85.03%	0.4220

Table 2. For comparison, we fixed the *Sp* values of GPS 3.0 algorithm so as to be similar or identical to GPS 1.0 and 2.0 algorithms, and compared the *Sn* values.

4. Conclusion

During the past several decades, accumulated experimental studies have made slow but steady contributions toward understanding molecular mechanisms and regulatory roles of various PTMs (Mann and Jensen, 2003; Walsh, 2005; Walsh and Jefferis, 2006). Recently, rapid progresses in the state-of-the-art HTP-MS techniques have boomed an explosion of modification data for systematically analyzing PTM regulation in a proteomic level (Choudhary and Mann, 2010). However, the biological functions of PTMs are still far from fully elucidated. In this regard, more efforts remain to be carried out.

In contrast with expensive and error-prone experimental methods, *in silico* prediction of PTM-specific substrates with their sites has emerged as a popular alternative approach. In this field, two questions should be addressed: 1) How to predict modification sites in a given protein sequence? 2) How to predict regulatory enzyme information of modification sites in a given protein sequence? The importance of the two questions is different for distinct types of PTMs. For example, a phosphoproteomics analysis can detect thousands of phosphorylation sites in a single experiment (Olsen et al., 2006; Villen et al., 2007). In this regard, the prediction of general or non-specific phosphorylation sites is not much useful at the current stage. However, there are only ~3,500 phosphorylation sites with known upstream PK information in the public databases (Xue et al., 2008; Xue et al., 2011). In this regard, the prediction of kinase-specific phosphorylation sites is still a great challenge, while the results can be a help for further experimental consideration. For sumoylation and palmitoylation, accurately large-scale identification of their substrates and sites is not easy to be performed. In this regard, the prediction of general sumoylation and palmitoylation in proteins is useful for guiding further experimental verifications. Also, since the experimentally identified enzyme-specific information for both sumoylation and palmitoylation is quite limited, the prediction of enzyme-specific sumoylation or palmitoylation is still not available due to data limitation.

Intuitively, the prediction of PTM sites seems to be a trivial job. Assume that one may easily obtain experimentally identified PTMs from one or two review articles as the training data set, casually select a machine learning algorithm such as PSSM, SVMs or ANN, carry out several validations to evaluate the performance, and develop a web server for the prediction. Then the manuscript can be written with a cup of coffee in hand. Previously, a number of researchers asked us by personal communications that why we did not used a simple existed algorithm to develop an integrate tool for the prediction of all types of PTMs sites. Is the prediction of PTMs sites really simple? From our research experience, the answer is "not at all". First, most of widely-used machine learning algorithms are derived from the fields of informatics or statistics and originally designed for general propose, but not specifically for PTMs sites prediction. Second, different types of PTMs can have distinct sequence features. One algorithm can generate promising performance for a specific PTM but exhibit poor accuracy for other types of PTMs. For example, the prediction of PKA-specific phosphorylation sites with any algorithm can generate satisfying performance (Xue et al., 2008). However, for PTMs with strong motifs, the scenario is different. For example, the sumoylation has a strong consensus motif of ψKXE, which about 77% of all known sumoylation sites follow this pattern (Xue et al., 2006b). Since the simple strong motif can generate great accuracy, development of new algorithms will not be necessary if the performance can not be significantly improved. For palmitoylation, two weak motifs can be obtained (Ren et al., 2008). Prediction of palmitoylation with weak motifs will generate poor performance. But it's also difficult for computational algorithm to retrieve informative features for prediction. In addition, for S-nitrosylation and nitration, even weak motifs are not available (Liu et al., 2011; Xue et al., 2010b). In this regard, development a novel and useful algorithm specifically for PTMs site prediction is an urgent demand. Also, great attention needs to be paid since different PTMs have different properties.

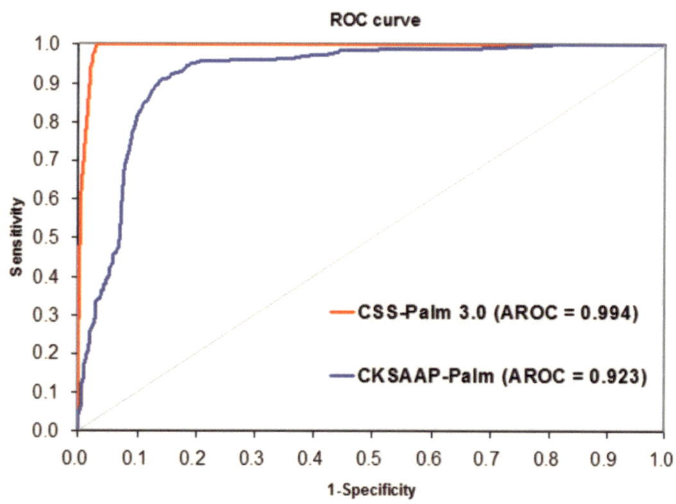

Fig. 6. Comparison of CSS-Palm 3.0 and CKSAAP-Palm (Wang et al., 2009).

During the past several years, our group take great efforts on designing and improve the GPS series algorithms, which were specifically designed for the prediction of PTM sites in proteins. Although the preliminary GPS 1.x algorithms could be only comparative to other approaches, the GPS 2.x exhibited superior performance against analogous predictors. The latest version of GPS 3.0 algorithm has been further improved and much better than our GPS 2.x algorithms. As an application, we used this algorithm to construct CSS-Palm 3.0 for

the prediction of palmitoylation sites. By comparison of a recently released predictor of CKSAAP-Palm (Wang et al., 2009), the performance of CSS-Palm 3.0 is significantly better (Fig. 6).
Finally, we do not propose that the GPS 3.0 will be the final version, while more strategies will be developed and included in GPS series algorithms. We anticipated that the combination of computational predictions and experimental verifications will become the foundation of systematically understanding the mechanisms and the dynamics of PTMs.

5. Acknowledgment

This work was supported by grants from the National Basic Research Program (973 project) (2010CB945400, 2011CB910400), National Natural Science Foundation of China (90919001, 31071154, 30900835, 30830036, 91019020, 21075045), and Fundamental Research Funds for the Central Universities (HUST: 2010JC049, 2010ZD018; SYSU: 11lgzd11, 11lgjc09).

6. References

Ackermann, B.L., & Berna, M.J. (2007). Coupling immunoaffinity techniques with MS for quantitative analysis of low-abundance protein biomarkers. *Expert Review of Proteomics*, Vol.4, No.2, (April 2007), pp. 175-186, ISSN 1744-8387

Balter, M. & Vogel, G. (2001). Nobel prize in physiology or medicine. Cycling toward Stockholm. *Science*, Vol.294, No.5542, (October 2001), pp. 502-503, ISSN 0036-8075

Blom, N., Sicheritz-Ponten, T., Gupta, R., Gammeltoft, S. & Brunak, S. (2004). Prediction of post-translational glycosylation and phosphorylation of proteins from the amino acid sequence. *Proteomics*, Vol.4, No.6, (June 2004), pp. 1633-1649, ISSN 1615-9853

Boschetti, E. & Righetti, P.G. (2009). The art of observing rare protein species in proteomes with peptide ligand libraries. *Proteomics*, Vol.9, No.6, (March 2009), pp. 1492-1510, ISSN 1615-9853

Caenepeel, S., Charydczak, G., Sudarsanam, S., Hunter, T. & Manning, G. (2004). The mouse kinome: discovery and comparative genomics of all mouse protein kinases. *Proceedings of the National Academy of Sciences of the United States of America*, Vol.101, No.32, (August 2004), pp. 11707-11712, ISSN 0027-8424

Choudhary, C., Kumar, C., Gnad, F., Nielsen, M.L., Rehman, M., Walther, T.C., Olsen, J.V. & Mann, M. (2009). Lysine acetylation targets protein complexes and co-regulates major cellular functions. *Science*, Vol.325, No.5942, (August 2009), pp. 834-840, ISSN 0036-8075

Choudhary, C. & Mann, M. (2010). Decoding signalling networks by mass spectrometry-based proteomics. *Nature Reviews Molecular Cell Biology*, Vol.11, No.6, (June 2010), pp. 427-439, ISSN 1471-0080

Dang, T.H., Van Leemput, K., Verschoren, A. & Laukens, K. (2008). Prediction of kinase-specific phosphorylation sites using conditional random fields. *Bioinformatics*, Vol.24, No.24, (December 2008), pp. 2857-2864, ISSN 1367-4811

Fukata, Y. & Fukata, M. (2010). Protein palmitoylation in neuronal development and synaptic plasticity. *Nature Reviews Neuroscience*, Vol.11, No.3, (March 2010), pp. 161-175, ISSN 1471-0048

Herwig, R., Poustka, A.J., Muller, C., Bull, C., Lehrach, H. & O'Brien, J. (1999). Large-scale clustering of cDNA-fingerprinting data. *Genome Research*, Vol.9, No.11, (November 1999), pp. 1093-1105, ISSN 1088-9051

Huang, H.D., Lee, T.Y., Tzeng, S.W. & Horng, J.T. (2005). KinasePhos: a web tool for identifying protein kinase-specific phosphorylation sites. *Nucleic Acid Research*, Vol.33, Web Server issue, (July 2005), pp. W226- W229, ISSN 1362-4962

Hunter, T. (2007). The age of crosstalk: phosphorylation, ubiquitination & beyond. *Molecular Cell*, Vol.28, No.5, (December 2007), pp. 730-738, ISSN 1097-2765

Jenuwein, T. & Allis, C.D. (2001). Translating the histone code. *Science*, Vol.293, No.5532, (August 2001), pp. 1074-1080, ISSN 0036-8075

Johnson, E.S. (2004). Protein modification by SUMO. *Annual Review of Biochemistry*, Vol.73, (June 2004), pp. 355-382, ISSN 0066-4154

Johnson, M., Zaretskaya, I., Raytselis, Y., Merezhuk, Y., McGinnis, S. & Madden, T.L. (2008). NCBI BLAST: a better web interface. *Nucleic Acid Research*, Vol.36, Web Server issue, (July 2008), pp. W5-W9, ISSN 1362-4962

Kim, J.H., Lee, J., Oh, B., Kimm, K. & Koh, I. (2004). Prediction of phosphorylation sites using SVMs. *Bioinformatics*, Vol.20, No.17, (November 2004), pp. 3179-3184, ISSN 1367-4803

Kresge, N., Simoni, R.D. & Hill, R.L. (2011). The process of reversible phosphorylation: the work of Edmond H. Fischer. *The Journal of Biological Chemistry*, Vol.286, No.3, (January 2011), pp. e1-e2, ISSN 0021-9258

Kumar, N. & Mohanty, D. (2010). Identification of substrates for Ser/Thr kinases using residue-based statistical pair potentials. *Bioinformatics*, Vol.26, No.2, (January 2010), pp. 189-197, ISSN 1367-4811

Lahiry, P., Torkamani, A., Schork, N.J. & Hegele, R.A. (2010). Kinase mutations in human disease: interpreting genotype-phenotype relationships. *Nature Reviews Genetics*, Vol.11, No.1, (January 2010), pp. 60-74, ISSN 1471-0064

Li, W. & Godzik, A. (2006). Cd-hit: a fast program for clustering and comparing large sets of protein or nucleotide sequences. *Bioinformatics*, Vol.22, No.13, (July 2006), pp. 1658-1659, ISSN 1367-4811

Linder, M.E. & Deschenes, R.J. (2007). Palmitoylation: policing protein stability and traffic. *Nature Reviews Molecular Cell Biology*, Vol.8, No.1, (January 2007), pp. 74-84, ISSN 1471-0080

Liu, Z., Cao, J., Ma, Q., Gao, X., Ren, J. & Xue, Y. (2011). GPS-YNO2: computational prediction of tyrosine nitration sites in proteins. *Molecular BioSystems*, Vol.7, No.4, (January 2011), pp. 1197-1204, ISSN 1742-2051

Mann, M. & Jensen, O.N. (2003). Proteomic analysis of post-translational modifications. *Nature Biotechnology*, Vol.21, No.3, (March 2003), pp. 255-261, ISSN 1087-0156

Manning, G., Whyte, D.B., Martinez, R., Hunter, T. & Sudarsanam, S. (2002). The protein kinase complement of the human genome. *Science*, Vol.298, No.5600, (December 2002), pp. 1912-1934, ISSN 1095-9203

Matsuoka, S., Ballif, B.A., Smogorzewska, A., McDonald, E.R., 3rd, Hurov, K.E., Luo, J., Bakalarski, C.E., Zhao, Z., Solimini, N., Lerenthal, Y., Shiloh, Y., Gygi, S. P. & Elledge, S. J. (2007). ATM and ATR substrate analysis reveals extensive protein networks responsive to DNA damage. *Science*, Vol.316, No.5828, (May 2007), pp. 1160-1166, ISSN 1095-9203

Norvell, A. & McMahon, S.B. (2010). Cell biology. Rise of the rival. *Science*, Vol.327, No.5968, (February 2010), pp. 964-965, ISSN 1095-9203

Obenauer, J.C., Cantley, L.C. & Yaffe, M.B. (2003). Scansite 2.0: Proteome-wide prediction of cell signaling interactions using short sequence motifs. *Nucleic Acid Research*, Vol.31, No.13, (July 2003), pp. 3635- 3641, ISSN 1362-4962

Olsen, J.V., Blagoev, B., Gnad, F., Macek, B., Kumar, C., Mortensen, P. & Mann, M. (2006). Global, in vivo & site-specific phosphorylation dynamics in signaling networks. *Cell*, Vol.127, No.3, (November 2006), pp. 635-648, ISSN 0092-8674

Ren, J., Gao, X., Jin, C., Zhu, M., Wang, X., Shaw, A., Wen, L., Yao, X. & Xue, Y. (2009). Systematic study of protein sumoylation: Development of a site-specific predictor of SUMOsp 2.0. *Proteomics*, Vol.9, No.12, (June 2009), pp. 3409-3412, ISSN 1615-9853

Ren, J., Wen, L., Gao, X., Jin, C., Xue, Y. & Yao, X. (2008). CSS-Palm 2.0: an updated software for palmitoylation sites prediction. *Protein Engineering, Design and Selection*, Vol.21, No.11, (November 2008), pp. 639-644, ISSN 1741-0134

Schwartz, D. & Gygi, S.P. (2005). An iterative statistical approach to the identification of protein phosphorylation motifs from large-scale data sets. *Nature Biotechnology*, Vol.23, No.11, (November 2005), pp. 1391-1398, ISSN 1087-0156

The UniProt Consortium (2010). The Universal Protein Resource (UniProt) in 2010. *Nucleic Acid Research*, Vol.38, Database issue, (January 2010), pp. D142-D148, ISSN 1362-4962

Villen, J., Beausoleil, S.A., Gerber, S.A. & Gygi, S.P. (2007). Large-scale phosphorylation analysis of mouse liver. *Proceedings of the National Academy of Sciences of the United States of America*, Vol.104, No.5, (January 2007), pp. 1488-1493, ISSN 0027-8424

Vogel, G. (2004). Nobel Prizes. Gold medal from cellular trash. *Science*, Vol.306, No.5695, (October 2004), pp. 400-401, ISSN 1095-9203

Walsh, C. (2005). *Posttranslational Modification of Proteins: Expanding Nature's Inventory*, Roberts and Co. Publishers, ISBN 978-097-4707-73-0, Colorado, USA

Walsh, G. & Jefferis, R. (2006). Post-translational modifications in the context of therapeutic proteins. *Nature Biotechnology*, Vol.24, No.10, (October 2006), pp. 1241-1252, ISSN 1087-0156

Wang, Q., Zhang, Y., Yang, C., Xiong, H., Lin, Y., Yao, J., Li, H., Xie, L., Zhao, W., Yao, Y. Ning, Z. B. Zeng, R. Xiong, Y. Guan, K. L. Zhao, S. & Zhao, G. P. (2010). Acetylation of metabolic enzymes coordinates carbon source utilization and metabolic flux. *Science*, Vol.327, No.5968, (February 2010), pp. 1004-1007, ISSN 1095-9203

Wang, X.B., Wu, L.Y., Wang, Y.C. & Deng, N.Y. (2009). Prediction of palmitoylation sites using the composition of k-spaced amino acid pairs. *Protein Engineering, Design and Selection*, Vol.22, No.11, (November 2009), pp. 707-712, ISSN 1741-0134

Xue, Y., Gao, X., Cao, J., Liu, Z., Jin, C., Wen, L., Yao, X. & Ren, J. (2010a). A summary of computational resources for protein phosphorylation. *Current Protein & Peptide Science*, Vol.11, No.6, (September 2010), pp. 485-496, ISSN 1875-5550

Xue, Y., Zhou, F., Zhu, M., Ahmed, K., Chen, G. & Yao, X. (2005). GPS: a comprehensive www server for phosphorylation sites prediction. *Nucleic Acid Research*, Vol.33, Web Server issue, (July 2005), pp. W184-W187, ISSN 1362-4962

Xue, Y., Li, A., Wang, L., Feng, H. & Yao, X. (2006a). PPSP: prediction of PK-specific phosphorylation site with Bayesian decision theory. *BMC Bioinformatics*, Vol.7, (March 2006), pp. 163, ISSN 1471-2105

Xue, Y., Zhou, F., Fu, C., Xu, Y. & Yao, X. (2006b). SUMOsp: a web server for sumoylation site prediction. *Nucleic Acid Research*, Vol.34, Web Server issue, (July 2006), pp. W254-W257, ISSN 1362-4962

Xue, Y., Ren, J., Gao, X., Jin, C., Wen, L. & Yao, X. (2008). GPS 2.0, a tool to predict kinase-specific phosphorylation sites in hierarchy. *Molecular & Cellular Proteomics*, Vol.7, No.9, (September 2008), pp. 1598-1608, ISSN 1535-9484

Xue, Y., Liu, Z., Gao, X., Jin, C., Wen, L., Yao, X. & Ren, J. (2010b). GPS-SNO: Computational Prediction of Protein S-Nitrosylation Sites with a Modified GPS Algorithm. *PLoS ONE*, Vol.5, No.6, (June 2010), pp. e11290, ISSN 1932-6203

Xue, Y., Liu, Z., Cao, J., Ma, Q., Gao, X., Wang, Q., Jin, C., Zhou, Y., Wen, L. & Ren, J. (2011). GPS 2.1: enhanced prediction of kinase-specific phosphorylation sites with an algorithm of motif length selection. *Protein Engineering, Design and Selection*, Vol.24, No.3, (March 2011), pp. 255-260, ISSN 1741-0134

Yates, J.R., Ruse, C.I. & Nakorchevsky, A. (2009). Proteomics by mass spectrometry: approaches, advances & applications. *Annual Review of Biomedical Engineering*, Vol.11, (April 2009), pp. 49-79, ISSN 1545-4274

Yoon, S., Ebert, J.C., Chung, E.Y., De Micheli, G. & Altman, R.B. (2007). Clustering protein environments for function prediction: finding PROSITE motifs in 3D. *BMC Bioinformatics*, Vol. 8, Suppl. 4, (June 2007), pp. S10, ISSN 1471-2105

Young, N.L., Plazas-Mayorca, M.D. & Garcia, B.A. (2010). Systems-wide proteomic characterization of combinatorial post-translational modification patterns. *Expert Review of Proteomics*, Vol.7, No.1, (February 2010), pp. 79-92, ISSN 1744-8387

Zhao, S., Xu, W., Jiang, W., Yu, W., Lin, Y., Zhang, T., Yao, J., Zhou, L., Zeng, Y., Li, H., Li, Y., Shi, J., An, W., Hancock, S. M., He, F., Qin, L., Chin, J., Yang, P., Chen, X., Lei, Q., Xiong, Y. & Guan, K. L. (2010). Regulation of cellular metabolism by protein lysine acetylation. *Science*, Vol.327, No.5968, (February 2010), pp. 1000-1004, ISSN 1095-9203

Zhou, F., Xue, Y., Chen, G.L. & Yao, X. (2004). GPS: a novel group-based phosphorylation predicting and scoring method. *Biochemical and Biophysical Research Communications*, Vol.325, No.4, (December 2004), pp. 1443-1448, ISSN 0006-291X

Zhou, F., Xue, Y., Yao, X. & Xu, Y. (2006a). CSS-Palm: palmitoylation site prediction with a clustering and scoring strategy (CSS). *Bioinformatics*, Vol.22, No.7, (April 2006), pp. 894-896, ISSN 1367-4811

Zhou, F., Xue, Y., Yao, X. & Xu, Y. (2006b). A general user interface for prediction servers of proteins' post-translational modification sites. *Nature Protocols*, Vol.1, No.3, (April 2007), pp. 1318-1321, ISSN 1750-2799

Zielinska, D.F., Gnad, F., Wisniewski, J.R. & Mann, M. (2010). Precision mapping of an in vivo N-glycoproteome reveals rigid topological and sequence constraints. *Cell*, Vol.141, No.5, (May 2010), pp. 897-907, ISSN 1097-2765

7

Protein Networks: Generation, Structural Analysis and Exploitation

Enrico M. Bucci[1,2], Massimo Natale[1] and Alice Poli[1]
[1]*Biodigitalvalley Srl – Pont Saint Martin- AO-*
[2]*CNR - IBB – Naples*
Italy

1. Introduction

The scientific community is well aware of the fact that the very presence of internet profoundly affects the way research is done. Internet is the major infrastructure through which results and data are communicated and shared, computing is parallelized, collaborations are started and enlarged, papers are published, and a myriad of other transactions are performed, so to involve nearly all aspects of everyday researcher's life.
Perhaps surprisingly, scientists paid little attention to the theoretical study of internet structure (i.e. topology) and of its dynamical behavior, until quite recently.
Even more surprisingly, however, when they did it, it quickly emerged some very rarely found scientific truth, of such a general kind, to directly reverberate from Internet studies into the field of molecular and cellular biology, with very small (if any) changes. Consequences manifested immediately, so that, at the turning of the millennium, in a now classical Nature paper (Vogelstein B. et al., 2000), Vogelstein, Lane and Levine, the discoverers of p53 and of its role as tumor-suppressor, wrote that "The cell, like the Internet, appears to be a 'scale-free network'." To let the reader fully appreciate this revolution, it is useful to recall that in 1999, just one year before the appearance of this particular paper, more than 15.000 independent articles have been already published on p53 and its role in cancer biology, making this protein one of the most studied topics ever. Yet, after 20 years of research, despite the enormous amount of available data, some aspects of p53 biology were still missing, and were so crucial to understand why this protein is found mutated in about 50% of cancer patients, to let its discoverer write "One way to understand the p53 network is to compare it to the Internet" (Vogelstein B. et al., 2000).
What had it happened? To understand this, we must look only two years before. In 1998, Zoltan Oltvai, a molecular pathologist, and Lazlo Barabási, a physicist studying Internet topology, were both working at the Northeastern University of Chicago. They both were Hungarians –actually Barabási was born Romanian, but lived and studied in Hungary - have small kids, and were home neighbors; thus it is hardly surprising that, as recalled by Barabási himself (Barabási AL, 2002), they usually met for dinner. By that time, Barabási had already found out that the Internet structure is a peculiar one: he had collected evidence that it is far from random. This was a non trivial result, given the fact that all large networks were modeled at the time as random. To understand this point, we have to think of

networks made of a huge number of nodes, like human social networks, communication networks and alike. To study how they work, i.e. how information spreads through the network, or whether the network is sensitive to external attacks, or how to find the best pathway from one node to another, it is not possible to perform a direct experiment, given the size of the object under study; instead, one has to model the network, then find out a proper set of equations, and simulate the behavior of the network varying the equation parameters. Since the 60s, the model of choice for large networks was that of Erdős and Rényi, which assumes that each node in a network is randomly connected to a fixed, average number of other nodes. As Barabási explains (Barabási AL, 2002), Erdős and Rényi "acknowledged for the first time that real graphs, from social networks to phone lines, are not nice and regular. They are hopelessly complicated. Humbled by their complexity, [they] assumed that these networks are random." In a random network, each node is equivalent to every else. Were the real network random, this would have several practical implications. For example, removing one server from Internet, was it a random network, would have on average the same effect of removing every other node, so that to protect Internet from hackers one should only care that, on average, a sufficiently high number of servers is shielded, wherever they are.

However, you could have already guessed that removing 100 servers from the Google facilities would have a larger impact on Internet than shutting down 100 servers in rural China (at least presently). This is due to the fact that Google machines are Internet *hubs*, i.e. they are continuously connected to an enormous number of other machines, and mediate a big amount of Internet data exchange. Barabási and its group were the first to notice the presence of hubs in the web (for example, the New York Times web site has an immense number of links, whereas an obscure blogger may have none), and recognized that the classical network theory of Erdős and Rényi was totally unable to deal with them.

To see whether this was a peculiarity of Internet or a general finding, they began to map the topology of other networks as well. It turned that 1) most real networks are different from both regular lattices and random structures and 2) they all exhibit a common underlying organization, based on few hubs and many poorly connected nodes. This last point is evident if one plots the number of nodes having a defined amount of connections, which is called node *degree* or *connectivity*, versus the degree itself. One gets a curve (the *degree distribution*), which smoothly descends from a maximum (many nodes with very low connectivity) to a minimum (few nodes with very high connectivity); since this curve is exponential, the obtained degree distribution obeys a power-law and is said to be *scale-free*.

Having already obtained the first evidence for the generality of its finding on network structure, Barabási met Oltvai, who, like the majority of the biologists, was very well aware of the intricacy of metabolic connections between the molecular constituents of a living cell. Indeed, the complicate diagrams on the walls of biochemistry labs represent complex networks, were the nodes are biomolecules of any sort, and the links are biological interactions (let us keep this description vague for the moment). The two researchers wondered if these biological networks were also scale free as those made by man. By 2002, Barabási and Oltvai had published their results obtained from 43 different organisms (Ravasz E et al., 2002): the metabolic networks connecting the main metabolites have essentially the same large-scale structure of complex, non-biological networks. They are all scale-free, with hubs and poorly connected nodes, despite significant differences in the particular biochemical pathways included, so that each cell of every examined organism resembles a tiny Internet and can be studied in pretty much the same way.

With a perfect timing and a great deal of intuition, Vogelstein, Lane and Levine realized that, when looking to cancer, p53 is indeed a crucial hub, sitting in the center of a complex protein network, and, very much like internet hacking, cellular hijacking by cancer proceeds by attacking hubs. This is why, *ex post*, one finds p53 mutated so many times: touching a cellular hub causes a great deal of effects, while mutating less prominent "client" proteins passes nearly unnoticed. This is also what brought the three researchers to publish the paper which changed p53 science forever.

Since these early observations, a lot of progresses have been made, to the point that protein networks are useful tools in the hands of molecular biologists. The rest of this chapter is devoted to a simple introduction to their structure and properties, in a way that purportedly simplify mathematical descriptions, keeping an eye on the biological meaning of protein networks.

2. The structure of protein networks: Scale freeness

Before entering in some details about protein networks, we want to point out some special characteristics of this type of networks.

First of all, one should keep in mind that a protein network is an abstract representation of the real world, instead of a physical entity like the Internet infrastructure or a phone line web. In particular, while for the latter the links between the nodes are physical (cables in both cases), in the former case the links represent only a *potential* interaction between two proteins. With the notable exception of macromolecular complexes, which can be thought of as networks of interacting proteins, one would never be able to visualize a protein web under a microscope.

Secondly, with the same exception mentioned before, molecular biologists do usually deal with a special type of protein network, one where the nodes represent all of the protein copies coded by a single gene, instead of all the individual proteins which are floating around in a cell. The protein network we will refer to in the following, thus, is a graph which resumes all the known interactions (the links) occurring between the product of every gene out of some list (the nodes); in this respect, such a network is more like a map of our current knowledge about some specific ensemble of proteins then a representation of a real molecular web.

As a third point, a simplification is usually made, by considering only one type of interaction between the proteins composing a given network. The links connecting the nodes thus correspond to one out of a number of possible biological interactions, ranging from very specific types -such as a network where a link between two proteins represents a physical interaction in a molecular complex- to broader concepts -such as a network where a link between two proteins occurs if they are found co-expressed in a given condition. Correspondingly, different protein networks can be obtained, joining the nodes according to different types of interaction: protein-protein interaction networks, transcription networks, enzymatic networks, signaling networks, co-expression networks and so on. However, one should not consider this simplification as an absolute constraint: software does exist, for example, which is able to color the links of a network according to the type of interaction, filtering them as wanted.

We can now start examining an example. Let us consider the human protein-protein interaction network, which can be downloaded from the Reactome organization institutional site (http://www.reactome.org/download/index.html). At the time when this chapter was written, data were available for more than 5000 proteins, summing up to 120661 unique physical interactions (including homomeric interactions, i.e. interactions between identical proteins). These links do not imply any *directed* action from a given protein to its connected neighbors; thus, as opposite to other types of protein networks, such as transcriptional networks, signaling networks and alike, the protein-protein interaction network is said to be *undirected*. The network, on the left of figure 1, is represented as a sphere of densely connected dots (the proteins).

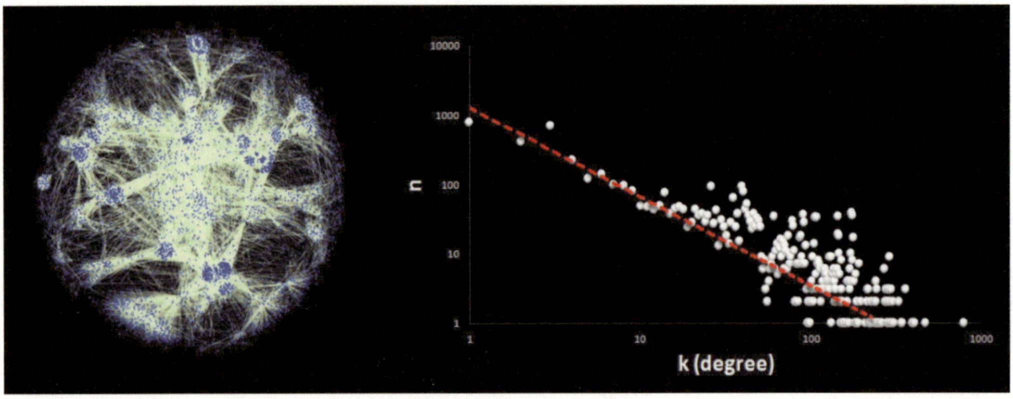

Fig. 1. Human protein-protein interaction network

On the right of figure 1, the degree distribution for this network is plotted on a log-log scale; as evident by the linearity of the obtained graph (red line), this distribution follows a power law, of the form:

$$n \sim k^{-\gamma} \quad (1)$$

and the network is scale-free. This fact in turn implies that there are few protein hubs, connected to most of all other proteins in the network, and a vast majority of proteins able to bind only few partners (poorly connected nodes). Hubs will be treated extensively in the next paragraph; instead, we will focus here on the properties imparted to a protein network by its scale-free nature.

First of all, let us consider the effects of removing some nodes from a scale-free protein network. To this aim, consider the network in figure 2.

On the left, upper corner, low-degree nodes were selected for removal (and are colored in yellow); the resulting network is on the left, lower corner. As evident, the effects are quite limited, and all remaining nodes are still fully connected. Consider now the removal of the same number of high-degree nodes, as shown in the upper, right corner of figure 2. The network falls apart (right, lower corner) and among the remaining nodes, there are examples of disconnected proteins, like the ones pointed by red arrows (which are not the only ones, if you look carefully). Since high-degree nodes are very rare in a scale free network, if compared to low-degree ones, you may have already guessed that protein networks are very resistant to random removal of nodes: if we really selected nodes by chance, than they will nearly always be peripheral, poorly connected proteins. As a consequence, to have even a slight probability of hitting a well-connected protein by random selection, we have to remove a disproportionately high number of nodes, or to repeat the selection process several times. This property of scale-free networks is called *robustness*. In the case of protein networks, robustness has some interesting biological consequences. First of all, it implies that low-frequency, random events such as protein mutations will really affect the cell - which relies for its functioning on several different types of proteins and biochemical networks - with an exceedingly low probability. Thus, we can say that, even before our immune system takes action against malfunctioning cells, we are protected by potentially dangerous random insults - which could give rise to serious diseases such as cancer - by the very architecture of the cell protein networks. On the opposite side, scale-free determined robustness means that the cell has some true Achille's heels - the few highly connected proteins - which can be exploited by selective attacks. Once again, we can refer to Internet for an useful comparison: selective, non random attacks to central routers are preferred by hackers, which purportedly aim to take control over the attacked networks. At cellular level, viruses can be considered hackers, which divert the protein network operations toward an illegitimate scope. Indeed, it has been found by several independent groups that viruses selectively target central proteins, causing large effects on the host protein network (de Chassey B. et al., 2008; Navratil V. et al., 2011; Zou X. et al., 2010).

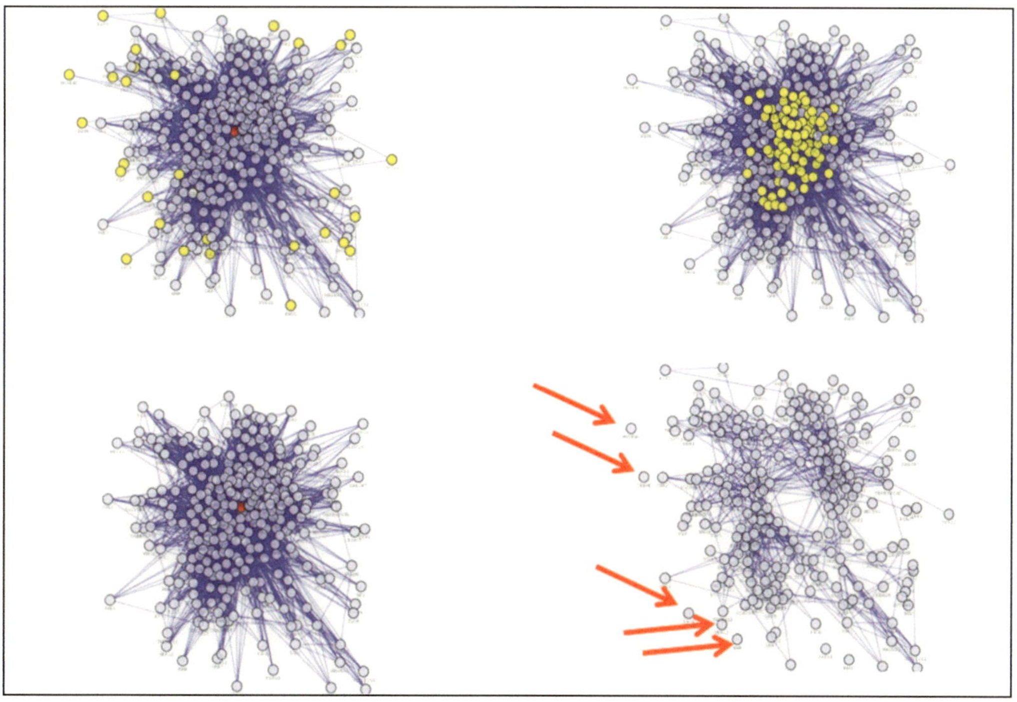

Fig. 2. The effects of removing nodes from a protein network: random removal (left) versus hub removal (right)

Beside robustness, scale-free protein networks show another interesting feature. They exhibit a *small-world* behavior: hopping from one node to a neighbor, any node can be reached from any other in few steps. The distance L between any couple of nodes, in particular, grows roughly proportional to the logarithm of the number of nodes N which are included in the network:

$$L \sim Log(N) \quad (2)$$

For protein-protein interaction networks, L grows even slower with N, and the average distance between two nodes is so small that they have been defined ultrasmall networks (Cohen R & Havlin S. 2003). The small-world property, shared by many complex networks, is particular relevant to biology when the considered network is made of proteins which can influence the neighbors through the links. This is typically the case of transcriptional networks, were two proteins are connected if one influences the expression of the other, signaling network, were proteins are coupled through phosphorylation and other post-translational modifications, and in general holds true for any protein network where the activity of one protein affects its connected partners. In these cases, the small-world structure implies that whatever stimulus changes the status or activity of a protein, its effects will rapidly propagate to the entire network, since on average only few proteins will separate the starting node from every other in the net. This in turn has the consequence that a cell, once its protein network has been stimulated at a single, peripheral node, may quickly change the status of a large number of proteins in response, so that the original signal propagates to a vast number of different proteins, synchronizing their status to the variation of the external stimulus. In other words, small-world protein networks, like any other small-world, display enhanced signal-propagation speed, computational power, and synchronizability (Watts DJ & Strogatz SH. 1998).

In the case of scale-free networks, the observed robustness and the small-world effect are mutually connected. In particular, since protein hubs are linked to the vast majority of all other nodes, most of the pathways connecting any couple of nodes pass through hubs, so that the average distance between any two nodes in the network does not change much if nodes are removed randomly: this is a formulation of network robustness equivalent to the one we mentioned before. At the same time, the more an hub is prominent, i.e. it is connected to an higher amount of protein nodes, the more the distance between any two uncoupled nodes will tend to a single hop through the hub. Hubs are thus key features of scale-free network, mediating both robustness and small-world properties of protein networks; we will dedicate the next paragraph to examine their properties.

3. The structure of protein networks: Hubs

Since the early times of protein network studies, the few, always present hubs attracted a lot of attention: quite naturally, it was thought that since highly connected proteins have a lot of different molecular partners, they should also be implied in the majority of the cellular processes. In case of a protein interaction network like the one depicted in figure 2, this reasoning goes as follows: proteins found in many different macromolecular complexes, represented as hubs in the interaction network, should be either core components of a single molecular complex, or elements conserved in many different molecular complexes, which works as switches and are used by the cell to coordinate the activation or repression of different molecular machineries. As a consequence, any alteration of the hubs of a protein-protein interaction network is predicted to have large effects on the cell biology. This assumption, which has been dubbed as *"centrality-lethality rule"*, has been extensively explored by experimentally knocking-down protein interaction hubs and quantitatively assessing the effects in different models (Jeong H. et al., 2011). Going a step further in the reasoning, it has been hypothesized that mutations affecting these proteins should be particularly related to the insurgence of diseases. Some experimental validation of this prediction has been indeed obtained: Rambaldi and his group (Rambaldi D. et al., 2008) provided evidence that virtually all proteins having a degree higher than 80 in the human protein-protein interaction network are target of known cancer-related mutations. Similarly, Ortutay and Vihinen (Ortutay C. & Vihinen M. et al., 2009), after building an interaction network comprising all human proteins involved in immune response, found that the network hubs include known disease-causing genes as well as 26 new genes related to primary immunodeficiency. In a further example, Chang and colleagues (Chang W. et al., 2009), found new gastric cancer candidate markers by looking to hubs in a protein-protein interaction network build from genes differentially expressed in the patient tissues.

So far, we looked to hubs in protein-protein interaction networks. However, hubs are a common characteristic of any complex web, albeit their biological meaning and relevance change according to the particular type of protein network considered. To understand this point, let us compare the three different human protein networks reported in figure 3.

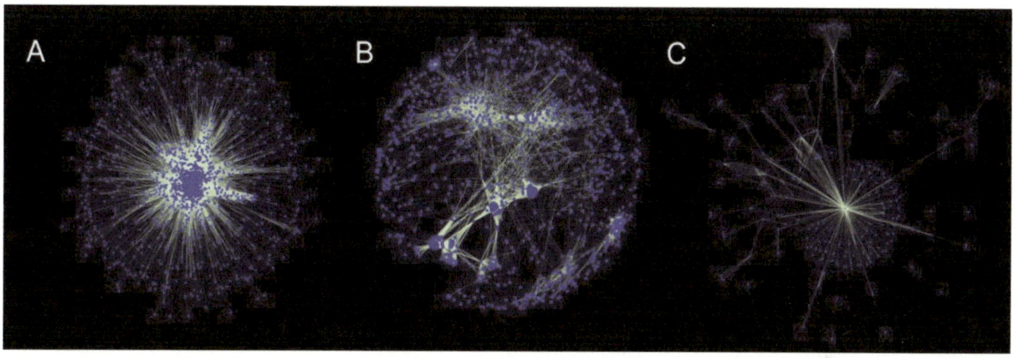

Fig. 3. A) Co-expression network; B) Biochemical/metabolic network; C) Process specific network

The first network on the left (network A) is obtained by considering all proteins differentially expressed in breast cancer patients. Proteins are connected if they are found co-expressed in at least 2 different and independent experiments, and the resulting network is a *co-expression network*. The central network (network B) is build by considering all the proteins which are linked to the p53 human protein by some known biochemical pathway, and is then a *biochemical/metabolic network*. The last network on the right (network C) is a network reporting all those proteins controlled by p53 or controlling it during the unfolding of the apoptotic process, and can be seen as a *process specific network*. You may have already noticed that hubs (and scale-freeness) are present in all three webs, despite the fact that the network size decreases from left to right. However, the biological relevance of hubs is very different in the three network.

In network A, hubs are cancer markers which are found co-expressed with nearly any other cancer protein. Hubs in this network are not granted to be very relevant for the pathogenesis of breast cancer: they can be proteins which are deregulated by the inflammation accompanying cancer, as well as cytoskeletal proteins altered due to the hyperproliferation of cancer cells or other type of very abundant proteins, with no specific role in cancer progression and insurgence. While these proteins are indeed dysregulated in breast cancer, and thus are useful for diagnosing it, their status of hubs do not privilege them with respect to other, less connected nodes, since their co-occurrence with many partners does not imply that their expression level changes more than that of any other node in the network. Indeed, the list of hubs of network A includes useful diagnostic proteins, such as ERBB2, ESR1 and BCRA1, as well as proteins with little meaning for cancer diagnosis, such as complement proteins. Thus, for co-expression networks like the one depicted in figure 3A, being an hub is of no particular merit for a protein.

In network B, since by construction each neighbor of a specific hub is connected to p53 by some biochemical chain, hubs are at the crossroad of several cellular pathway involving p53. In this respect, hubs of this network are in a prominent position to act as checkpoints for controlling the (very redundant) flow of biochemical information from and toward p53, and thus we can expect them to be important controllers and mediators of p53 activity. For example, we found in network B that the 15 hubs with the highest degree are all different subunits of all the three mammalian RNA polymerase, but two, which are important transcription factors (TF2A and TF2B); this is hardly surprising, since p53 in the very end exerts its prominent and multiple actions regulating the transcriptional process, so that all p53 pathways converge into the regulation of the RNA polymerase machinery. By considering hubs with a lower degree, we find the mitosis controlling kinase NEK-2, the nuclear cap-binding protein 1 and 2, and several other proteins which have prominent roles in regulating the cellular status. As a general rule, although there are exceptions, the lower is the degree, the more specific is the position of the protein in the p53 network (or the lesser is known about it). For example, among the proteins having k=1, we find the liprin alpha 4, a protein which binds to the intracellular membrane-distal phosphatase domain of tyrosine phosphatase LAR, and appears to localize LAR for regulating the disassembly of focal adhesion and orchestrating cell-matrix interactions; or E2F-3, a transcription factor which binds specifically to the protein RB1 , in a cell-cycle dependent manner; or MDB4, the Methyl-CpG-binding domain protein 4, which is a mismatch-specific DNA N-glycosylase involved in DNA repair, specific for G:T mismatches within CpG sites. Thus, for biochemical/metabolic networks like the one depicted in figure 3B, hubs are checkpoints for most of the pathways considered in building the network (in the presented case, p53-related pathways), acting as crucial mediators of biological activity and behaving like switches for several biochemical pathways. On the opposite site, if interested to specific, less studied biochemical players, one should concentrate on low-degree nodes of the network, a group which is enriched in proteins involved in few, specific metabolic modules.

In network C, starting from p53, nodes are attached if they co-occur in at least one biochemical pathway and are involved in apoptosis. The fact that two proteins co-occur in more than one pathway is represented by multiple links. This network can be considered as extracted from network B, by filtering out those proteins not involved in apoptosis. As for network B, hubs of this network are to be considered prominent biochemical regulators; however, since we are restricted to a single, specific biological process, there is no special role for low-degree proteins, which are simply peripheral players in a specific apoptotic pathway, among the many redundant possibilities. Hubs are thus the only targets for the analysis of network C: they are important mediators of p53-related apoptosis, controlling most of the network, and their knocking out can be expected to perturb largely the apoptotic control of the cell. As a matter of fact, ordering by degree the nodes of this network, after p53, which is trivially an hub, we find MDM2, possibly the most important regulator of p53 mediated apoptosis, and the apoptosis-stimulating protein of p53 ASPP2, which influences the apoptotic response of cells without affecting p53-induced cell cycle arrest. On the opposite side, we find cdc42, an important cellular protein, which nonetheless mediates only one of the apoptotic pathways controlled by p53 (Thomas A. et al., 2000). Thus, for networks like that depicted in figure 3C, hubs may be considered the most relevant proteins to be found involved in the selected biological process, and they can be safely assumed as targets for further analysis.

After the preceding discussion, it should be clear at this point that protein hubs are extremely variable in their relevance, and that before considering the degree of a node as a topological guide to prioritize protein lists, one must carefully select the type of network to be used, i.e. the rule to generate links between nodes. However, even having the best network may be not enough. To understand why, let us first make a general consideration and then go on with an example.

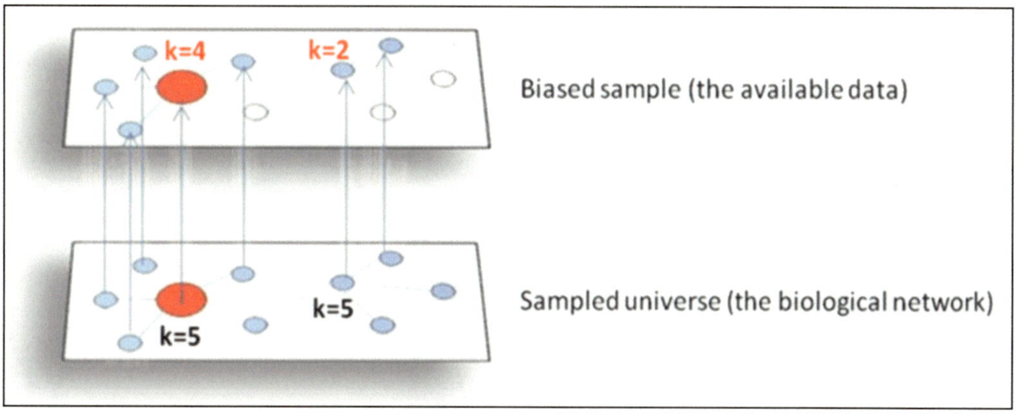

Fig. 4. Sampling bias in biological networks

As shown in figure 4, we must face a sampling problem. “Sampling”, in this context, means to accumulate knowledge on a specific node or part of the network, useful to define its connectivity. In facts, whatever biological network we are exploring, we are only getting an incomplete representation of the real thing, one which was produced by a finite number of experiments interrogating a biological entity. If sampling of the real network (sitting on the lower plane in figure 4) is non-random, i.e. it is concentrated around some “hot” protein (represented in red), then we get a skewed representation (sitting on the upper plane in figure 4) where nodes originally having the same degree are represented as very different in the reconstructed network. Besides being biased, the representation we have can also be error-prone. For example, in figure 4 many links are missing on the upper plane as compared to the lower one (negative error); the opposite situation, were extra links are erroneously added to the representation - for example due to non-specific binding in protein interaction experiments - is also common. Both errors and biases obviously affect the definition of hubs in a network. However, while simple tests exist to check whether an identified hub is a genuine one in an error-prone web (Vallabhajosyula RR. et al., 2009), bias may have subtler effects, much more difficult to deal with. To see this last point, let us consider a further example.

On the left of figure 5, there is a co-expression network which includes all proteins studied in breast cancer. Two proteins are connected if, by any experimental method, they were found to be co-expressed in a breast cancer human sample, whatever the stage or the provenance of the sample. Proteins which are known targets for drug currently used to treat breast cancer or under development are highlighted in red. On the right of the same figure, there is a box-plot which shows the degree distribution for nodes which have been never entered the drug development process (first box on the left), are in preclinical development (second box), are in clinical development phases (phase I, II and III corresponding to the third, fourth and fifth box respectively) and are already on the market (last box on the right). A clear trend may be seen, with the degree regularly increasing as the clinical development of a target proceeds. Is this a genuine trend to be used for drug target identification, i.e. is it true that the more a protein is an hub, the better is to target it from a pharmacological perspective? Quite the opposite. If we consider the same network in a temporal perspective, we will see why. Have a look at figure 6. For the sake of simplicity, we will focus on three exemplary pharmaceutical targets in breast cancer (the vascular endothelial growth factor VEGF, the tymidilate synthase TYMS and the clusterin CLU).

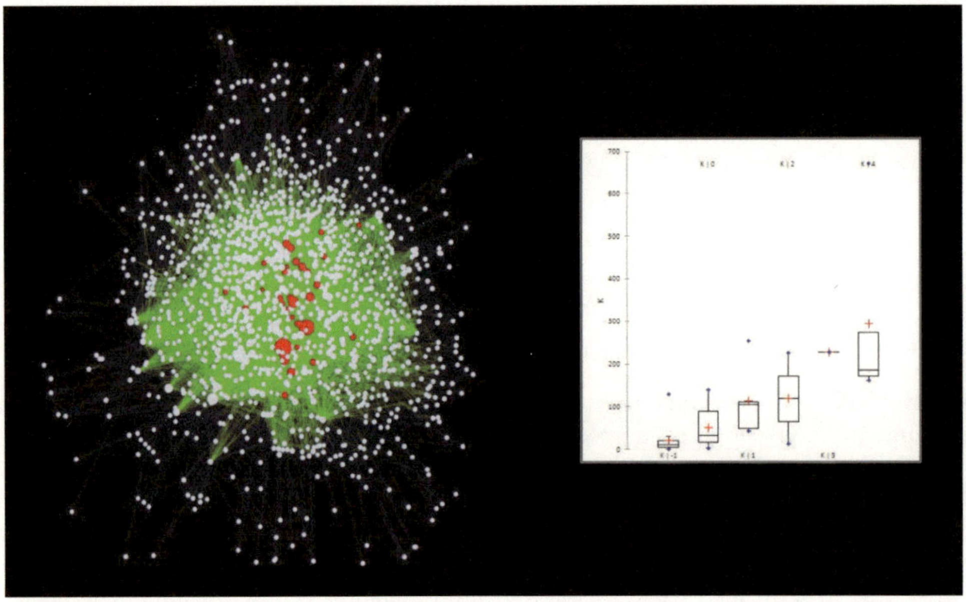

Fig. 5. Degree distribution for pharmacological targets in a breast cancer co-expression network

We want to study their position as hub during time, to see whether it is constant or changes, as new experiments are performed and new network nodes are added. Since the network grows in time, instead of the degree we will consider the ratio between the degree and the total number of nodes; this is the fraction of network nodes connected to the considered protein, and we will refer to this quantity as to "net occupancy". You may have already noticed that this quantity varies in an unpredictable manner. Something connected to about 6.5% of all network nodes in 1993 (TYMS), a true hub for the network, became connected to less than 1% in 2002, to go back to about 3% in 2009. VEGF, which was an important hub in 2009, was barely connected to the network before 1997, and was certainly not an hub by that time. From the graph, we can notice three temporal points associated to an abrupt trend change for all the selected proteins: 2002 for TYMS, 1997 for VEGF and 2005 for CLU. What happened at the time? In 2002, pemetrexed, a drug targeting TYMS, was introduced for breast cancer therapy; in 1997, the antiangiogenic therapy was hypothesized as an option to treat breast cancer; in 2002, the experimental drug OTX-111, targeting CLU, was shifted to prostate cancer, due to mixed results in breast cancer trials. We can thus directly observe that, in the selected cases, the industrial interest immediately precedes a topological change of a protein in the network, promoting to hubs those proteins which are under industrial development, and downsizing those proteins which were not up to the standard in clinical trials. Such kind of an effect may also be caused by interests different from the industrial ones. For example, it is probable that strong academic groups tend to produce a lot of data on their "pet" proteins; moreover, most studied proteins tend understandably to be of human origin, well soluble, stable and easily detected. Large scale "unbiased" experiments, such those using microarrays, two-yeast hybrid or proteomic techniques, produce data which are also biased toward detectable proteins (Ivanic J. et al., 2009), and are still very often affected by the interest of the experimenter (think to the study of knock-out models). Some possible solutions which have the potential to mitigate biases as well as errors in reconstructing protein networks have been recently proposed. These approaches make use of network alignment between different organisms (Tan CS. Et al., 2009). In particular, evidence has been recently produced demonstrating that even though the present protein network data are strongly biased by the experimental methods used to produce them, they still exhibit species-specific similarity and reproducibility (Fernandes LP. Et al., 2010). While intra-species conservation approaches tend to contribute "core" networks, i.e. networks made of conserved proteins and conserved topologies which do not account for inter-species variability, they have the indubitable advantage to average biases (because the networks used for the alignment come from different scientific communities, and are less vexed by pharmaceutical industry interests) and errors (because more large scale experiments are taken into accounts). Moreover, hubs conserved among different species are likely to be very relevant for the basic biology of the cell, as shown by the fact that they tend to be duplicated so to increase the mutational robustness of the corresponding biological network (Kafri R. et al., 2008)

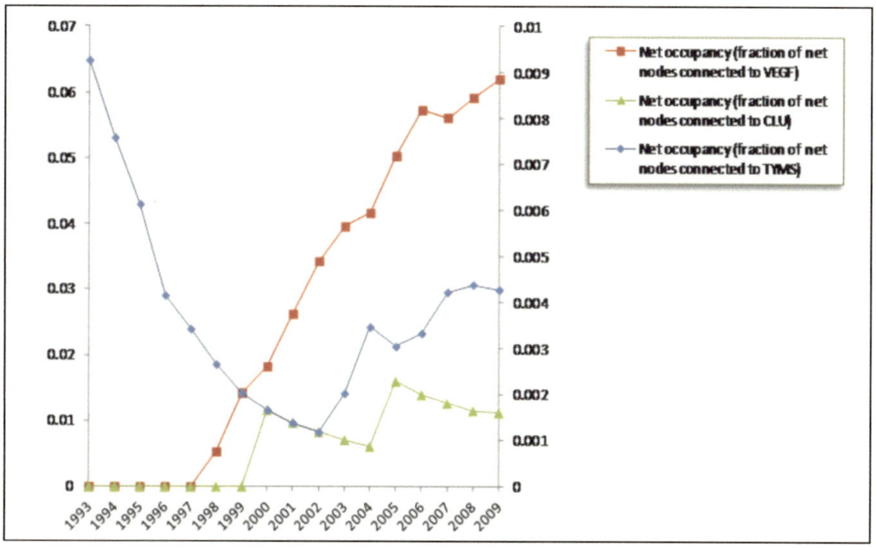

Fig. 6. VEGF, TYMS, and CLU network occupancy

We want to conclude this paragraph by the following message: when properly taking into account biases and errors, the topological prominence of hubs is indeed informative and useful for protein/gene prioritization; however, the real biological meaning of "hubbiness" is strictly dependent on the linking rule applied for building a specific network, as shown in this paragraph for co-expression networks, biochemical/metabolic networks and process specific networks.

4. The structure of protein networks: Neighborhoods

In a graph, the neighbors of a given node consist in all those other nodes that are connected to it up to a certain distance. Distance in this context is intended as the minimal number of steps connecting the source node to any other. In other words, for a particular protein x in a network (which we will call the seed), we define the neighborhood of x, N(x), to be the subgraph of the network whose vertex set consists of all of x's interaction neighbors and the edges between them, up to a preselected distance *D*.

According to the type of graph, neighborhoods can be used to derive useful biological information.

We will try to illustrate this by showing how:

1. in a network, the biological roles of the neighbors can be used to infer the unknown functions of a seed;
2. in protein-protein interaction networks, a group of highly interconnected neighbors sharing a given biological function likely coincides with a macromolecolar complex or part of it.

As for the first example, it is useful to remember that traditionally the function of a protein is inferred from its sequence and/or structure by homology modeling. Unsurprisingly, this approach performs poorly for those proteins which have unusual sequences and unknown structures. In this particular circumstances, an analysis of the biological functions of the network neighbors of the protein can be decisive. In particular, it has been proved that in protein networks the probability that a certain biological function is shared between two proteins is higher if the 2 considered proteins are proximal neighbors, and then decreases as the distance *D* increases (Shamir R. et al., 2007). This is true in many different network types, such as protein-protein interaction networks, metabolic/biochemical protein networks, genetic interaction networks etc. Moreover, if a given protein with an unknown function is at short distance (usually *D=1*) from several proteins sharing a given function, the probability that it too shares that particular function is obviously even higher. On this basis, a neighborhood-guided labeling strategy is possible to assign biological functions to virtually any protein in a network, providing that at least a fraction of the nodes in its neighborhood has a known biological role. The process is exemplified in figure 7, were functional annotation is symbolized by node coloring.

As can be intuitively understood by looking at figure 7, the functional annotation of a given node is guided by several factors, including distance and number of neighbors with a given biological function, their own connectivity and their heterogeneity (which led to the lack of propagation for the red and the blue colors in the example). Mathematical modeling of the labeling procedure basically consists in weighting all these factors in a single probability function, so to obtain a score for the assignment of a given biological role to all the network nodes. While the details of the proposed methods are out of the scope of this introductory text, we would like to stress here that the procedure depends always on the local topology, which affects the label propagation by determining the number of neighbors a given node communicate with, and on the type of network considered, which limits the distance and the direction of propagation of a label along the edges.

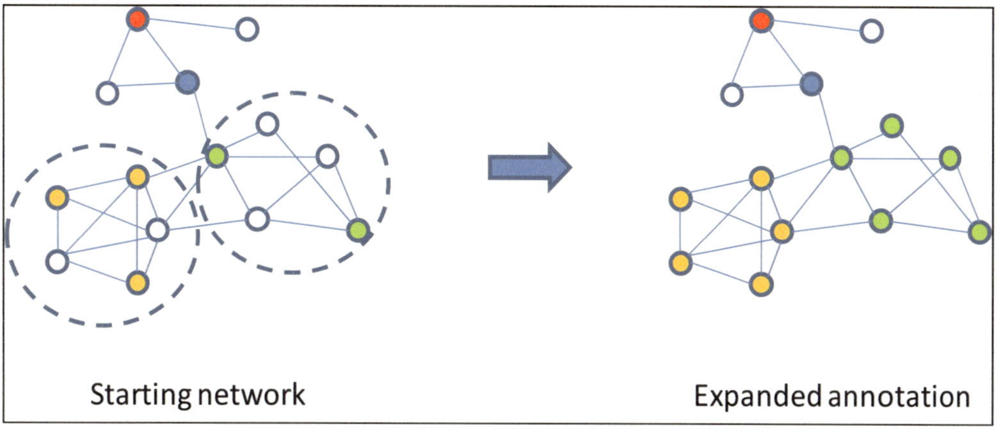

Fig. 7. Functional annotation of a given node

As for the second example, we will refer to a recent work of Fox et al. (Fox AD et al., 2011) on protein-protein interaction networks. Consider in particular the two alternative situations illustrated in figure 8.

In A, the neighborhood for *D=1* of the selected seed (shown in blue) is made of two groups of nodes, which are not directly connected; on the opposite, in B the neighborhood is highly interconnected in a single cluster.

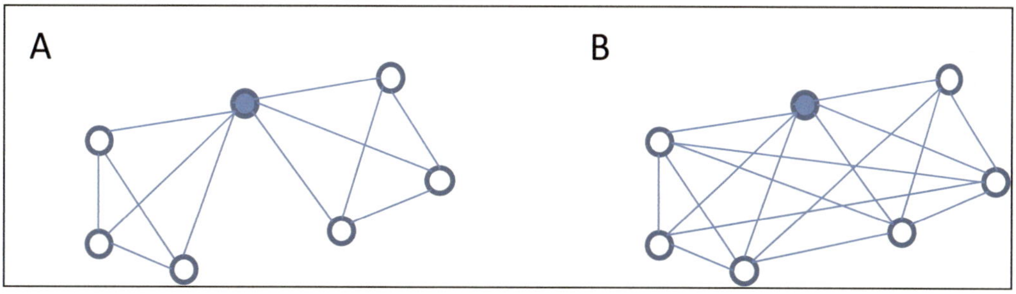

Fig. 8. A) Two disconnected neighborhood; B) Highly interconnected neighborhood

As reported by the authors, the structure observed in A suggests the possibility that the two groups of neighbors might be active under different conditions, as opposite to B. Indeed, it was found that single-component neighborhoods like the one represented in B are enriched in protein sharing similar functions and participating to molecular complexes, and are thus more likely to represent a single, defined protein complex, while multiple-components neighborhoods like the one represented in A tend to represent different molecular complexes, sharing a single component. Interestingly, we found that this concept can be extended beside protein-protein interaction networks. Let us consider, for example, all those proteins, which are reported as changed in expression by at least two different papers on Parkinson's Disease. We will consider two proteins connected, if they co-occur at least 2 times, i.e. if they are reported together by at least two papers. The obtained co-expression network is shown in figure 9A.

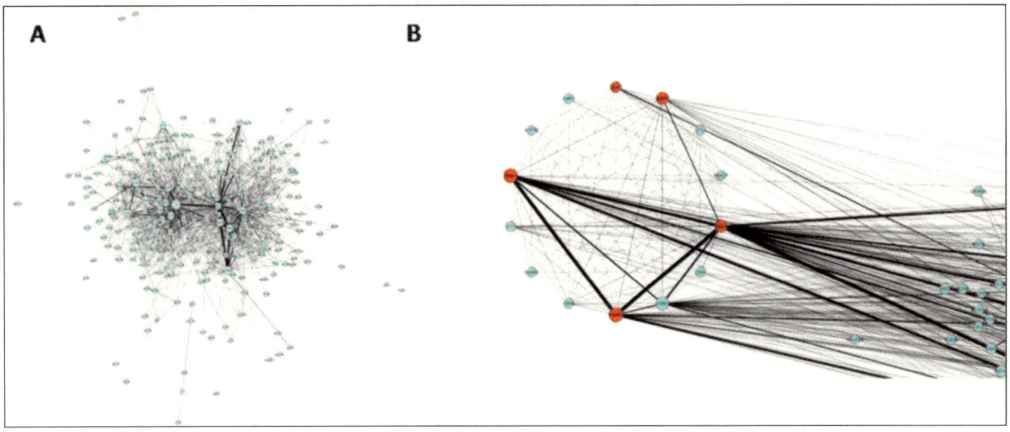

Fig. 9. A) Parkinson's Disease co-expression network; B) A clique from the same network (red nodes have many connections outside the clique).

On the right, in figure 9B, a neighborhood of 14 proteins is extracted from the network, which are all fully interconnected (meaning that each protein is connected to any other). Among these 14 proteins, the red ones are those which have at least as many bonds outside the neighborhood as they have inside it (i.e. at least 13 bonds outside the network). These 14 proteins are arranged in a way similar to that exemplified in figure 8B: a single cluster of highly connected nodes. Much in the same way predicted for protein-protein interaction networks, the cluster is enriched in proteins sharing some functional aspect: in particular, it turns out that 13 out of the 14 components are found in inclusion bodies, a hallmark of neurodegeneration in Parkinson's Disease. Intriguingly, in a sense they represent once again a macromolecular complex - albeit a non-specific one, being a structurally random aggregate, which may vary in its particular composition from case to case. Thus, while the starting network is a co-expression network, where edges do not represent physical interactions among proteins, also in this case proteins in well connected neighborhoods tend to share biological functions and to be involved in the formation of complexes.

5. The structure of protein networks: Graphlet degree signatures

Until now, we have examined pretty simple topological features of the nodes in a protein network. Recently, however, more complex metrics have been introduced, which have several advantages over the older ones. In particular, many of these sophisticated parameters are useful because they recapitulate a larger amount of information with respect to simpler ones. One of such parameter is the "graphlet degree signature" of a node, first introduced by Milenković T. & Przulj N. (2008). To understand what is it, let us consider figure 10.

Imagine that we want to study the local topology around the two colored nodes shown in figure 10A. A possible way would be to count all the graphlets of a certain type which pass through the nodes. Graphlets are small connected network subgraphs with a predetermined number of nodes. In figure 10B, we reported all the possible graphlets with 2 nodes and 3 nodes, with the designation G_0, G_1 and G_2 originally introduced by Pržulj. As evident by figure 10C , node 1 is touched by 3, 3 and 1 G_0, G_1 and G_2 graphlets respectively. Node 2 is touched by 5, 5 and 2 G_0, G_1 and G_2 graphlets respectively. You can check the number of G_0 and G_1 graphlets on the left part of figure 10C, and the number of G_2 graphlets (triangles) on the right; these numbers are called G_0, G_1 and G_2 graphlet degree of a node. Thus, with respect to two- and three nodes graphlets, it is possible to define an ordered vector of the type $<g_0, g_1, g_2>$, which will describe for each node how many graphlets of any possible type actually pass through the node. For node 1 and node 2, this vector assumes the values of <3,3,1> and <5,5,2> respectively. The vector obtained considering all the 29 possible graphlets having from 2 to 5 nodes has been originally dubbed "graphlet degree signature" or simply "signature" of a node.

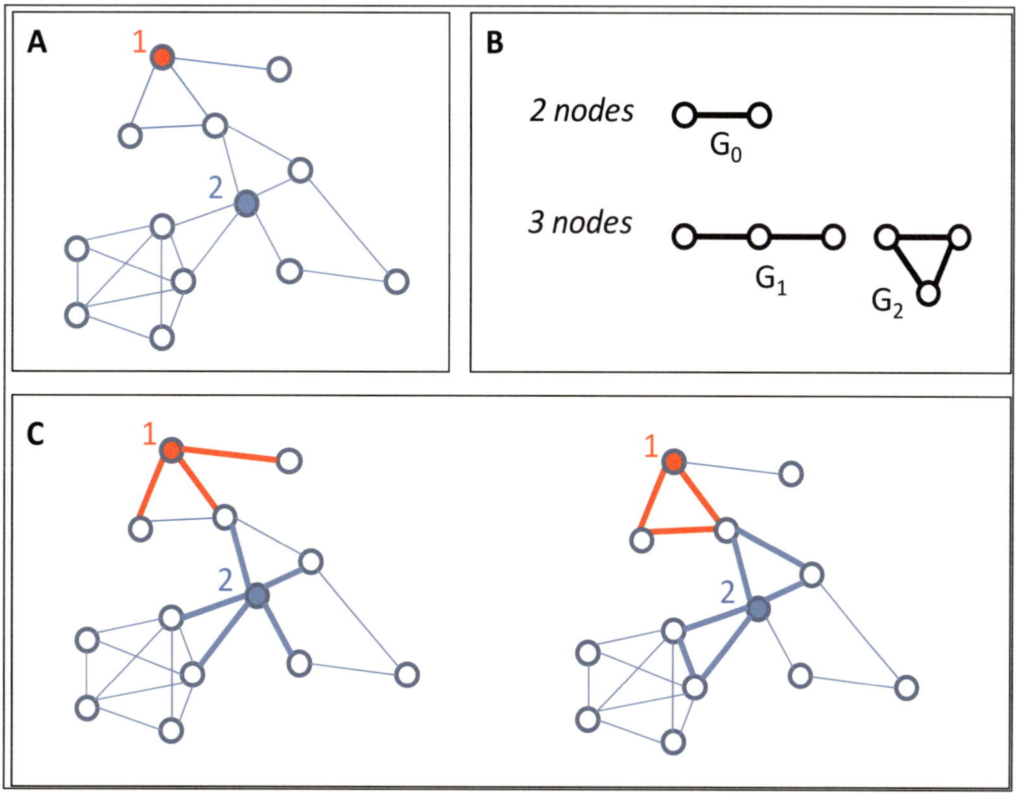

Fig. 10. A) Two nodes with a distinct topology; B) All the possible connection arrangement (graphlets) for groups of 2 or 3 nodes; C) Left, prevalence of G0 and G1 graphlets passing by nodes 1 and 2; Right, prevalence of G2 graphlets passing by nodes 1 and 2.

Before going further, it is important to note the following:

1. the G_0 degree is equivalent to the node degree we saw in the preceding paragraphs; in this respect, the graphlet degree signature can be seen as a generalization of the node degree, which is not limited to the count of only a single type of graphlet;
2. By considering the type and number of graphlet connected to a node, the graphlet degree signature captures in a single metric both the degree of the node, the neighborhood abundance and its topology, recapitulating in a single measure the complementary aspects we introduced in the previous paragraphs.

Once defined in the way we have seen, the graphlet degree signature can be used to cluster all the nodes of a network according to their similarity. We will not enter into the details of the method, which is fully described elsewhere (Milenković T. & Przulj N. 2008); to our purposes, it is enough to understand that nodes with a graphlet degree signature similar above a certain threshold (which implies a similar centrality and a topologically equivalent neighborhood) can be grouped together. The resulting groups, however, may contain nodes which are quite far in the original network, so that nodes in the same cluster are in general scattered all over the network. Mostly relevant to the biologist, it has been shown how these clusters contain proteins of similar biological role and functioning (Milenković T. & Przulj N. 2008). This means that, at least in principle, if one selects a node with known biological features, it is possible to calculate its graphlet degree signature, search for nodes with a similar signature and transpose the biological features to what has been found, without considering the distance of the newly identified nodes from the starting point, as opposite to what we have seen in the previous paragraph.

6. A brief review of successful applications

We will see at this point how the topological analysis of biological networks has been already applied to achieve interesting results. Due to the limited space, we will restrict ourselves to few examples; however, the literature describing successful applications of network analysis in biology is growing at exponential rate, as evident by comparing the papers produced yearly and indexed by Pubmed for "network biology"in 2000 (372) to the corresponding figure for 2010 (2324).

A first, obvious application of topological network analysis consists in illuminating new aspects of the cell biology, which are evident only when looking to the full puzzle represented by a molecular net, instead then to the single pieces of it. The instruments used for such an analysis are many; however, even the simplest topological descriptors we have introduced in this chapter, such as the cliques, may be very useful.

To illustrate this point, let us refer initially to the classic work of Spirin and Mirny on the yeast protein-protein interaction network (Spirin V. & Mirny LA. 2003). These two authors were the first to describe the presence of densely connected modules in protein-protein interaction network, i.e. neighborhoods whose internal connectivity is very high compared to the average network connectivity. As we already know, in extreme cases - i.e. in case all the neighborood's components are fully connected - these protein groups are cliques. As discovered by the aforementioned authors, cliques and very connected neighborhoods represent molecular complexes and/or functional modules. Thanks to this fact, the authors were able to identify a full wealth of new functional modules, including several previously

unknown molecular machineries, such as an eight-member module of cyclin-dependent kinases, cyclins and their inhibitors regulating the cell cycle, a six-member module of proteins involved in bud emergence and polarity establishment and a six-member module of CDCs, septins, and Ser/Thr protein kinases involved in mitotic control. From their starting seminal work, which for the first time shifted the network analysis from single node centrality to community of nodes, a deluge of research followed. This trend culminated in several complex applications of clique analysis, such as a recent work which nicely illustrated how the mitotic spindle functioning is regulated by a cascade of events which involves cliques (i.e. molecular complexes) instead of single proteins (Chen TC. Et al., 2009).
With regard to more complex topological parameters, such as the graphlet degree signature introduced in the previous paragraph, there are obviously fewer examples, given the fact that they have been introduced much later. However, being refined instruments, the results obtained by their systematic application are somehow superior in generality, and uncover the real potency of the topological approach in molecular network analysis. To understand this, is sufficient to read a recent paper by Milenkovic T. et al., 2010. The authors describe how in a human protein-protein interaction network oncogenes do have a very similar graphlet degree signature, which is different from that of genes unrelated to cancer, at a point that they are able to use this signature to identify new oncogenes. If this finding will be confirmed by others, we will be forced to admit that the detailed topology around a node in a global protein-protein interaction map is important in determining the function of the corresponding protein at least at the same level as its sequence and three-dimensional structure - a somehow unexpected result, given the fact that protein-protein interaction networks are only a very abstract map of all the interactions which have been observed, without spatial and temporal resolution, and do not corresponds to any physical entity. However, we want to conclude this paragraph by stressing the fact that, albeit this and similar fundamental problems rest to be solved, and are matter of current and future research in the field, we are seeing already the first applications of network analysis in human therapy. In particular, although network science is still in its infancy, it is currently shifting from a better understanding of why a given drug works or not to the identification of new therapeutic interventions. As an example, consider the case of multi-drug therapy, which is a very active field of research and experimental work, due to its high potential in overcoming several obstacle to the effective pharmacological treatment of different conditions. As opposed to the classical "magic bullet" pharmacological paradigm, aiming to the ultra-specific targeting of a single protein, a new kind of approach to the design of a therapy is emerging, which relies on simultaneously targeting several molecular processes. The topological analysis of the molecular network underlying a specific disease is the only way to rational implement such an approach, allowing the quest for modulators acting on different network areas, so to attack different cellular pathways. This way to proceed was recently validated by some groups, which could identify the right combination of drugs to be used in a number of oncological conditions, such as incurable pancreatic adenocarcinoma (Azmi AS. Et al., 2010), as well as head and neck chemoresistant cancer (Ratushny V. et al., 2009).
From the point of view of the network topology, the approaches described in these papers can be seen as the targeting of control hubs within neighborhoods with quite distinct compositions and cellular functions (i.e. separated neighborhoods enriched in proteins with different functional annotation), a practical strategy which relies on the concepts discussed previously in this chapter and which wait to be extended to several other cases.

7. Conclusion: A concept-map for the analysis of network topology

Having listed some few examples, we would like to recall to the attention of the reader those elements which allow a successful analysis: a proper selection of the data set to start with, a correct identification of the rules used to build the network (i.e. the type of network to be analyzed), few general assumptions on the relationships between the topology and the biological properties of the proteins to be found, and a correctly chosen null-hypothesis for the minimization of false positives (which, if possible, should also take into account bias and errors).

Let us discuss briefly the first point, i.e. the selection of a proper data set to derive the nodes of the network. This step is crucially influenced by the scope of the network. For example, if the aim is to find potential drug targets for a given condition, a literature-derived dataset, including all the proteins known to be related to a certain disease – irrespectively of the type of relation they have with the studied condition - might be useful. A protein expression data set, containing data on differential protein expression, would be equally useful. On the contrary, taking into account a complete protein-protein interaction data set may be both misleading – given the fact that there is no guarantee that the proteins contained in it are expressed in the selected condition – and useless, because this type of database lacks information on those proteins which have strong activity and expression in the selected condition, but do not have any identified molecular partner.

As for the second point, usually people select the type of network (and thus the node linking rule) they want to build at the very first step – i.e., they use protein interaction databases to build protein-protein interaction networks, expression databases for co-expression networks and so on. However, there are certain cases were this passage is not automatic. For example, if the data source for the node list is the scientific literature, instead of building a literature co-occurrence network one can derive the linking rule from a different source, like a microarray experiment database. By combining a literature-derived list of nodes with microarray information for linking them, one would obtain a network, whose nodes are selected on the basis of a specific scientific topic, and are bound by co-expression, without the need to perform an actual experiment in the condition of interest.

As for the third point, it is true that, in general, the topology of a node is correlated to the relevance of the role that the corresponding protein plays in the particular condition the network refers to. However, one has to recall that:

1. The meaning of "topologically relevant proteins" varies with the type of network – for example, hubs in co-expression networks are usually housekeeping proteins, while in protein-protein interaction networks they may be core constituent of molecular complexes;
2. the specific meaning varies also with the network dimension – so that in a network including the full yeast proteome, topologically prominent proteins are heterogeneous in function, while in a network made of proteins involved in apoptosis the hubs are key apoptosis regulators;
3. obviously protein prioritization is affected by the particular topological quantity one is measuring –a protein may be an hub, yet may have no clique including it;
4. the relevance of a protein for the cell may be in gross contrast with what is perceived as relevant by the investigator – housekeeping proteins are very relevant to the functioning of the cell, but not so to someone wanting to find new drug targets.

Finally, coming to the forth point, we want to stress here that control models to be used for underpinning significant topological properties should vary, depending on the topological quantity under study. Thus, to get a control network for testing the relevance of some topological characteristic of a certain group of node, one may compare the results obtained on the actual network with those obtained in:

a. a random network, i.e. a network made of the same number of nodes and edges, with fully random connection between the nodes- this is enough to test for the global distribution of topological quantities, such as the degree distribution or the existence of statistically relevant neighborhoods;
b. a degree-preserving random network, i.e. a network made of the same number of nodes and edges, with the degree of each node preserved, but a completely different wiring – this is the proper control, when one want to test the association between some topological parameter and a specific biological attribute, which depends on the particular nodes considered;
c. a set of random network (degree preserving type or not) – this is the proper control, when one want to test the probability of the emergence of the observed topology in a network

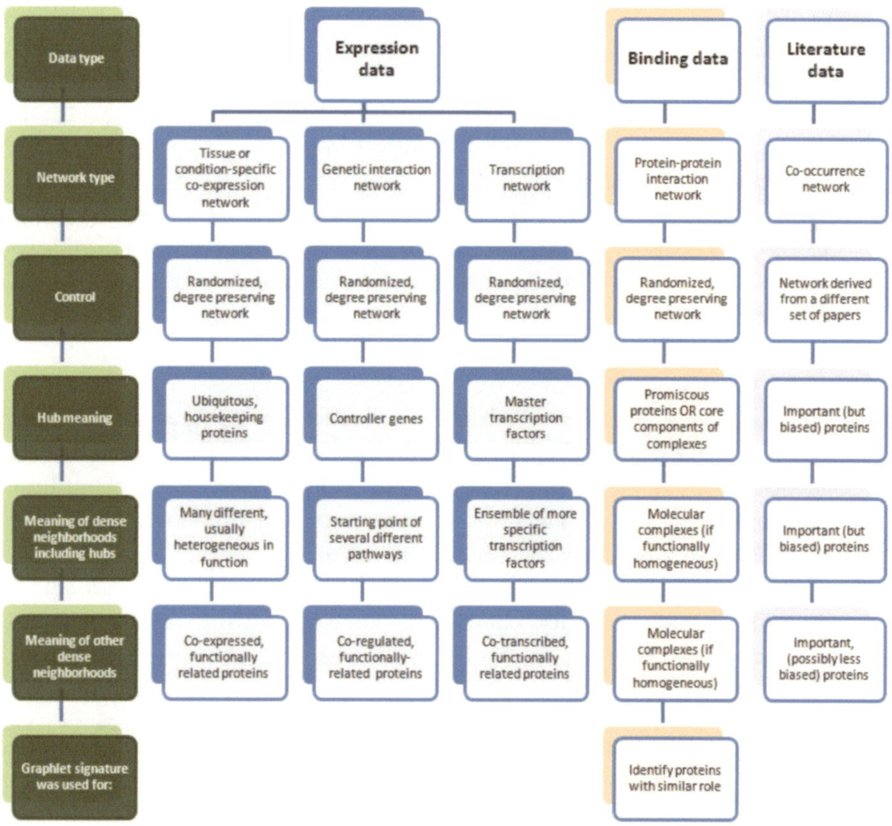

Fig. 11. Concept map for network topology analysis

After these considerations, we will conclude this chapter by outlining a general concept-map, reported in figure 11, which we feel can be useful in analyzing the topology of protein networks. This map should be regarded as a contribution to avoid common misinterpretations of the meaning of topological parameters in different contexts, not as an all-inclusive description of the possible applications and types of protein networks.

8. Acknowledgment

We would like to thanks all the BioDigitalValley team involved into developing ProteinQuesttm, the tool which we used to explore the wonderful world of biological networks and to outline the concepts described in this chapter.

9. References

Azmi AS, Wang Z, Philip PA, Mohammad RM. & Sarkar FH. (2010). Proof of concept: network and systems biology approaches aid in the discovery of potent anticancer drug combinations. *Mol Cancer Ther.*, Vol. 9, No. 12, (December 2010), pp. 3137-44

Barabási AL. (2002). Linked: How Everything Is Connected to Everything Else and What it Means for Business, Science, and Everyday Life. ISBN 0-452-28439-2

Chang W, Ma L, Lin L, Gu L, Liu X, Cai H, Yu Y, Tan X, Zhai Y, Xu X, Zhang M, Wu L, Zhang H, Hou J, Wang H. & Cao G. (2009). Identification of novel hub genes associated with liver metastasis of gastric cancer. *Int J Cancer,* Vol. 125, No.12, (December 2009), pp. 2844-53

Chen TC, Lee SA, Chan CH, Juang YL, Hong YR, Huang YH, Lai JM, Kao CY. & Huang CY. (2009). Cliques in mitotic spindle network bring kinetochore-associated complexes to form dependence pathway. *Proteomics*, , Vol. 9, No. 16, (August 2009), pp. 4048-62.

Cohen R& Havlin S. (2003). Scale-free networks are ultrasmall. *Phys Rev Lett.*, Vol. 90, No. 5, (February 2003), 058701

de Chassey B, Navratil V, Tafforeau L, Hiet MS, Aublin-Gex A, Agaugué S, Meiffren G, Pradezynski F, Faria BF, Chantier T, Le Breton M, Pellet J, Davoust N, Mangeot PE, Chaboud A, Penin F, Jacob Y, Vidalain PO, Vidal M, André P, Rabourdin-Combe C. & Lotteau V. (2008). Hepatitis C virus infection protein network. *Molecular Systems Biology*, Vol. 4, No. 230, (November 2008)

Fernandes LP, Annibale A, Kleinjung J, Coolen AC. & Fraternali F. (2010). Protein networks reveal detection bias and species consistency when analysed by information-theoretic methods. *PLoS One*, Vol. 5, No. 8, (August 2010), e12083

Fox AD, Hescott BJ, Blumer AC. & Slonim DK. (2011). Connectedness of PPI network neighborhoods identifies regulatory hub proteins. *Bioinformatics*, Vol. 27, No. 8, (Apr 2011), pp.1135-42

Ivanic J, Yu X, Wallqvist A. & Reifman J. (2009). Influence of protein abundance on high-throughput protein-protein interaction detection. *PLoS One*, Vol. 4, No. 6, (June 2009), e5815

Jeong H, Mason SP, Barabási AL, Oltvai ZN. (2001). Lethality and centrality in protein networks. Nature, Vol. 411, No. 6833, (May 2001), pp.41-2

Kafri R, Dahan O, Levy J. & Pilpel Y. (2008). Preferential protection of protein interaction network hubs in yeast: evolved functionality of genetic redundancy. *Proc Natl Acad Sci U S A*, Vol. 105, No. 4, (January 2008), pp. 1243-8

Milenkovic T, Memisevic V, Ganesan AK. & Przulj N. (2010). Systems-level cancer gene identification from protein interaction network topology applied to melanogenesis-related functional genomics data. *J R Soc Interface*, Vol. 7, No. 44, (March 2010), pp. 423-37

Milenković T. & Przulj N. (2008). Uncovering biological network function via graphlet degree signatures. *Cancer Inform*, Vol. 6, (Apr 2008), pp. 257-73

Navratil V, de Chassey B, Combe CR. & Lotteau V. (2011). When the human viral infectome and diseasome networks collide: towards a systems biology platform for the aetiology of human diseases. *BMC Syst Biol*, Vol. 21, (January 2011), pp. 5-13

Ortutay C. & Vihinen M. (2009). Identification of candidate disease genes by integrating Gene Ontologies and protein-interaction networks: case study of primary immunodeficiencies. *Nucleic Acids Res.*, Vol. 37, No. 2, (February 2009), pp. 622-8

Rambaldi D, Giorgi FM, Capuani F, Ciliberto A. & Ciccarelli FD. (2008). Low duplicability and network fragility of cancer genes. *Trends Genet.*, Vol. 24, No. 9, (September 2008), pp. 427-30

Ratushny V, Astsaturov I, Burtness BA, Golemis EA. & Silverman JS. (2009). Targeting EGFR resistance networks in head and neck cancer. *Cell Signal*, Vol. 21, No. 8, (August 2009), pp. 1255-68

Ravasz E, Somera AL, Mongru DA, Oltvai ZN. & Barabási AL. (2002). Hierarchical organization of modularity in metabolic networks. *Science*, Vol. 297, No. 5586, (August 2002), pp. 1551-5

Sharan R, Ulitsky I. & Shamir R. (2007). Network-based prediction of protein function. *Mol Syst Biol.*, Vol. 3, No. 88, (March 2007)

Spirin V. & Mirny LA. (2003). Protein complexes and functional modules in molecular networks. *Proc Natl Acad Sci U S A*, Vol. 100, No. 21, (October 2003), pp.12123-8

Tan CS, Bodenmiller B, Pasculescu A, Jovanovic M, Hengartner MO, Jørgensen C, Bader GD, Aebersold R, Pawson T. & Linding R. (2009). Comparative analysis reveals conserved protein phosphorylation networks implicated in multiple diseases. *Sci Signal.*, Vol. 2, No. 81, (July 2009), ra39

Thomas A, Giesler T. & White E. (2000). p53 mediates bcl-2 phosphorylation and apoptosis via activation of the Cdc42/JNK1 pathway. Oncogene, Vol. 19, No. 46, (November 2000), pp. 5259-69

Vallabhajosyula RR, Chakravarti D, Lutfeali S, Ray A. & Raval A. (2009). Identifying hubs in protein interaction networks. *PLoS One*, Vol. 4, No. 4, (April 2009), e5344

Vogelstein B, Lane D, Levine AJ. (2000) Surfing the p53 network. *Nature*, Vol. 408, No. 6810, (November 2000), pp. 307-10.

Watts DJ & Strogatz SH. (1998). Collective dynamics of 'small-world' networks. *Nature*, Vol. 393, No. 6684, (June 1998), pp. 440-2

Zou X. (2010). The Topological Properties of Virus-Human Protein Interaction Networks, *The Fourth International Conference on Computational Systems Biology*, ISB 2010, Suzhou, China, September, 2010

Part 2

Gene Regulation, Networking and Signaling in and Between Genomes

8

Prediction and Analysis of Gene Regulatory Networks in Prokaryotic Genomes

Richard Münch, Johannes Klein and Dieter Jahn
Institute of Microbiology, Technische Universität Braunschweig, Braunschweig Germany

1. Introduction

The availability of over 1500 completely sequenced and annotated prokaryotic genomes offers a variety of comparative and predictive approaches on genome-scale. The results of such analyses strongly rely on the quality of the employed data and the computational strategy of their interpretation. Today, comparative genomics allows for the quick and accurate assignment of genes and often their corresponding functions. The resulting list of classified genes provides information about the overall genomic arrangement, of metabolic capabilities, general and unique cellular functions, however, almost nothing about the underlying complex regulatory networks. Transcriptional regulation of gene expression is a central part of these networks in all organisms. It determines the actual RNA, protein and as a consequence metabolite composition of a cell. Moreover, it allows cells to adapt these parameters in response to changing environmental conditions. An integral part of transcriptional regulation is the specific interaction of transcription factors (TFs) with their corresponding DNA targets, the transcription factor binding sites (TFBSs) or motifs. Recent advances in extensive data mining using various high-throughput techniques provided first insights into the complex regulatory networks and their interconnections. However, the computational prediction of regulatory interactions in the promoter regions of identified genes remains to be difficult. Consequently, there is a high demand for the *in silico* identification and analysis of involved regulatory DNA sequences and the development of software tools for the accurate prediction of TFBSs.

In this chapter we focus on methods for the prediction of TFBSs in whole prokaryotic genomes (regulons). Although, many studies were sucessfully performed in eukaryotes they are often not transferable to the special features of bacterial gene regulation. In particular the prokaryotic genome organization concerning clusters of co-transcribed polycistronic genes, the lack of introns and the shortness of promoter sequences necessitates adapted computational approaches. Besides the genomic structure there are also differences in the regulatory control logic. Prokaryotic promoters often possess one or few regulatory interactions while the repertoire of regulators consists of only a couple of global TFs but many local TFs (Price et al., 2008). On the other hand, eukaryotic promoters and enhancers involve the concerted binding of multiple regulators, so called cis-regulatory modules (CRMs) or composite elements (Loo & Marynen, 2009). Many excellent reviews in the field prokaryotic gene regulation were recently published with focus on the broad spectrum of approaches for the experimental and theoretical reconstruction of gene regulatory networks and their interspecies transfer (Baumbach, 2010; Rodionov, 2007; van Hijum et al., 2009; Zhou & Yang, 2006). Here, we focus on practical aspects how to detect new members of a regulon for genes or genomes of interest. We will summarize useful bioinformatics databases, methods and algorithms available for unraveling bacterial gene regulatory networks from whole genome sequences. Finally, we want to indicate the limitations and technical problems of such approaches and give a survey on recent improvements in this field.

2. Strategies for the prediction of transcripion factor binding sites

Basically, today exist at least two general approaches to recognize regulatory sequence patterns. One challenging approach called **pattern discovery** relies on a statistical overrepresentation of DNA sequence motifs present in promoters of structurally and funktionally related or co-regulated genes. In that case it is a *de-novo* prediction where the binding site and the corresponding regulator are unknown. The list of investigated genes can be derived from clusters of co-expressed genes available in microarray experiments, from ChIP-on-chip experiments or from orthologous genes of related organisms. In the latter case this method is called phylogenetic footprinting (McCue et al., 2001). Pattern discovery algorithms are top-down approaches that use various learning principles with different degrees of performance (Sandve et al., 2007; Su et al., 2010; Tompa et al., 2005). The advantage of this method is the detection of potential regulatory DNA sequences even if there is little known about the corresponding regulation. A recent study in prokaryotes appling a pattern discovery approach revealed that the predicted patterns matched up to 81% of known individual TFBSs (Zhang et al., 2009). However, this approach has limitted value in getting a clue about what specific regulator is involved in a predicted TFBS.

An alternative approach on which we focus in this chapter is called **pattern matching**. It makes use of prior knowledge in form of a predetermined pattern that can be assigned to a specific regulator. The pattern is usually build based on a profile of known TFBSs for which experimental evidence is available (Fig. 1 A). Using this set of DNA sequences a probabilistic model describing the pattern degeneracy is constructed. Application of the model on a given sequence results in a score for the likelihood that the investigated sequence belongs to the same sequence family. The application of pattern matching involves the availability of a reliable training set of TFBSs. For that purpose, several specalized databases provide collections and patterns of prokaryotic TBFSs supplemented with various related information like promoter and operon structures. A limited list of important data sources is shown in table 1.

In the following examples a data set of 40 experimentally proven TFBSs from the anaerobic regulator Anr of *Pseudomonas aeruginosa* is used (Trunk et al., 2010). There are different ways of pattern representation. Traditionally, the usage of IUPAC code for base ambiguities is a straightforward way to describe a binding motif (NC-IUB, 1985). In this approach, combinations of certain bases are assigned to an extended alphabet of specific letters (Fig. 1 B). IUPAC code can be easily converted into a regular expression (Fig. 1 C). A regular expression is a formal language for pattern matching, that can be used to scan for ambiguous IUPAC strings in order to predict new TFBSs (Betel & Hogue, 2002). (Fig. 1 B). Allthough the IUPAC letter code is very concise and still widely used among biologists it does not describe a proper weighting of bases. Additionally, the majority rules how to generate a consensus sequences are to some extent arbitrary (Day & McMorris, 1992). However, in the case that the training set consists of only a few sequences the usage of IUPAC code can still make sense.

Name	Year	Data content	URL	References
CoryneRegNet	2006	*Coynebacerium* TFBSs, regulatory networks, predictions	http://www.coryneregnet.de	Baumbach et al. (2009)
DBTBS	2001	*B. subtilis* TFBSs, operons, predictions	http://dbtbs.hgc.jp	Sierro et al. (2008)
DPInteract	1998	*E. coli* TFBSs, PWMs	http://arep.med.harvard.edu /dpinteract	Robison et al. (1998)
PRODORIC	2003	prokaryotic TFBSs, PWMs, promoters, expression data	http://www.prodoric.de	Grote et al. (2009)
PromEC	2001	*E. coli* promoters	http://margalit.huji.ac.il /promec	Hershberg et al. (2001)
RegPrecise	2010	predicted TFBSs	http://regprecise.lbl.go	Novichkov et al. (2010)
RegTransBase	2007	prokaryotic TFBSs, PWMs	http://regtransbase.lbl.gov	Kazakov et al. (2007)
RegulonDB	1998	*E. coli* TFBSs, PWMs, operons,	http://regulondb.ccg.unam.mx	Gama-Castro et al. (2011)
Tractor_DB	2004	predicted TFBSs of γ-proteobacteria	http://www.tractor.lncc.br	Pérez et al. (2007)

Table 1. List of important public databases about bacterial gene regulation. The table shows the name, year of establishment, data content, the internet address and the latest reference of the respective database.

A more accurate description of a binding pattern is achieved by probabilistic models like a frequency matrix (or alignment matrix) (Staden, 1984). Instead of considering only the most common bases at each position a matrix comprises the frequencies for each nucleotide at each position (Fig. 1 D). Based on frequency matrices many models for the calculation of weights were proposed. Such a model is broadly called position weight matrix (PWM) or position specific scoring matrix (PSSM). PWMs can be considered as simplified profile hidden Markov models (HMM) that do not allow insertion and deletion states (Durbin et al., 1998). Formally, a PWM is an array M of weights w where each column corresponds to the position of the TFBS motif of the length l and each row represents the letter of the sequence alphabet $\mathcal{A}$. In case of DNA $\mathcal{A} \in \{A, C, G, T\}$ (equation 1).

$$M = \begin{vmatrix} w_{A,1} & w_{A,2} & \cdots & w_{A,l} \\ w_{C,1} & w_{C,2} & \cdots & w_{C,l} \\ w_{G,1} & w_{G,2} & \cdots & w_{G,l} \\ w_{T,1} & w_{T,2} & \cdots & w_{T,l} \end{vmatrix} \tag{1}$$

Many very related examples for the calculation of individual weights were proposed in the literaure (Berg & von Hippel, 1987; Fickett, 1996; Schneider et al., 1986; Staden, 1984; Stormo, 2000). The information theoretical approach and modifications of it ((Schneider et al., 1986)) are widely used and some of the most successful methods for both the modeling and the prediction of potential TFBSs. Information is a measure of uncertainty which means that a highly conserved position with the exclusive occurence of one specific nucleotide gets the highest information value of 2 bits. In other words there is a maximum certainty of finding this nucleotide at this position. In contrast, an information value of 0 bits represents a highly degenerated position and the highest uncertainty of finding a specific nucleotide. The information vector $R(l)$ represents the total information content of a profile of aligned sequences at the position l with $f(b,l)$ indicating the frequency of the base b at position l.

$$R(l) = 2 + \sum_{b=A}^{T} f(b,l) \log_2 f(b,l) \quad (2)$$

An information PWM $m(b,l)$ is generated by multiplying the base frequencies $f(b,l)$ with the total information content $R(l)$ (Fig. 1 E).

$$m(b,l) = f(b,l) \cdot R(l) \quad (3)$$

For pattern matching applications a PWM is used by summing up the corresponding weights of a candidate sequence to a score. Afterwards, these scores are compared to a predefined cut-off (or threshold) to filter out potential predictions. The derived score is often correlated to the binding affinity of a TF thus the information score can be interpreted as an rough estimate to the specific bindung energy. However, this is only possible under the simplifying assumption that each position of a pattern contributes independently to the TF-TFBS interaction. This additivity assumption is controversially discussed but is was shown that it is in fact a reasonable approximation (Benos et al., 2002). The graphical representation of an information PWM is called sequence logo (Schneider & Stephens, 1990). In a sequence logo each PWM weight is equivalent to the individual letter size so the total height of the stack of letters represents the information content $R(l)$ at this position. Sequence logos allow an illustrative visualization of the sequence conservation and binding preference of a regulator (Fig. 1 F).

3. Statistical significance of pattern matching

Regulatory sequences are commonly short (usually 6-18 bp), the sample size of experimentally proven sites is often limited and in many cases the observed level of sequence conservation is low. Consequently, the genome-wide statistically occurance frequency of derived patterns is often unrealistically high. In such cases, searches generally generate increasing numbers of false-predictions the lower the threshold score is set. This is demonstrated in Fig. 2 showing the score distributions of true and false predictions of a genome wide search in *P. aeruginosa* using the PWM of the Anr regulator (Fig. 1 E). In the shown example matches in coding regions were considered as false-predictions (false-positives) and matches that are part of the training set were naturally ranked as true-predictions (true-positives). Score distributions are also important indicators to evaluate the predictive capacity of a PWM (Medina-Rivera et al., 2011).

In order to improve the predictive power of pattern matching, commonly a cut-off score is set in a way, that improves the ratio of true- and false-predictions. However, thereby the total number of hits will still contain to some extent false-positives while some true matches become lost (false-negatives). From this it follows that matches of TFBS predictions can not be classified in a binary manner like a dignostic test, since true-positives and false-positives are always coexisting. Alternatively, they can be grouped into a classification schema consisting

A) Excerpt of 40 Sample sequences (training set)

1	T	T	G	A	T	T	C	C	G	G	T	C	A	A
2	T	T	G	A	C	T	T	T	C	A	T	C	A	A
3	T	T	G	A	T	T	G	C	C	A	T	C	A	A
4	T	T	G	A	C	C	G	G	A	A	T	C	A	A
⋮								⋮						
40	T	T	G	A	T	G	T	C	G	A	T	C	A	A

B) IUPAC consensus

Y T G H Y N B N B V K C A R

C) Regular Expression

[CT]TG[ACT][CT][ACGT][CGT][ACGT][CGT][ACG][TG]CA[AG]

D) Frequency Matrix

	1	2	3	4	5	6	7	8	9	10	11	12	13	14
A	2	0	0	29	0	6	0	10	6	20	0	0	40	35
C	5	0	0	7	17	17	15	10	15	3	1	40	0	0
G	0	0	40	0	3	10	9	18	16	17	2	0	0	5
T	33	40	0	4	20	7	16	2	3	0	37	0	0	0

E) Position Weight Matrix

	1	2	3	4	5	6	7	8	9	10	11	12	13	14
A	0.06	0.00	0.00	0.65	0.00	0.02	0.00	0.07	0.04	0.35	0.00	0.00	2.00	1.27
C	0.15	0.00	0.00	0.16	0.30	0.05	0.17	0.07	0.09	0.05	0.04	2.00	0.00	0.00
G	0.00	0.00	2.00	0.00	0.05	0.03	0.10	0.12	0.10	0.30	0.08	0.00	0.00	0.18
T	0.97	2.00	0.00	0.09	0.35	0.02	0.18	0.01	0.02	0.00	1.43	0.00	0.00	0.00

F) Sequence Logo

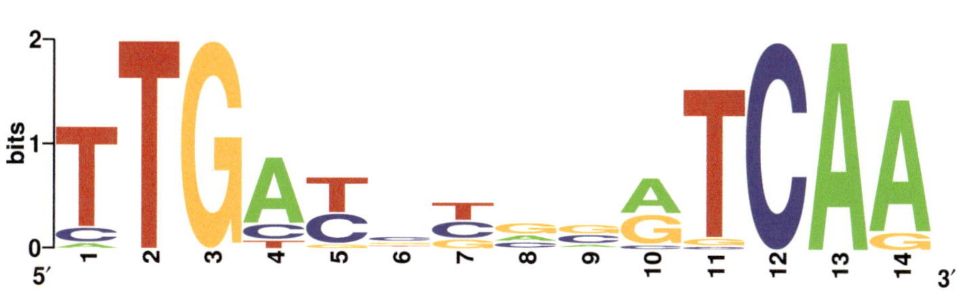

Fig. 1. Various pattern representations for a taining set 40 Anr binding sites from *Pseudomonas aeruginosa* (Trunk et al., 2010). The deduced IUPAC consensus (B), regular expression (C), frequency matrix (D), position weight matrix (E) and sequence logo (F) are shown.

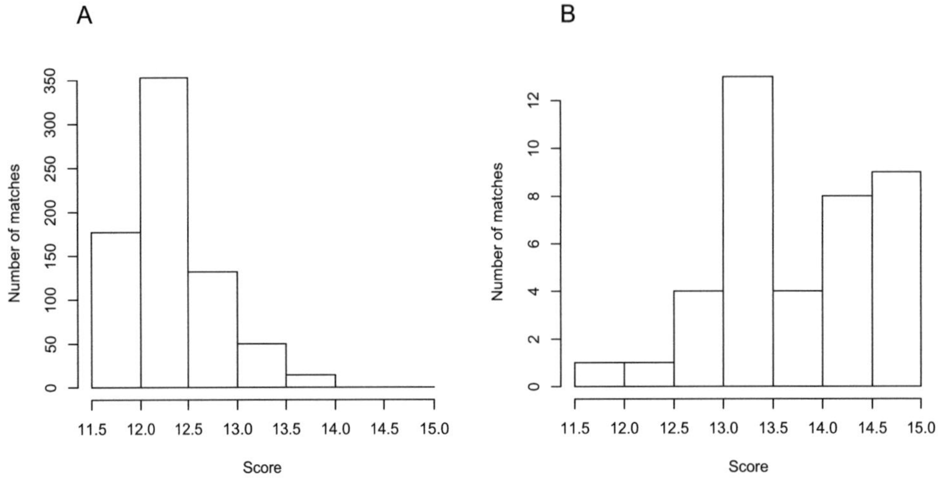

Fig. 2. Score distributions of false-positive matches (A) and true-positive matches (B) from a genome wide search in *P. aeruginosa* using the Anr PWM.

of four different classes (Fig. 3) which is called a two-by-two confusion matrix or contingency table (Fawcett, 2004).

	Dataset	
	Positive	Negative
Match	True-Positive	False-Positive
No Match	False-Negative	True-Negative

Fig. 3. A two-by-two confusion matrix illustrates all four possible outcomes of matches in the positive and in the negative dataset.

Thus, setting a cut-off score can be considered as important decision-making process. Instead of setting an arbitrary cut-off value it is possible to determine an optimized threshold. For that purpose, a number of statistical performance measurements for binary classification are available. Sensitivity Sn (or true-positive rate) measures the proportion of positive matches which are correctly identified at a given cut-off score c. Hereby, the positive matches include both the number of true-positives TP and false-negatives FN.

$$Sn(c) = \frac{TP}{TP + FN} \tag{4}$$

Similarly, specificity Sp (or true-negative rate) measures the proportion of correctly identified negative matches at a given cut-off score c where the amount of negative matches is the sum of true-negatives TN and false-positive FP.

$$Sp(c) = \frac{TN}{TN + FP} \tag{5}$$

This definition involves that the sensitivity and specificity plots as a function of the cut-off show opposite behaviour which results in an increase of specificity (get less false-positives) at the cost of sensitivity (find less true-positives) and vice versa (Fig. 4 A). A receiver operating characteristics (ROC) curve summarizes the classification performance in a plot of sensitivity versus (1-specificity). ROC curves are fundamental tools for the evaluation of the classification models. An optimal ROC curve would cross the upper left corner or coordinate (0,1) representing 100% sensitivity and specificity whereas a random guess would produce a point along the diagonal line (Fig. 4 A). Thus, the diagonal line divides the ROC space: points above the digonal represent good classification results, points below the line indicate poor results (Fawcett, 2004).

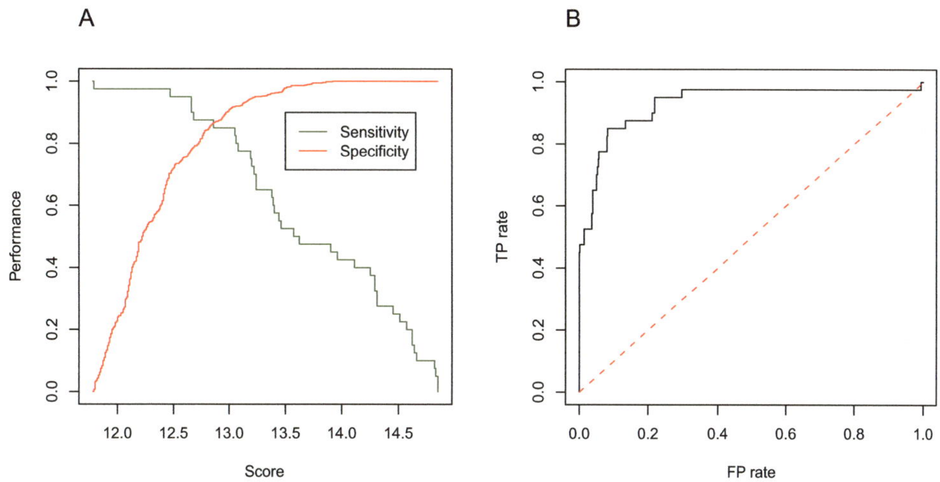

Fig. 4. Performance measurements for the prediction of the Anr regulon in *Pseudomonas aeruginosa*. (A) Sensitivity (green) and specificity (red) plot. (B) ROC graph.

An alternative way to optimize the performance of pattern matching and to produce stastistically significant results is the calculation of a *p*-value. A *p*-value depicts the likelihood to find a score that is as least as good by change. *P*-values can be either determined by simulation or estimated via a compound importance sampling approach (Oberto, 2010).
Finally, appropriate thresholds for pattern searches are determined as a tradeoff between sensitivity and specificity to maximize both values. Despite optimized cut-off values this approach can results in a poor sensitivity and a loss of 40-60% of known functional sites (Benítez-Bellón et al., 2002). In addition, the fact that false-predictions commonly exeeds true-predictions by several orders of magnitude (Fig. 2 B) was called **'futility theorem'** (Wasserman & Sandelin, 2004). Fortunately, there are many sophisticated approaches to overcome this problem in a reasonable way (see section 4).

4. Improvements to increase the accuracy of TFBS predictions

4.1 Modifications of the score

In several studies the information score was modified in different ways. One of the most critical points of equation 2 is that it postulates an equal nucleotide distribtuion of the target genome which is the case e.g. for *Escherichia coli* with a GC content of 51.8%. For this reason, the calculation of the information content of motifs in genomes with highly biased nucleotide composition is likely to be over- or underestimated. A more generalized form that considers the background frequencies P_b is given in equation 6.

$$R(l) = -\sum_{b=A}^{T} f(b,l) \log_2 \frac{f(b,l)}{P_b} \quad (6)$$

This new term turned out to be the relative entropy or Kullback-Leibler distance (Stormo, 2000). An other promising approach deals with biased genome as a discrete channel of noise to discriminate a motif from its background (Schreiber & Brown, 2002). However, it was recently demonstrated, that the unmodified information score performs on average better than other alternatives (Erill & O'Neill, 2009). One reason might be, that binding sites shift towards the genome skew in a co-evolutionary process between TFs and its corresponding TFBSs.

Other modifications concern the way the score is computationally calculated. Since the information vector usually peeks at certain well conserved positions it is possible to get overestimated matches by forming the overall sum. For that purpose, it is useful to define a core region consisting of the highly conserved positions. Using this approach it is possible to realize the computation of the score in two steps. Potential matches have to pass first the core cut-off before they are evaluated by the overall cut-off score (Münch et al., 2005; Quandt et al., 1995).

Finally, it is possible to enhance the accuracy by combining multiple (independent) criterions. Apart from the pure sequence information, DNA exhibits distinct structural properties caused by interactions from neighboring nucleotides. This includes for example DNA curvature, flexibility and stability, amongst others. Structural DNA features are available as di- and trinucleotide scale values assigning a particular value to each possible nucleotide combination (Baldi & BaisnÃ©e, 2000). These values are derived from empirical measurements or theoretical approaches. The calculation of structural features within a DNA sequence stretch is usually performed by summing up and averaging the corresponding di- or trinucleotide scales. Prokaryotic promoters usually exhibit distinct structural features which imply that these DNA sequences are more curved and less flexible in comparision to coding regions. This feature is necessary in order to enable the melting of the DNA strands for the onset of transcription. In most bacterial promoters structural peaks are present around the position -40 upstream of the transcriptional start point (Pedersen et al., 2000). Structural features can provide distinct scores independent from PWM based sequence similarity scores. Recently, pattern matching was combined with a binding site model that was trained using 12 different structural properties (Meysman et al., 2011). In this approach, based on conditional random fields, it was shown, that the classification of matches was significantly improved. In a similar way, structural and chemical features of DNA decreased the number of false-positives in a supervised learning approach (Bauer et al., 2010).

4.2 Positional preference of TFBSs

Prokaryotic genomes usually consist of 6-14% non-coding DNA (Rogozin et al., 2002). In contrast to eukaryotes, the evolvement of non-coding regions appears to be determined primarily by the selective pressure to minimize the amount of non-functional DNA, while maintaining the essential TFBSs. Additionally, it was demonstrated in *Escherichia coli*, that many PWMs show a strong preference for matches in non-coding regions (Robison et al., 1998). Figure 5 A shows the distance of 1741 genomic TFBSs relative to the translational start site of the target gene. Only 3.6% of all TFBSs are located after the start codon within the coding region. However, the largest amount of TFBSs is accumulated directly upstream. This is also demonstrated in the cumulative percentage of TFBSs against the distance to the translational start (Fig. 5 B). According to this result, a total of 75.3% and 87.9% of all TFBSs are located 200bp and 300bp upstream, respectively. Thus, prokaryotic promoters are usually short and it is reasonable to constrain searches to non-coding regions with a limit of a few hundred bp upstream to the translational start.

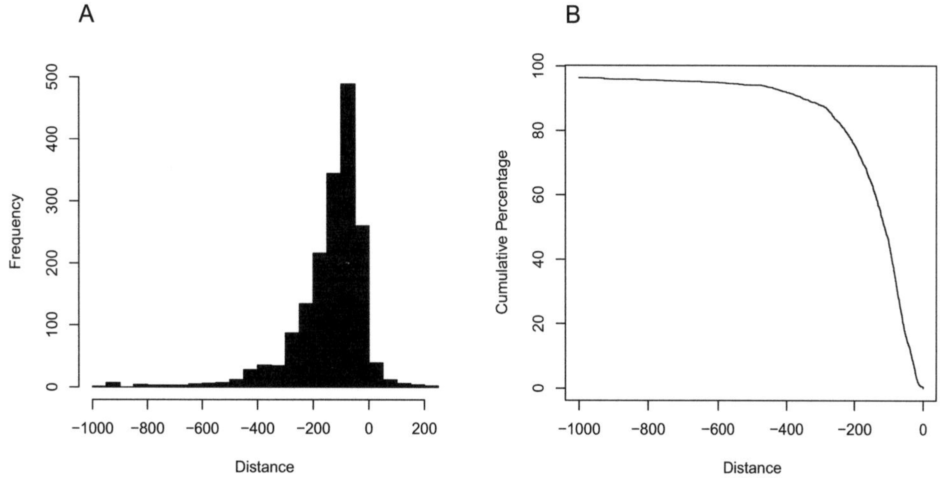

Fig. 5. Histogram of TFBS distances to the translational start site. The used dataset consisted of 1741 genomic TFBSs from various bacterial species taken from the PRODORIC database

4.3 Phylogenetic conservation of regulatory interactions

The large number of sequenced bacterial genomes offers comparative genomics approaches to predict and to analyze regulatory interactions. Similar to phylogenetic footprinting, highly conserved matches in promoter regions of paralogous genes are more likely to be functional targets than non-conserved matches (McCue et al., 2001). This is particulary important for the interspecies transfer of gene regulatory networks (Babu et al., 2006; Baumbach, 2010) but also for the scanning of new regulon members (Pérez et al., 2007). The utilization of pattern matching methods in combination with phylogenetic conservation is also called regulog analysis (Alkema et al., 2004). During a regulog analysis the relativ conservation score RCS is defined by the fraction of orthologs, that share the same potential TFBS.

$$RCS = \frac{orthologs_{observed}}{orthologs_{expected}} \tag{7}$$

In the first step of this and related approaches, the orthologous regulators and the corresponding target gene set are determined. This is often realized by bi-directional best BLAST hits (BBH) (Mushegian & Koonin, 1996). In the second step, conserved TFBSs are extracted via pattern matching or pattern discovery approaches. Predicted TFBSs with phylogenetic conservation can also be used to extend or to build new PWMs. Huge datasets based on phylogenetic reconstruction were generated in various groups of bacteria (Baumbach et al., 2009; Novichkov et al., 2010; Pérez et al., 2007). Further investigetion of regulon evolution revealed the availability of a core set of genes that is widely conserved across related species and a variable set of target genes reflecting the degree of specialization (Browne et al., 2010; Dufour et al., 2010). However, it was shown, that the outlined approach is commonly only feasible between closely related clades which is due to the fact that TFs evolve rapidly and independently of their target genes (Babu et al., 2006). Morover, orthologous TFs in bacteria often have different functions and regulate different sets of genes (Price et al., 2007). In summary, a high *RCS* value for a TFBS match represents an independent score for the validation for a real functional targets while a low *RCS* does not necessarily rule out false-positive matches. The phylogenetic conservation approach represents a powerful approach to predict gene regulatory networks in highly related organisms and to get insights into the evolution of regulons.

5. Conclusion and outlook

In summary the genome-wide recognition of DNA patterns by computational methods is still a challanging task. However, major improvements in this field allow for reliable predictions in many cases. Especially the rising number of sequenced bacterial genomes in combination with data from high-throughput technologies offers many posibilities for the development of more sophisticated methods in comparative genomics approaches. Nevertheless, computational methods for TFBSs prediction can not replace wet-lab experiments but they can help to find new hypotheses that can be verified in an iterative process.

6. References

Alkema, W. B. L., Lenhard, B. & Wasserman, W. W. (2004). Regulog analysis: detection of conserved regulatory networks across bacteria: application to *Staphylococcus aureus.*, *Genome Res.* 14(7): 1362–1373.
URL: *http://dx.doi.org/10.1101/gr.2242604*

Babu, M. M., Teichmann, S. A. & Aravind, L. (2006). Evolutionary dynamics of prokaryotic transcriptional regulatory networks., *J Mol Biol* 358(2): 614–633.
URL: *http://dx.doi.org/10.1016/j.jmb.2006.02.019*

Baldi, P. & BaisnÃ©e, P. F. (2000). Sequence analysis by additive scales: DNA structure for sequences and repeats of all lengths., *Bioinformatics* 16(10): 865–889.

Bauer, A. L., Hlavacek, W. S., Unkefer, P. J. & Mu, F. (2010). Using sequence-specific chemical and structural properties of dna to predict transcription factor binding sites., *PLoS Comput Biol* 6(11): e1001007.
URL: *http://dx.doi.org/10.1371/journal.pcbi.1001007*

Baumbach, J. (2010). On the power and limits of evolutionary conservation–unraveling bacterial gene regulatory networks., *Nucleic Acids Res* .
URL: *http://dx.doi.org/10.1093/nar/gkq699*

Baumbach, J., Wittkop, T., Kleindt, C. K. & Tauch, A. (2009). Integrated analysis and reconstruction of microbial transcriptional gene regulatory networks using coryneregnet., *Nat Protoc* 4(6): 992–1005.
URL: *http://dx.doi.org/10.1038/nprot.2009.81*

Benos, P. V., Bulyk, M. L. & Stormo, G. D. (2002). Additivity in protein-DNA interactions: how good an approximation is it?, *Nucleic Acids Res* 30(20): 4442–4451.

Benítez-Bellón, E., Moreno-Hagelsieb, G. & Collado-Vides, J. (2002). Evaluation of thresholds for the detection of binding sites for regulatory proteins in *Escherichia coli* K12 DNA., *Genome Biol* 3(3): 13.

Berg, O. G. & von Hippel, P. H. (1987). Selection of DNA binding sites by regulatory proteins. Statistical-mechanical theory and application to operators and promoters., *J Mol Biol* 193(4): 723–750.

Betel, D. & Hogue, C. W. V. (2002). Kangaroo–a pattern-matching program for biological sequences., *BMC Bioinformatics* 3(1): 20.

Browne, P., Barret, M., O'Gara, F. & Morrissey, J. P. (2010). Computational prediction of the crc regulon identifies genus-wide and species-specific targets of catabolite repression control in *Pseudomonas* bacteria., *BMC Microbiol* 10: 300.
URL: *http://dx.doi.org/10.1186/1471-2180-10-300*

Day, W. H. & McMorris, F. R. (1992). Critical comparison of consensus methods for molecular sequences., *Nucleic Acids Res* 20(5): 1093–1099.

Dufour, Y. S., Kiley, P. J. & Donohue, T. J. (2010). Reconstruction of the core and extended regulons of global transcription factors., *PLoS Genet* 6(7): e1001027.
URL: *http://dx.doi.org/10.1371/journal.pgen.1001027*

Durbin, R., Eddy, S., Krogh, A. & Mitchison, G. (1998). *Biological sequence analysis*, Cambridge University Press.

Erill, I. & O'Neill, M. C. (2009). A reexamination of information theory-based methods for dna-binding site identification., *BMC Bioinformatics* 10: 57.
URL: *http://dx.doi.org/10.1186/1471-2105-10-57*

Fawcett, T. (2004). ROC graphs: Notes and practical considerations for researchers, *Technical report*, HP Laboratories.
URL: *http://www.hpl.hp.com/techreports/2003/HPL-2003-4.pdf*

Fickett, J. W. (1996). Quantitative discrimination of MEF2 sites., *Mol Cell Biol* 16(1): 437–441.

Gama-Castro, S., Salgado, H., Peralta-Gil, M., Santos-Zavaleta, A., Muñiz-Rascado, L., Solano-Lira, H., Jimenez-Jacinto, V., Weiss, V., García-Sotelo, J. S., López-Fuentes, A., Porrón-Sotelo, L., Alquicira-Hernández, S., Medina-Rivera, A., Martínez-Flores, I., Alquicira-Hernández, K., Martínez-Adame, R., Bonavides-Martínez, C., Miranda-Ríos, J., Huerta, A. M., Mendoza-Vargas, A., Collado-Torres, L., Taboada, B., Vega-Alvarado, L., Olvera, M., Olvera, L., Grande, R., Morett, E. & Collado-Vides, J. (2011). Regulondb version 7.0: transcriptional regulation of escherichia coli k-12 integrated within genetic sensory response units (gensor units)., *Nucleic Acids Res* 39(Database issue): D98–105.
URL: *http://dx.doi.org/10.1093/nar/gkq1110*

Grote, A., Klein, J., Retter, I., Haddad, I., Behling, S., Bunk, B., Biegler, I., Yarmolinetz, S., Jahn, D. & Münch, R. (2009). PRODORIC (release 2009): a database and tool platform for the analysis of gene regulation in prokaryotes., *Nucleic Acids Res* 37(Database issue): D61–D65.
URL: *http://dx.doi.org/10.1093/nar/gkn837*

Hershberg, R., Bejerano, G., Santos-Zavaleta, A. & Margalit, H. (2001). PromEC: An updated database of *Escherichia coli* mRNA promoters with experimentally identified transcriptional start sites., *Nucleic Acids Res* 29(1): 277.

Kazakov, A. E., Cipriano, M. J., Novichkov, P. S., Minovitsky, S., Vinogradov, D. V., Arkin, A., Mironov, A. A., Gelfand, M. S. & Dubchak, I. (2007). RegTransBase–a database of regulatory sequences and interactions in a wide range of prokaryotic genomes., *Nucleic Acids Res* 35(Database issue): D407–D412.
URL: *http://dx.doi.org/10.1093/nar/gkl865*

Loo, P. V. & Marynen, P. (2009). Computational methods for the detection of cis-regulatory modules., *Brief Bioinform* 10(5): 509–524.
URL: *http://dx.doi.org/10.1093/bib/bbp025*

McCue, L., Thompson, W., Carmack, C., Ryan, M. P., Liu, J. S., Derbyshire, V. & Lawrence, C. E. (2001). Phylogenetic footprinting of transcription factor binding sites in proteobacterial genomes., *Nucleic Acids Res* 29(3): 774–782.

Medina-Rivera, A., Abreu-Goodger, C., Thomas-Chollier, M., Salgado, H., Collado-Vides, J. & van Helden, J. (2011). Theoretical and empirical quality assessment of transcription factor-binding motifs., *Nucleic Acids Res* 39(3): 808–824.
URL: *http://dx.doi.org/10.1093/nar/gkq710*

Meysman, P., Dang, T. H., Laukens, K., Smet, R. D., Wu, Y., Marchal, K. & Engelen, K. (2011). Use of structural dna properties for the prediction of transcription-factor binding sites in *Escherichia coli.*, *Nucleic Acids Res* 39(2): e6.
URL: *http://dx.doi.org/10.1093/nar/gkq1071*

Münch, R., Hiller, K., Grote, A., Scheer, M., Klein, J., Schobert, M. & Jahn, D. (2005). Virtual Footprint and PRODORIC: an integrative framework for regulon prediction in prokaryotes., *Bioinformatics* 21(22): 4187–4189.
URL: *http://dx.doi.org/10.1093/bioinformatics/bti635*

Mushegian, A. R. & Koonin, E. V. (1996). A minimal gene set for cellular life derived by comparison of complete bacterial genomes., *Proc Natl Acad Sci U S A* 93(19): 10268–10273.

NC-IUB (1985). Nomenclature Committee of the International Union of Biochemistry (NC-IUB). Nomenclature for incompletely specified bases in nucleic acid sequences. Recommendations 1984., *Eur J Biochem* 150(1): 1–5.

Novichkov, P. S., Laikova, O. N., Novichkova, E. S., Gelfand, M. S., Arkin, A. P., Dubchak, I. & Rodionov, D. A. (2010). RegPrecise: a database of curated genomic inferences of transcriptional regulatory interactions in prokaryotes., *Nucleic Acids Res* 38(Database issue): D111–D118.
URL: *http://dx.doi.org/10.1093/nar/gkp894*

Oberto, J. (2010). Fitbar: a web tool for the robust prediction of prokaryotic regulons., *BMC Bioinformatics* 11: 554.
URL: *http://dx.doi.org/10.1186/1471-2105-11-554*

Pedersen, A. G., Jensen, L. J., Brunak, S., Staerfeldt, H. H. & Ussery, D. W. (2000). A DNA structural atlas for *Escherichia coli.*, *J Mol Biol* 299(4): 907–930.
URL: *http://dx.doi.org/10.1006/jmbi.2000.3787*

Pérez, A. G., Angarica, V. E., Vasconcelos, A. T. R. & Collado-Vides, J. (2007). Tractor_DB (version 2.0): a database of regulatory interactions in gamma-proteobacterial genomes., *Nucleic Acids Res* 35(Database issue): D132–D136.
URL: *http://dx.doi.org/10.1093/nar/gkl800*

Price, M., Dehal, P. & Arkin, A. (2008). Horizontal gene transfer and the evolution of transcriptional regulation in *Escherichia coli.*, *Genome Biol* 9(1): R4.
URL: *http://dx.doi.org/10.1186/gb-2008-9-1-r4*

Price, M. N., Dehal, P. S. & Arkin, A. P. (2007). Orthologous transcription factors in bacteria have different functions and regulate different genes., *PLoS Comput Biol* 3(9): 1739–1750.
URL: *http://dx.doi.org/10.1371/journal.pcbi.0030175*

Quandt, K., Frech, K., Karas, H., Wingender, E. & Werner, T. (1995). MatInd and MatInspector: new fast and versatile tools for detection of consensus matches in nucleotide sequence data., *Nucleic Acids Res* 23(23): 4878–4884.

Robison, K., McGuire, A. M. & Church, G. M. (1998). A comprehensive library of DNA-binding site matrices for 55 proteins applied to the complete *Escherichia coli* K-12 genome., *J. Mol. Biol.* 284(2): 241–254.

Rodionov, D. A. (2007). Comparative genomic reconstruction of transcriptional regulatory networks in bacteria., *Chem Rev* 107(8): 3467–3497.
URL: *http://dx.doi.org/10.1021/cr068309+*

Rogozin, I. B., Makarova, K. S., Natale, D. A., Spiridonov, A. N., Tatusov, R. L., Wolf, Y. I., Yin, J. & Koonin, E. V. (2002). Congruent evolution of different classes of non-coding DNA in prokaryotic genomes., *Nucleic Acids Res* 30(19): 4264–4271.

Sandve, G. K., Abul, O., Walseng, V. & DrablÃ¸s, F. (2007). Improved benchmarks for computational motif discovery., *BMC Bioinformatics* 8: 193.
URL: *http://dx.doi.org/10.1186/1471-2105-8-193*

Schneider, T. D. & Stephens, R. M. (1990). Sequence logos: a new way to display consensus sequences., *Nucleic Acids Res* 18(20): 6097–6100.

Schneider, T. D., Stormo, G. D., Gold, L. & Ehrenfeucht, A. (1986). Information content of binding sites on nucleotide sequences., *J Mol Biol* 188(3): 415–431.

Schreiber, M. & Brown, C. (2002). Compensation for nucleotide bias in a genome by representation as a discrete channel with noise., *Bioinformatics* 18(4): 507–512.

Sierro, N., Makita, Y., de Hoon, M. & Nakai, K. (2008). Dbtbs: a database of transcriptional regulation in bacillus subtilis containing upstream intergenic conservation information., *Nucleic Acids Res* 36(Database issue): D93–D96.
URL: *http://dx.doi.org/10.1093/nar/gkm910*

Staden, R. (1984). Computer methods to locate signals in nucleic acid sequences., *Nucleic Acids Res* 12(1 Pt 2): 505–519.

Stormo, G. D. (2000). DNA binding sites: representation and discovery., *Bioinformatics* 16(1): 16–23.

Su, J., Teichmann, S. A. & Down, T. A. (2010). Assessing computational methods of cis-regulatory module prediction., *PLoS Comput Biol* 6(12): e1001020.
URL: *http://dx.doi.org/10.1371/journal.pcbi.1001020*

Tompa, M., Li, N., Bailey, T. L., Church, G. M., Moor, B. D., Eskin, E., Favorov, A. V., Frith, M. C., Fu, Y., Kent, W. J., Makeev, V. J., Mironov, A. A., Noble, W. S., Pavesi, G., Pesole, G., RÃ©gnier, M., Simonis, N., Sinha, S., Thijs, G., van Helden, J., Vandenbogaert, M., Weng, Z., Workman, C., Ye, C. & Zhu, Z. (2005). Assessing computational tools for the discovery of transcription factor binding sites., *Nat Biotechnol* 23(1): 137–144.
URL: *http://dx.doi.org/10.1038/nbt1053*

Trunk, K., Benkert, B., Quäck, N., Münch, R., Scheer, M., Garbe, J., Jänsch, L., Trost, M., Wehland, J., Buer, J., Jahn, M., Schobert, M. & Jahn, D. (2010). Anaerobic adaptation in *Pseudomonas aeruginosa*: definition of the Anr and Dnr regulons., *Environ Microbiol* 12(6): 1719–1733.
URL: *http://dx.doi.org/10.1111/j.1462-2920.2010.02252.x*

van Hijum, S. A. F. T., Medema, M. H. & Kuipers, O. P. (2009). Mechanisms and evolution of control logic in prokaryotic transcriptional regulation., *Microbiol Mol Biol Rev* 73(3): 481–509, Table of Contents.
URL: *http://dx.doi.org/10.1128/MMBR.00037-08*

Wasserman, W. W. & Sandelin, A. (2004). Applied bioinformatics for the identification of regulatory elements., *Nat Rev Genet* 5(4): 276–287.
URL: *http://dx.doi.org/10.1038/nrg1315*

Zhang, S., Xu, M., Li, S. & Su, Z. (2009). Genome-wide de novo prediction of cis-regulatory binding sites in prokaryotes., *Nucleic Acids Res* 37(10): e72.
URL: *http://dx.doi.org/10.1093/nar/gkp248*

Zhou, D. & Yang, R. (2006). Global analysis of gene transcription regulation in prokaryotes., *Cell Mol Life Sci* 63(19-20): 2260–2290.
URL: *http://dx.doi.org/10.1007/s00018-006-6184-6*

9

Mining Host-Pathogen Interactions

Dmitry Korkin, Thanh Thieu, Sneha Joshi and Samantha Warren
University of Missouri, Columbia, USA

1. Introduction

Infections are caused by a vast variety of pathogenic agents including viruses, bacteria, fungi, protozoa, multicellular parasites, and even proteins (Anderson and May 1979; Morse 1995; Bartlett 1997; Mandell and Townsend 1998) that target host organisms from virtually all kingdoms of life (Daszak, Cunningham et al. 2000; Williams, Yuill et al. 2002). Infectious diseases in humans account for 170 thousand deaths in the United States and 14,7 million deaths world-wide (2004; Rossi and Walker 2005). "Neglected diseases", a group of tropical diseases that are spread among the poorest segment of the world's population, account for a large portion of human infections (Ayoola 1987; Trouiller, Olliaro et al. 2002). With the reluctance of the pharmaceutical industry to invest in the development of drugs for neglected diseases, there is an increasing pressure on the scientific community in academia and non-profit organizations to obtain a fast and inexpensive cure (Trouiller, Torreele et al. 2001; Maurer, Rai et al. 2004; Fehr, Thurmann et al. 2006). In addition to human infections, infections in plant and animals have a multibillion dollar economic impact each year (Bowers, Bailey et al. 2001; Whitby 2001). Expanding the studies to the whole animal kingdom allows scientists to study the host-pathogen evolution of virulence mechanisms that are common among plant and animals, such as type III secretion system (T3SS), an elaborate protein-delivery system (Espinosa and Alfano 2004; Abramovitch, Anderson et al. 2006). Moreover, studying interactions between pathogens and simpler model organisms, such as drosophila, has led to important findings in mammalian systems and is critical for understanding human infections (Cherry and Silverman 2006). Recently another threat has come to scientists' attention: the potential use of some pathogens as bioweapons (Whitby 2001; Moran, Talan et al. 2008). The attacks can target population directly, or they can target strategic resources such as the world's most consumed crops. Studying HPIs may provide critical knowledge for the development of infection diagnosis and treatment for disaster planning in case of a bioterrorism event.

A pathogen causing an infectious disease generally exhibits extensive interactions with the host (Munter, Way et al. 2006). These complex crosstalks between a host and a pathogen may assist the pathogen in successfully invading the host organism, breaching its immune defence, as well as replicating and persisting within the organism. Systematic determination and analysis of HPIs is a challenging task from both experimental and computational approaches, and is critically dependent on the previously obtained knowledge about these interactions. The molecular mechanisms of host-pathogen interactions (HPIs) include interactions between proteins, nucleotide sequences, and small ligands (Lengeling, Pfeffer et al. 2001; Kahn, Fu et al. 2002; Stebbins 2005; Forst 2006). The interactions between the pathogen and host proteins are one of the most important and therefore widely studied group of HPIs (Stebbins 2005). During the last decade, an increasing amount of experimental data on virulence factors, their structures, and their functions has become available (Sansonetti 2002; Stebbins 2005). The first steps towards large-scale systematic determination and analysis of molecular HPIs have recently emerged for important pathogens (Shapira, Gat-Viks et al. 2009; Dyer, Neff et al. 2010). Recent progress in data mining and bioinformatics allows scientists to accurately predict novel protein-protein interactions, structurally characterize individual proteins and protein complexes, and predict protein functions on a scale of an entire proteome (Thornton 2001; Russell, Alber et al. 2004; Shoemaker and Panchenko 2007). Unfortunately, there have been only a handful of methods designed to address the protein interactions between pathogenic agents and their hosts (Cherkasov and Jones 2004; Davis, Barkan et al. 2007; Dyer, Murali et al. 2007; Lee, Chan et al. 2008; Evans, Dampier et al. 2009; Tyagi, Krishnadev et al. 2009; Doolittle and Gomez 2011). As it is the case for many bioinformatics areas, collecting HPI data into a centralized repository is instrumental in developing accurate predictive methods. Recently, several such HPI repositories have been introduced, some are manually curated, while others are reliant on the existing databases (Winnenburg, Urban et al. 2008; Driscoll, Dyer et al. 2009; Kumar and Nanduri 2010). While this is a promising first step towards a large-scale HPI data collection, one of the largest and most comprehensive sources of experimentally verified HPI data remains largely underexplored: PubMed, a database of peer-reviewed biomedical literature, which includes abstracts of more than 20 million research papers and books (http://www.ncbi.nlm.nih.gov/pubmed/). Unfortunately, the comprehensive manual identification and data extraction of the abstracts containing HPI information from PubMed is not feasible due to the size of PubMed. Furthermore, no informatics approach currently available to do this automatically.

In this chapter, we discuss several possible solutions to the problem of automated HPI data collection from the publicly available literature. The chapter is organized as follows. First, we describe some of the popular HPI databases that are currently available publicly. Second, we discuss the state-of-the-art approaches to a related problem of mining general protein-protein interactions from the literature. Third, we propose three approaches to mine HPIs and discuss the advantages and disadvantages of these approaches. In conclusion, we discuss the future steps in the area of HPI text mining by highlighting factors that are critical for its successful development.

2. Host-pathogen interaction databases

During the last several years, a number of resources collecting HPI data have emerged (Snyder, Kampanya et al. 2007; Winnenburg, Urban et al. 2008; Driscoll, Dyer et al. 2009; Kumar and Nanduri 2010). Many resources rely on the automated post-processing of the large-scale databases for general protein-protein interactions, while some other obtain the HPI data by manually curating the biomedical literature. Often the resources focus on the human-pathogen interactions. Next, we will briefly describe some of the popular databases that include HPI data.

HPIDB - Host-Pathogen Interaction DataBase. One of the most recent HPI database, HPIDB (Kumar and Nanduri 2010) integrates the information from other HPI database, PIG

(Driscoll, Dyer et al. 2009), and more general protein-protein interaction databases, BIND (Gilbert 2005), GeneRIF(Mitchell, Aronson et al. 2003; Pruitt, Tatusova et al. 2003), IntAct (Aranda, Achuthan et al. 2010), MINT (Zanzoni, Montecchi-Palazzi et al. 2002), and Reactome (Matthews, Gopinath et al. 2009). Currently, the database has 22,841 protein-protein interactions between 49 host and 319 pathogen species (Kumar and Nanduri 2010). HPIDB is searchable via a keyword search, a BLAST search, or a homologous HPI search. For each query, the following output information is obtained: UniProt accession numbers of both host and pathogen proteins, host and pathogen names, detection method, author name, PubMed publication ID (PMID), interaction type, source database, and comments. The homologous HPI search option allows the user to do one or both of the following: search for a set of homologous host proteins, and search for a set of homologous pathogen proteins.

PATRIC - PAThosystems Resource Integration Center. PATRIC is a resource that integrates genomics, proteomics, and interactomics data on a comprehensive set of bacterial species as well as a set of data mining and comparative genomics tools (Snyder, Kampanya et al. 2007; Sullivan, Gabbard et al. 2010). The human-pathogen interaction data for 30 bacterial pathogens are also a part of the resource. Similar to HPIDB, the data are extracted and post-processed from a number of general protein-protein interaction databases including BIND (Gilbert 2005), DIP (Xenarios, Fernandez et al. 2001), IntAct (Aranda, Achuthan et al. 2010), and MINT (Zanzoni, Montecchi-Palazzi et al. 2002). With PATRIC a user selects a pathogen from the home page. The search can be refined by selecting specific interaction types (*e.g.*, "direct interaction", "colocalization"), detection methods (*e.g.*, "coimmunoprecipitation", "two hybrid"), or source databases. The results can be visualized as a network of interacting proteins with the colour nodes representing different species and weighted edges representing the number of independent experimental sources supporting the interaction. The Pathogen Interaction Gateway (PIG) is a part of PATRIC that is focused on collecting and analysing exclusively the protein-protein human-pathogen interactions and the corresponding interaction networks (Driscoll, Dyer et al. 2009). The PIG web interface allows mining the data using two query types: the BLAST search and text keyword search. PIG also has a utility that allows the user to visualize the network of protein-protein HPIs followed by the network comparison between the HPI networks extracted for two different pathogen genes.

PHI-base - the Pathogen-Host Interaction dataBASE. PHI-base collects information on experimentally verified pathogenicity, virulence and effector genes from bacterial, fungal, and Oomycete pathogens and includes a variety of infected hosts from plants, mammals, fungus, and insects (Winnenburg, Urban et al. 2008). All database entries are manually curated and are supported by experimental evidence and literature citations. The current version has a total of 1,065 gene entries participating in 1,335 interactions between 97 pathogens and 76 hosts, supported by 720 literature references. The interaction between a host and pathogen organism is considered in this database in a more general sense and often is not associated with any physical interaction between the host and pathogen proteins. Using the PHI-base web interface, a user can do either a simple quick search or an advanced search, where the user selects one or many of the following search terms: gene, disease (caused by pathogen), host, pathogen, anti-infective, phenotype, and experimental evidence. The search output is a list of interactions and their details including PHI-base accession number, gene name, EMBL accession number, phenotype of the mutant, pathogen species, disease name, and experimental host. The user can also obtain additional information on nucleotide and amino acid sequences of the pathogen gene, experimental evidence of the interaction, gene ontology (pathogenesis, molecular function, and biological process), and a publication reference.

3. Current approaches for mining protein-protein interactions

Rapid growth of published biomedical research has resulted in the development of a number of methods for biomedical literature mining over the last decade (Krallinger and Valencia 2005; Rodriguez-Esteban 2009). The methods dealing with the biomolecular information can be generally divided into three categories based on the domain of biomedical knowledge they target: (i) automated protein or gene name identification in a text (Mika and Rost 2004; Seki and Mostafa 2005; Tanabe, Xie et al. 2005), (ii) literature-based functional annotation of genes and proteins (Chiang and Yu 2003; Jaeger, Gaudan et al. 2008), and (iii) extracting the information on the relationships between biological molecules, such as proteins and RNAs, or genes (Hu, Narayanaswamy et al. 2005; Shatkay, H^glund et al. 2007; Lee, Yi et al. 2008). The relationships detected by the third group of methods range from a co-occurrence of the genes and proteins in a text (Hoffmann and Valencia 2005) to detecting the protein-protein interactions (PPIs) (Blaschke and Valencia 2001; Marcotte, Xenarios et al. 2001; Donaldson, Martin et al. 2003) and identification of signal transduction networks and metabolic pathways (Friedman, Kra et al. 2001; Hoffmann, Krallinger et al. 2005; Santos and Eggle 2005). Being a special case of protein-protein interactions, HPIs could directly benefit from the advancements of the currently existing text mining methods.

Extraction of protein-protein interactions from the text has been one of the three main tasks for the recent BioCreAtIvE (Critical Assessment of Information Extraction systems in Biology) challenges, a community-wide effort for evaluating biological text mining and information retrieval systems (Hirschman, Yeh et al. 2005; Krallinger, Leitner et al. 2008). Three subtasks have been specified: (i) detection of protein-protein interactions relevant documents (interaction article subtask, IAS), (ii) identification of sentences with protein-protein interactions (interaction sentences subtask, ISS), and (iii) identification of interacting protein pairs (interaction pair subtask, IPS). A relevant problem, the protein interaction method subtask (IMS), is concerned with identification of the type of experimental data used to determine an interaction. Approaches that address these subtasks vary from supervised machine learning classifiers, to address the first subtask, to statistical language processing and grammar-based methods to address the second and third subtasks.

A simple approach to extract protein-protein interactions is to determine the co-existence of proteins in the same sentence (Stephens, Palakal et al. 2001; Hoffmann and Valencia 2005). However, this approach is insufficient to handle structured information of biomedical sentences. Therefore, pattern matching methods have been proposed that rely on either manually defined patterns (Leroy and Chen 2002; Corney, Buxton et al. 2004) or patterns that are automatically generated using dynamic programming (Huang, Zhu et al. 2004; Hao, Zhu et al. 2005). Another popular group of methods employs the natural language processing parsers. A basic approach, called shallow parsing, decomposes sentences into non-overlapping fragments and chunks, and defines the dependencies between the chunks without extracting their internal structure (Thomas, Milward et al. 2000; Leroy, Chen et al. 2003). Many shallow parsing approaches employ finite-state automata to recognize the interaction relationships between proteins or genes (Thomas, Milward et al. 2000; Leroy, Chen et al. 2003). One of the most prominent approaches relies on the deep parsing techniques, where the entire structure of a sentence is extracted (Park, Kim et al. 2001; Ding, Berleant et al. 2003; Daraselia, Yuryev et al. 2004; Pyysalo, Ginter et al. 2004; Kim, Shin et al. 2008; Miyao, Sagae et al. 2009). Many deep parsing approaches have successfully employed link grammars (Sleator and Temperley 1995), context-free grammars that rely on a dictionary of rules (linking requirements) to connect, or "link", pairs of related words (Ahmed, Chidambaram et al. 2005; Seoud, Youssef et al. 2008; Yang, Lin et al. 2009).

Each of the above methods, while directly addressing the second and the third subtasks, can also solve the abstract classification problem from the first subtask, based on whether or not the method is able to extract any protein-protein interactions. The accuracy of such classification, however, depends on the accuracy of a more difficult subtask of protein-protein interaction extraction. Thus, several methods have been developed to directly address the problem of binary classification of protein-protein interaction relevant publications (Marcotte, Xenarios et al. 2001; Calli 2009; Kolchinsky, Abi-Haidar et al. 2010). The methods primarily rely on supervised and unsupervised feature-based classification techniques. Recently, the first method for classification of HPI-relevant documents has been introduced, which employs a Support Vector Machines (SVM) supervised classifier (Yin, Xu et al. 2010).

4. New approaches to detection and mining host-pathogen interactions from biomedical abstracts

HPI literature mining is related to a general problem of protein-protein interaction literature mining. However, the additional requirement that the interaction occurs exclusively between the host and pathogen proteins makes the task more challenging. The accuracy of an HPI mining method will depend on additional factors, such as its ability to correctly assign a host or pathogen organism to the interacting protein. Similar to the way the BioCreAtIvE initiative defines three types of protein-protein interaction mining problems (Hirschman, Yeh et al. 2005), the problem of HPI mining can be split into three specific tasks:

HPI Mining Task 1: Given a biomedical publication (a paper or an abstract), determine whether or not it contains information on HPIs.

HPI Mining Task 2: Given a biomedical publication containing HPI information, determine specific sentences that contain this information.

HPI Mining Task 3: Given a biomedical publication that contain HPI information, determine specific pairs of host and pathogen proteins participating in the interactions and the corresponding organisms.

The first task can be formulated as a standard classification problem, which is often addressed by machine learning methods and for which a number of the method assessment protocols have been developed. Here we rely on the following five basic measures. The first measure, accuracy, is calculated as $f_{AC} = \left(N_{TP} + N_{TN}\right)/N$, where N_{TP} and N_{TN} are the number of true positives and negatives, correspondingly, and N is the number of classified interfaces. The other two related measures, precision and recall, are calculated as $f_{PR} = N_{TP}/\left(N_{TP} + N_{FP}\right)$ and $f_{RE} = N_{TP}/\left(N_{TP} + N_{FN}\right)$, correspondingly, where N_{FP} and N_{FN} are the number of false positives and negatives. F-score is calculated as $F = 2\frac{f_{PR} f_{RE}}{f_{PR} + f_{RE}}$. The last measure, the Matthew correlation coefficient is calculated as $MCC = \frac{N_{TP}N_{TN} - N_{FP}N_{FN}}{\sqrt{(N_{TP}+N_{FP})(N_{TP}+N_{FN})(N_{TN}+N_{FP})(N_{TN}+N_{FN})}}$. Similarly, performance on the last task can be easily assessed based on the available information about the host and pathogen proteins and their respective organisms. Specifically, we use four different measures. The first two measures, f_{ORG} and f_{PRT}, address the accuracy of detecting the pairs of interacting host and pathogen organisms as well as their proteins. Each measure is calculated as a percentage of the number of correctly detected pairs of organisms/proteins to the total number of pairs. The other two measures, g_{ORG} and g_{PRT}, account for the partial detection of HPI information, when at least one of the two organisms or proteins is detected. Both measures are defined as the percentage of the total number of detected organisms/proteins to the total number of organisms/proteins in all HPIs.

Unfortunately, evaluating a method's performance for the second task is more challenging, since the HPI data are often (i) scattered across multiple sentences and (ii) redundant (for instance, the same interaction between two proteins can be mentioned in several sentences). The method assessment for the second task becomes even more challenging when multiple HPIs are present in the same abstract.

We next introduce several strategies that address the above tasks for the PubMed biomedical abstracts (here and below, we will always consider an abstract of the biomedical publication together with the publication's title; the latter often provides important information on HPIs). One of the main reasons behind extracting HPI information from the abstracts rather than entire papers is the fact that for many papers, the abstract is the only information that is freely available in PubMed. The first strategy is to rely on the existing methods for mining protein-protein interactions followed by additional post-processing to filter out the intra-species interactions. Another approach employs the language-based methods traditionally used in protein-protein interaction literature mining. The last approach introduces a supervised-learning feature-based methodology, which has recently emerged in the area of biomedical literature mining. While each of the approaches is applicable to each of the three tasks, here we will focus on assessing their performance for the first and third tasks.

4.1 Data collection

Collecting accurate, unbiased, non-redundant data on HPIs is a critical step for efficient training of a supervised method as well as for an accurate assessment of any literature mining approach. Both the positive set (abstracts containing HPI information) and the negative set (abstracts that do not contain HPI information) were manually selected and annotated. To obtain the set of potential candidates for the positive and negative sets we have combined of both searching the existing HPI databases and the PubMed database. Our positive set consisted of 175 HPI containing abstracts that include human and non-human hosts. The abstracts containing human-pathogen interactions were collected by searching and manually curating abstracts from PIG, a database of host-pathogen interactions manually extracted from the literature (Driscoll, Dyer et al. 2009). For each abstract, we required the presence of organism and protein names for both the host and the pathogen, resulting in 89 abstracts. Unfortunately, in its current form, PIG only has the abstracts with annotated human-pathogen interactions. Therefore to obtain the list of interactions between non-human hosts and their pathogens, we searched using an extensive PubMed query. We required the presence in the same abstract of (i) at least one (non-human) host name, (ii) at least one pathogen name, (iii) and at least one interaction keyword. We then manually selected from the list another 86 abstracts that contained HPI information, adding them to the positive set.

To obtain candidates for the negative set, we performed an almost identical search strategy using the same PubMed query but including 'human' to the list of the host names. We again manually selected the abstracts to ensure that that they did not have any HPI information, even though they contained the important keywords. Note that it is significantly harder for a computational approach to distinguish between the abstracts from the obtained negative training set and those from the positive set, compared to a negative training set consisting of abstracts that were randomly chosen from PubMed. As a result, we selected 175 abstracts where no HPI information was found, although some of the abstracts included information on intra-species protein-protein interactions. The list of manually curated positive and negative sets of PubMed abstracts can be found at: http://korkinlab.org/datasets/philm/philm_data.html

4.2 A naïve approach based on literature mining of protein-protein interactions

In a simple naïve approach, we first establish whether an abstract contains any information on a protein-protein interaction using the existing state-of-the-art literature mining methods followed by extraction of the pair of interacting proteins (Fig. 1A). We rely on the PIE system, which integrates the natural language processing and machine learning methods to determine the sentences that contain protein-protein interactions in a PubMed abstract and extract the corresponding protein names and the interaction keywords (Kim, Shin et al. 2008). Next, for each interacting protein we identify its corresponding organism by applying NLProt protein/gene tagging software (Mika and Rost 2004). NLProt uses a number of techniques, such as the dictionary search, rule-based detection, and feature-based supervised learning, to extract the names of proteins and genes and tag them using SWISS-PROT or TrEMBL identifiers (Boeckmann, Bairoch et al. 2003). The method also predicts the most likely organisms associated with these proteins/genes. It was reported to have a precision of 75% and a recall of 76% on detecting protein/gene names (Mika and Rost 2004). Finally, for each sentence identified as containing a protein-protein interaction by the PIE system, we determine if this interaction is a HPI. Specifically, if each of the two proteins forming a protein-protein interaction belongs to a different organism, and these organisms can be assigned the host-pathogen roles, then the interaction is classified as an HPI. To assign the host-pathogen roles, we use our manually curated dictionaries of host and pathogen organism names (Table 1).

We assessed the naïve approach by applying it to our testing set of 88 abstracts, 44 positive and 44 negative examples. As a result in addressing Task 1, the obtained accuracy was 0.53, precision was 1.0, and recall was 0.07 for the classification of HPI-containing abstracts (Task 1); F-score and Matthews Correlation Coefficient were 0.13 and 0.19, correspondingly. We found that the method almost completely failed to detect the abstracts containing HPI information; the contribution to the accuracy came primarily from the true negative hits, containing 44 (out of 44) abstracts from the negative testing set. Interestingly, both high precision and low recall values could be attributed to the same property of the naïve approach: it failed to accurately detect the protein-protein interactions. Indeed, all 41 false negatives were not due to the approach's failure to assign the host and pathogen roles to the identified organisms, but due to its failure to identify a protein-protein interation in the abstract.

It is also not surprising that the naïve approach performed poorly when addressing Task 3: the method was able to detect only two proteins out of 44 protein pairs and none of the 44 pairs of organisms, resulting in the only non-zero score of g_{PRT} = 0.02; the other three scores, f_{ORG}, f_{PRT}, and g_{ORG} were equal to zero.

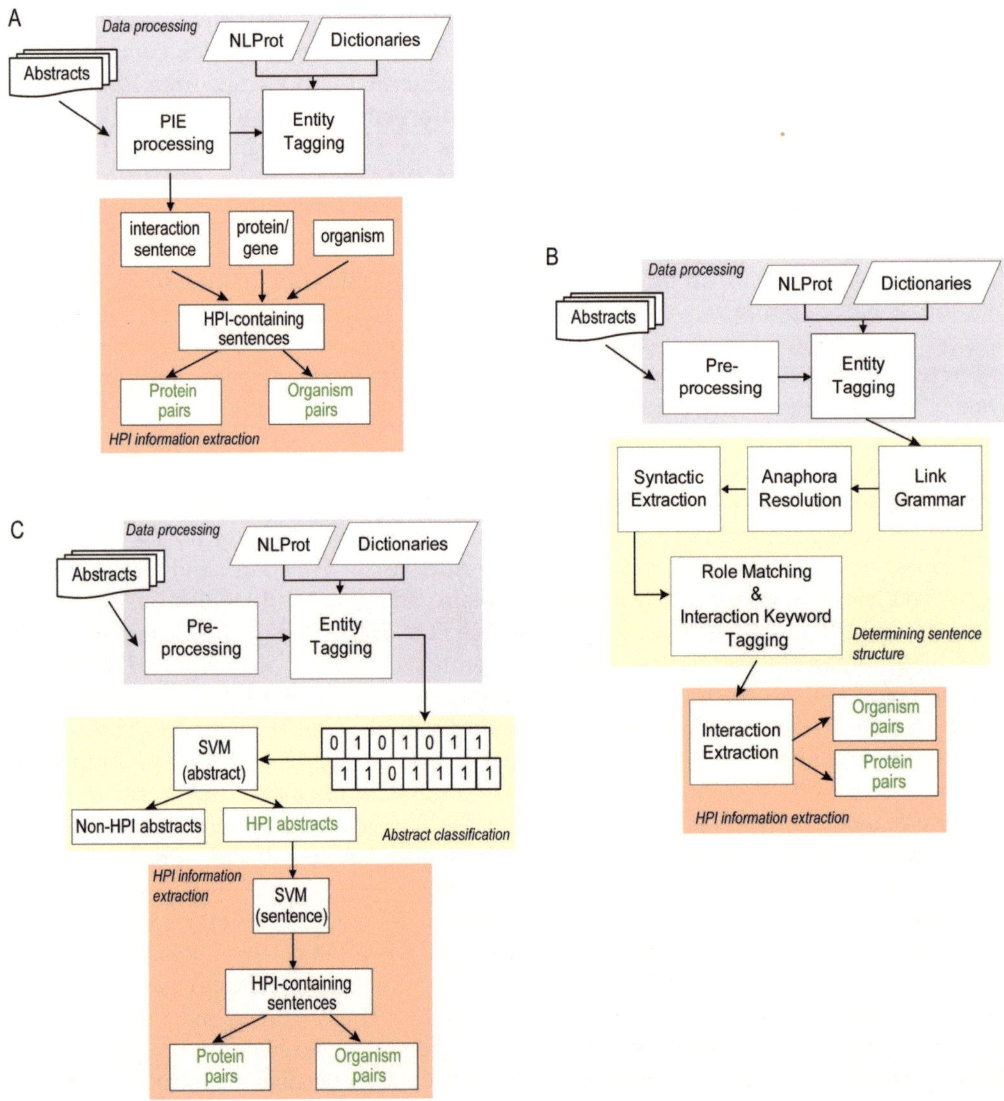

Fig. 1. Three HPI literature mining approaches. (A) Naïve approach. (B) Language-based approach (C) Feature-based supervised machine learning approach.

Dictionary name	N	Examples
Interaction keywords	54	*Interact, associate, bind*
Experimental keywords	28	*Yeast two-hybrid, chemical crosslinking*
Negation keywords	11	*Not, neither, inability*
HPI specific keywords	17	*Virulence, effectors, infection*
Host names	309	*Host, plant, human*
Pathogen names	349	*Listeria monocytogenes, Hepatitis virus*

Table 1. Dictionaries of keywords used by all three approaches. N is the number of unique entries for each dictionary.

4.3 A language-based approach

Our second approach is inspired by the language-based methods in biomedical text mining, which are also widely used in mining protein-protein interactions. In HPI text mining, we are faced with additional challenges such as correctly associating the organism name for each protein, ensuring that the extracted interaction is inter- and not intra-species interaction, and combining the information about an HPI from multiple sentences. As a result, these additional challenges necessitate adding new modules to the computational pipeline of our approach compared with a pipeline for extracting general protein-protein interactions. The HPI mining pipeline consists of the following 7 steps (Fig. 1B): (1) text preprocessing, (2) entity tagging, where we identify protein/gene and organism names, (3) grammar parsing, where we parse the input text into dependency structures (4) anaphora resolution, where we identity references to pronouns, (5) syntactic extraction, where we split a complex sentence into simple ones, (6) role matching, where we identify semantic roles in each simple sentence, (7) interaction keyword tagging, and (8) extraction of the actual HPI information. We note that this approach directly addresses Tasks 2 and 3 by finding the sentences containing HPI information and extracting the corresponding pairs of host and pathogen organisms and the interacting proteins/genes. Task 1 is addressed by classifying each abstract based on whether there was at least one HPI with the complete information extracted from the abstract's text.

Entity tagging. The entity tagging module identifies named entities in a abstract, such as protein/gene names and the corresponding organism names. For a language-based text mining approach, it is critical that all named entities are accurately identified. Thus, our language-based approach for HPI literature mining has the most elaborate entity tagging module of all three approaches introduced here. Specifically, the module includes three stages: (i) protein/gene name tagging using NLProt, (ii) host/pathogen organism dictionary match, and (iii) post-processing. First, we apply the NLProt tagger to identify the names of all proteins/genes occurring in the text and the corresponding organism names (Mika and Rost 2004). We note that in a case when a protein with the same name exists for multiple species, NLProt assigns the most likely organism for each entry of this protein. Second, we find a UniProt accession number (Bairoch, Apweiler et al. 2005) for each identified protein followed by grouping the proteins/genes with the same accession number into a protein/gene entity. Third we search for the organisms missed by NLProt using expanded versions of our host and pathogen organism dictionaries that include synonyms for each organism name and group the organisms under NCBI Taxonomy IDs (Wheeler, Barrett et al. 2006). Since NLProt may not identify all organisms in the abstract, our module rescans the abstract text again to find the remaining host and pathogen organisms. Finally, the system revisits the entity tagging module again after the next module, Link grammar parsing, provides the internal structure of the sentences in terms of its basic units, phrases. The idea is that we can use the internal sentence structure to (i) find additional host/pathogen information that is not present in the dictionary, and (ii) reassign protein/gene name to its correct organism, if needed. This stage plays an important role in the entity tagging module, since our host and pathogen dictionaries are potentially incomplete (not all organisms provided by NLProt may be covered); in addition, the dictionaries overlap with each other (the same organism can be both, a host and a pathogen). If an organism name suggested by NLProt for a protein is not found in our dictionary, the entity tagging module nevertheless tries to assign the organism's role as a host or pathogen. It does so by searching for generic keywords (such as "host", "pathogen", "pathogenic", "pathogenesis", *etc.*), in each phrase containing the organism name. Similarly, the module checks the organism name suggested by NLProt for a protein/gene by identifying the organism's name in the phrase that contains a protein/gene name. To do so the module relies on two search patterns:

1. Organism name + protein name (*e.g.*, "*Arabidopsis* RIN4 protein");
2. Protein name + preposition + organism name (*e.g.*, "RXLX of human").

The newly obtained information about the organism assignment then replaces the current suggestions provided by NLProt. For instance, in the phrase "the *Arabidopsis* RIN4 protein", NLProt associates RIN4 with a pathogenic organism, while the dictionary search matches *Arabidopsis* as a host organism and identifies this phrase as pattern P1. Therefore, *Arabidopsis* is assigned as the organism for RIN4 protein, followed by the correct assignment of RIN4 as a host protein.

Link grammar parsing. In our next module, we use natural language processing methods to determine the intrinsic structure of each sentence in the abstract. In our approach, all grammatical constructions are based on the link grammar, a context-free grammar that relies on the dependency structure of natural language (Sleator and Temperley 1995). In link grammar, every word has a linking requirement, which specifies which types of other words or phrases can link to it. Two words can only be linked if their linking requirements match. A link is represented as an arc above the two words (Fig. 2). The linking requirements are organized into a dictionary that the grammar parser refers to when analyzing a sentence. The principal structure in link grammar is the linkage, a set of links that completely connect all words in a sequence. Such a sequence of words is called a link grammar sentence if it satisfies three conditions: (i) the links do not cross (planarity), (ii) each word is connected to at least another word by a link (connectivity), and (iii) the linking requirements for each word in the sentence are not violated (satisfaction). For example, the linkage for the sentence "Avirulence protein B targets the Arabidopsis RIN4 protein" is shown in Fig. 2. In total, the link grammar has 107 main links, each of which can derive many sub-links. We implemented the module using an open source link grammar parser from AbiWord project (http://www.abisource.com/projects/link-grammar/). This project implements the original link grammar (Sleator and Temperley 1995), combining it with additional features such as adaptation of the parser to the biomedical sublanguage, BioLG (Pyysalo, Salakoski et al. 2006) and an English-language semantic dependency relationship extractor, RelEx (Fundel, Kuffner et al. 2007).

```
                            +--------------Os--------------+
                            |     +----------D*u----------+
+----AN---+--GN-+--Ss-+     |     +----G---+--AN--+
|         |     |     |     |     |        |      |
Avirulence protein.n B targets.v the Arabidopsis RIN4 protein.n
```

Fig. 2. Internal sentence structure annotated by a link grammar parser for an HPI relevant sentence. Words are labelled with the part-of-speech tags: .n (noun) and .v (verb). A link between two words can be formed to specify a dependency relation. Each dependency type has its own unique label: AN, GN, Ss, Os, D*u, G.

Anaphora resolution. In the anaphora resolution module, we determine semantic meaning for pronouns (it, they, he, she), and other language structures in the sentences. Unlike the case of intra-species protein-protein interactions, the information on HPIs often spans multiple sentences, with the pronouns often replacing the names of organisms or proteins/genes. Therefore, to extract the complete information on a HPI, it is critical to have an accurate anaphora resolution module. The module relies on the RelEx anaphora resolution method, which employs Hobbs' pronoun resolution algorithm (Hobbs 1978). For example, in the sentence "The *Pseudomonas syringae* type III effector protein avirulence protein B (AvrB) is delivered into plant cells, where it targets the *Arabidopsis* RIN4 protein", the anaphora resolution module resolves 'it' as 'The *Pseudomonas syringae* type III effector protein avirulence protein B (AvrB)'.

Syntactic extraction. Our syntactic extraction module splits each sentence into one or more simple sentences, where a simple sentence consists of four components organized into the following structure:

Subject (S) + Verb (V) + Object (O) + Modifying phrase of verb (M).

The module is built based on the automated extractor InTex (Ahmed, Chidambaram et al. 2005); it scans a sentence to find all links of the following four types. The first type, *S*-link, connects a subject to a verb, where the subject is located before the verb in the sentence. The second type, *RS*-link connects a verb to a subject, *i.e.*, the subject is located after the verb in the sentence. The third type, *O*-link, connects a verb to an object. Finally, the fourth type, *MV*-link, connects a verb to a modifying phrase. The module first determines the beginning of each simple sentence, which can be either an S-link or an RS-link. Following each verb from an *S*- or *RS*-link, the module determines the verb range by including all possible verb phrases, adverb phrases, or adjective phrase, before and after the verb. Finally, for each simple sentence the module determines the objects and modifying phrases for the verb in the corresponding verb range by identifying possible O-links and MV-links. For example, the modules split sentence "The Pseudomonas syringae type III effector protein avirulence protein B (AvrB) is delivered into plant cells, where it targets the Arabidopsis RIN4 protein" into two simple sentences: "The Pseudomonas syringae type III effector protein avirulence protein B (AvrB) is delivered into plant cells" and "The Pseudomonas syringae type III effector protein avirulence protein B (AvrB) targets the Arabidopsis RIN4 protein".

Interaction keyword tagging. In this module, the interaction keywords are tagged by searching (i) our manually curated dictionary of interaction keyword stems, to reduce the search time, and (ii) lexical database WordNet, which contains nouns, verbs, adjectives, and adverbs grouped by semantic concepts, and which uses a morphological function to infer the stem of a word (Fellbaum 1998). In the previous example, the module identifies interaction keywords that are found in our dictionary: "delivered" (the stem is "deliver") and "targets" (the stem is "target").

Role type matching. In this module, we specify the role of each syntactic component depending on whether the component contains complete information about an HPI. Here, we consider three types of roles: elementary, partial, and complete. A component of the elementary type is defined to be a host entity, a pathogen entity, or an interaction keyword. A component of the partial type includes any two distinct components of the elementary type. Finally, a syntactic component of complete type includes components of all three elementary types.

Interaction extraction. Once the role of each syntactic component is identified, the components are searched against a set of interaction patterns. We first select components of the complete type, since they contain complete information about an HPI occurring between two proteins/genes. Next, we combine the elementary and partial components such that they provide the complete HPI information.

An interaction pattern is defined as *LS*=*RS*. The left side (*LS*) is used to match the complete type from syntactic component(s), and the right side (*RS*) is used to extract the interaction information from each component. For example, the pattern S<E>V<E>O<E> = P<S>I<V>H<O> indicates that if a simple sentence includes three components, each of elementary type: subject, verb, and object, then the sentence contains (i) a pathogen entity in the subject, (ii) an interaction keyword in the verb, and (iii) a host entity in the object. Note that both sides include a matching part S-V-O. In this work, for our patterns we considered the following seven matching parts: S-V-O, S-O, S-V-M, S-M, S, O, and M (for abbreviations, see *Syntactic extraction* subsection). In addition to the above patterns, we use a set of three template-based filters that allows us to remove those simple sentences that although satisfy an interaction pattern, do not have a semantic connection between the host entity, pathogen entity, and interaction keyword. The introduced templates are similar to those employed by RelEx:

Pattern 1: A + interaction verb + B

Pattern 2: Interaction noun + 'between' + A + 'and' + B.

Pattern 3: Interaction noun + 'of' + A + 'by' + B,

where an interaction keyword can be either the interaction verb or interaction noun.

Interaction Normalization. When mining HPI information from literature, there are several sources for ambiguous information. First, there may be multiple HPIs in the same abstract. Second, the information about a single HPI may be spread over multiple sentences. Finally, the sentences may contain duplicate information about the same HPI. Our last module ensures that all sentences containing duplicate HPIs are accounted for and each HPI is reported only once. To do so, we first extract all HPIs and then determine the duplicate pairs. We define two HPIs as duplicate if they have the same host entity and the same pathogen entity. We note that two duplicate HPIs may still have different interaction keywords. To detect the duplication in HPIs, the module refers to the normalized protein/gene names (in terms of UniProt accession numbers) and organism names (in terms of taxonomy ids) obtained at the entity tagging module.

Performance of the language-based approach. To compare with the feature-based approach, the language-based approach was evaluated using the same testing set of 44 positive and 44 negative examples. We first assessed the method's performance in addressing Task 1. The method was able to classify the abstracts with 0.65 accuracy, 0.84 precision, and 0.36 recall. The F-score and Matthew correlation coefficient measures were 0.51 and 0.36, correspondingly. The performance of the approach on a more difficult Task 3 was significantly better than of the naïve approach, especially in partial predictions: f_{ORG} = 0.18, f_{PRT} = 0.14, g_{ORG} = 0.25, and g_{PRT} = 0.25. With the pre-calculated NLProt annotation, the average running time of the system on a single abstract was 36.3 sec. on a 2.4 Ghz Intel workstation. The computationally most expensive, link grammar parsing, module used 99.95% of the total running time.

4.4 A feature-based machine learning approach

The basic idea behind the feature-based approach introduced here is to extract a set of characteristic features that provide sufficient information for discriminating between an abstract containing HPI information and another abstract that does not. Using a training set of pre-annotated abstracts, the system can then learn how to efficiently discriminate between these two abstract types. Moreover, the same characteristic features can be calculated for the individual sentences in the abstract. Thus, we can use the same supervised-learning approach to solve Tasks 1 and 2. Finally, to solve Task 3 one can use a simple dictionary-based search for each sentence classified as containing HPI information. Our feature-based approach consists of four basic stages (Fig. 1C). First, each abstract is pre-processed to find each protein/gene in the abstract and identify its organism name. Second, for each abstract a feature vector is generated. Third, our supervised learning system is trained by providing the feature vectors generated from the positive and negative sets. Finally, the trained system is used on an independent testing set of HPI and non-HPI abstracts to assess the approach.

Text preprocessing. We first add the publication title to the abstract as its first sentence. The abstract is then further split into individual sentences by detecting the sentence termination patterns. A basic pattern of a period (.), followed by a space and capitalized letter can be directly used to distinguish sentences in a standard text. However, there are known challenges when preprocessing a biomedical (or any scientific) publication. For instance, the above simple approach is not always applicable, since the periods are often used in the names of proteins, abbreviations such as "*i.e.*", "*e.g.*", "*vs.*", and others. We first identify such cases using a predefined dictionary, replace periods in these words by spaces, and then apply the above basic pattern. The next steps of the preprocessing stage concerns with detecting the organism and protein/gene names using the entity tagging software NLProt (Mika and Rost 2004).

Support vector machines in text categorization. The problem of detecting whether an abstract contains HPI information can be formulated as a problem of supervised text categorization, with the goal of classifying abstracts into one of the selected categories. In our case, two categories can be naturally defined: (i) abstracts containing HPI information and (ii) abstracts without HPI information. Formally, given a training set of n objects, each represented as a vector of N numerical features, $\mathbf{x}^i = (x_1, x_2, \ldots, x_N)$, and their classification into one of the two classes $y \in \{-1,1\}$, the goal is to train a feature-based classifier based on the training set. After the training stage is completed, the classifier can assign a class label from y for any new abstract $\mathbf{x}$. In our approach, we use support vector machines (SVM) (Vapnik 1998), a supervised learning method, which is well established in bioinformatics and has been recently applied to identify abstracts containing host-bacteria interaction information (Yin, Xu et al. 2010). The basic type of support vector machine (SVM) that addresses this problem is a linear classifier defined by its discriminant function:

$$f(\boldsymbol{x}) = \langle \boldsymbol{w}, \boldsymbol{x} \rangle + b, \quad \langle \boldsymbol{w}, \boldsymbol{x} \rangle = \sum_{i=1}^{N} w_i x_i \,,$$

where w is a weight vector (Vapnik 1998). Geometrically, the problem can be described as finding the decision boundary, a hyperplane that separates two sets of points, corresponding to the sets of positive and negative examples. To do that, we maximize the margin defined by the closest to the hyperplane positive and negative examples. An optimal solution can be found by solving a related quadric optimization problem. The problem is further generalized by introducing soft margins, allowing the classifier to misclassify some points. The general optimization problem is often formulated in its dual form:

$$\underset{\alpha}{\text{minimize}} \left[-\sum_{i=1}^{n} \alpha_i + \frac{1}{2} \sum_{i=1}^{n} \sum_{j=1}^{n} y_i y_j \alpha_i \alpha_j \langle \boldsymbol{x}^i, \boldsymbol{x}^j \rangle \right]$$

$$\text{subject to: } \sum_{i=1}^{n} y_i \alpha_i = 0, \quad 0 \le \alpha_i \le C \text{ for } i = 1, 2, \ldots, n$$

and the discriminant function is defined as:

$$f(\boldsymbol{x}) = \sum_{i=1}^{n} y_i \alpha_i \langle \boldsymbol{x}^i, \boldsymbol{x} \rangle + b \,.$$

Examples from the training set for which $\alpha_i > 0$ are called support vectors. The formalism can be further extended by introducing non-linear classifiers defined using kernel functions, $K(\boldsymbol{x}, \boldsymbol{x}')$, similarity measures that replace the standard inner product $\langle x, x' \rangle$. In our approach, we applied and compared two widely used non-linear kernel functions: the polynomial kernel, $K^P(\boldsymbol{x}, \boldsymbol{x}') = (\langle \boldsymbol{x}, \boldsymbol{x}' \rangle + 1)^d$, where d is degree of the polynomial, and Gaussian radial basis function (RBF), $K^G(\boldsymbol{x}, \boldsymbol{x}') = \exp(-\|\boldsymbol{x} - \boldsymbol{x}'\|^2 / c)$. Both kernels are implemented using *libsvm* a freely available SVM software package (Chang and Lin 2001).

Feature vectors. One approach to generating a descriptive set of features for an abstract is to calculate the frequencies of occurrences of individual words (unigrams) as well as the word pairs (bigrams) from a biomedical text corpus (Yin, Xu et al. 2010). While these features can provide important information on the word usage, the number of features depends strongly on the size of the corpus and can easily reach thousands of features. In our approach, we propose to use a simpler 12-dimensional feature vector representation, $\boldsymbol{x} = (x_1, x_2, \ldots, x_{12})$, focusing on quantifying the information directly related to host-pathogen interaction. Features x_1 and x_2 quantify the presence of host and pathogen protein or gene names in the abstract and are calculated based on the protein/gene entity tagging obtained by NLProt (Mika and Rost 2004). Each protein is classified as a host or pathogen protein based on the source organisms extracted either from the NLProt tagging results or directly from the abstract by searching against our dictionary of host and pathogen organisms (Table 1). The dictionary was built using the set of organisms extracted from several databases (Winnenburg, Urban et al. 2008; Driscoll, Dyer et al. 2009; Kumar and Nanduri 2010) and by adding generic keywords, such as "pathogen", "host", "plant", etc. Similarly, features x_3

and x_4 specify the number of occurrences for the host and pathogen organism names. These features are defined using NLProt-based organism annotation and the dictionary of host and pathogen organisms. Binary feature x_5 specifies the presence or absence of the general protein-protein interaction keywords in the abstract. It is obtained by scanning the extended abstract against our interaction keyword dictionary (Table 1). Features x_6 and x_7 describe additional statistics on protein-protein interaction keyword occurrences. The former feature is defined as the percentage of interaction keywords in the total number of words in the abstract. The latter feature is defined as the percentage of sentences containing the interaction keywords in the total number of abstract sentences. Feature x_8 is calculated based on the cumulative keyword *typicality* for each abstract. We define the typicality of a keyword as the percentage of abstracts in the training set containing this keyword. Feature x_8 is calculated as a sum of typicalities for all protein-protein interactions keywords in a given abstract. Our next feature x_9 quantifies the amount of experimental evidence used to support the HPI and is defined as the total number of experimental keywords in the abstract, where each keyword is detected by scanning the abstract against our dictionary experimental keywords (Table 1). Some abstracts report the absence of an interaction between host and pathogen proteins. Determining the absence of interaction in an abstract by a feature-based approach is difficult, since such an abstract is likely to contain the information similar to an abstract describing a true HPI. One of the key differences between these abstracts is the presence of negation keywords present in the former abstract. Feature x_{10} accounts for such keywords and is defined as the percentage of negation keywords in the total number of words in the abstract. Similar to other keywords, these keywords are identified using our dictionary of collected negation keywords (Table XX3). A related feature, x_{11}, estimates whether a negation keyword is related specifically to the information on protein-protein interaction in the abstract. The feature is defined as the number of words between the negation keyword and the closest interaction keyword in a sentence. The last feature, x_{12}, accounts for the HPI-specific keywords, such as *virulence, effectors, factors, etc.* determined using the corresponding dictionary (Table 1). It is calculated as a percentage of such keywords in the total number of words in the abstract.

Supervised training and HPI detection using SVM. The trained SVM classifier is applied in our method twice. First, it is applied to the abstracts to identify those containing HPI information (Task 1). Second, it is applied to the individual sentences to determine those that contain this HPI information (Task 2). When applied to a sentence, we generate a 12-dimensional feature vector solely based on the information in this sentence and use it as an input to the SVM classifier. Once the sentences containing HPI are identified, we use the dictionaries of host and pathogen organisms combined with the protein/gene names to find the pairs of host and pathogen organisms and the corresponding proteins/genes (Task 3). The accuracy of an SVM-based classifier generally can be improved by optimizing a number of parameters during the training stage. The error cost parameter, C, controls the tradeoff between allowing training errors and forcing rigid margins. In our approach we select the cost parameter and another parameter, Gamma, by evaluating the accuracies of trained models for Task 1 using leave-one-out cross-validation. The values of C range from 2 to 20 and the values for Gamma range from 2^{-10} to 2^1. The set of parameters on which the SVM classifier reaches its maximum accuracy is selected as a final model. In addition, we optimize the degree of the polynomial when considering the polynomial kernel.

Assessment protocols. To assess the performance of the feature-based approach in abstract classification, we employ two benchmarking protocols. In the first protocol, the SVM model training is done on the training set and the assessment is performed exclusively on the testing set (Table 2). For the second protocol we use the leave-one-out and 10-fold cross validations on the training set.

Type	Training	Testing
Negative	131	44
Positive- Human	67	22
Positive-Non-Human	64	22
Total	262	88

Table 2. Testing and training sets of positive (HPI-relevant) and negative (HPI-irrelevant) abstracts. Testing data are used to evaluate all three approaches, and training data are used for SVM learning in the feature-based approach. The abstracts are extracted from then PubMed database and manually curated.

Performance of the feature based approach. During the leave-one-out cross-validation, an SVM model with the polynomial kernel of degree 3 and parameter values C=2 and Gamma=0.0175 was found to be the most accurate in the abstract classification problem (Table 3). The polynomial kernel was also the most accurate SVM model across both assessment protocols. In addition, this SVM model had the highest recall value, with the precision approaching its highest value. Overall, the performance of all three SVM kernels, across all evaluation protocols, was similar. The performance of the feature-based approach on Task 3 was slightly better than that of the language-based approach in partial predictions: g_{ORG} = 0.39 and g_{PRT} = 0.35. However the performance in complete pair predictions was worse: f_{ORG} = 0.07 and f_{PRT} = 0.07. The SVM classifier was efficient, taking only 0.003 sec. to classify 92 abstracts by an SVM classifier on a 2.66 Ghz Intel Xeon (Quad) workstation. However, the high efficiency of this approach was offset by a significantly slower protein tagging component that was done using NLProt and took ~18 min. on the same workstation to tag proteins in 262 abstracts from the dataset.

Protocol	f_{AC}	**PR**	**RE**	**AUC**	**F-score**
10-fold	72%	73%	71%	0.78	0.72
Test	66%	69%	60%	0.72	0.64
LOO	71%	72%	72%	0.78	0.71

Table 3. Evaluations of the feature-based classifier. LOO and 10-fold denote leave-one-out and 10-fold cross-validation protocols applied to the models that are trained on the set of 262 abstracts. The last protocol corresponds to the evaluation performed only on the testing set of 88 abstracts.

5. Conclusion

In this chapter, we discussed a new problem for biomedical literature mining that was concerned with mining molecular interactions between the host and pathogen organisms. Collecting HPI data is one of the very first steps towards studying and fighting infectious diseases. Creating an automated framework for extracting the HPI information from the biomedical literature, including millions of abstracts publicly available in PubMed database, is instrumental in completing this step. We formulated three key tasks of HPI literature mining and proposed three computational approaches that addressed these tasks: (i) a naïve approach, which was based on the existing protein-protein interaction mining methods, (ii) a language-based approach, which employed the link grammar, and (iii) a feature-based supervised learning approach, which relied on SVM methodology. Both, feature-based and language-based, approaches have been implemented in the PHILM (Pathogen-Host Interaction Literature Mining) web-server, accessible at http://korkinlab.org/philm.html.

Several important conclusions can be drawn from the comparative assessment of all three approaches. First, it became clear that being a new problem in biomedical literature mining (and a more difficult one than mining general protein-protein interactions), HPI text mining required development of new methods tailored to address the specifics of this problem. Indeed, for the first task the naïve approach performed with the disappointingly low accuracy of 53% and f-score of just 13%, while accuracy and f-score of the language-based approach were significantly higher, 65% and 51%, correspondingly; the feature-based method had even higher (10-fold) accuracy and f-score, 72% and 72%, correspondingly. We note that the performance accuracy of both language-based and feature-based approaches even at this early stage were comparable to the state-of-the-art protein-protein interactions mining methods (Krallinger, Leitner et al. 2008). In addition to its poor performance in the abstract classification task, the naïve approach completely failed to detect protein interaction pairs and organism pairs in the third task. The feature-based approach performed significantly better when detecting one of the interacting proteins or organisms, while still failing to accurately detect the complete pairs. It was not surprising that the highest accuracy of detecting both, host-pathogen organism pairs and protein pairs, was achieved by the most sophisticated language-based approach. Second, the analysis of incorrectly classified abstracts and identified pairs of proteins and organisms supported our conclusion that increasing the accuracy of the name tagging system is pivotal to increasing the classification accuracy in both approaches. Finally, both language-based and feature-based approaches demonstrated good performance but in different tasks, which suggests that by integrating these two approaches, one can obtain a system with a more accurate overall performance than either of the individual approaches.

6. Acknowledgment

We acknowledge funding from University of Missouri (Mizzou Advantage to DK), National Science Foundation (DBI-0845196 to DK), and Department of Education (GAANN Fellowship to SW).

7. References

(2004). WHO, The world health report 2004: changing history. Geneva, World Health Organization.

Abramovitch, R. B., J. C. Anderson, et al. (2006). "Bacterial elicitation and evasion of plant innate immunity." *Nat Rev Mol Cell Biol* 7(8): 601-611.

Ahmed, S. T., D. Chidambaram, et al. (2005). *IntEx: A Syntactic Role Driven Protein-Protein Interaction Extractor for Bio-Medical Text.* Proceedings of the ACL-ISMB Workshop on Linking Biological Literature. Ontologies and Databases: Mining Biological Semantics, Detroit, Association for Computational Linguistics.

Anderson, R. M. and R. M. May (1979). "Population biology of infectious diseases: Part I." *Nature* 280(5721): 361-367.

Aranda, B., P. Achuthan, et al. (2010). "The IntAct molecular interaction database in 2010." *Nucleic Acids Research* 38(Database issue): D525-531.

Ayoola, E. A. (1987). "Infectious diseases in Africa." *Infection* 15(3): 153-159.

Bairoch, A., R. Apweiler, et al. (2005). "The Universal Protein Resource (UniProt)." *Nucleic Acids Res* 33(Database issue): D154-159.

Bartlett, J. G. (1997). "Update in infectious diseases." *Annals of internal medicine* 126(1): 48-56.

Blaschke, C. and A. Valencia (2001). "The potential use of SUISEKI as a protein interaction discovery tool." *Genome informatics. International Conference on Genome Informatics* 12: 123-134.

Boeckmann, B., A. Bairoch, et al. (2003). "The SWISS-PROT protein knowledgebase and its supplement TrEMBL in 2003." *Nucleic Acids Research* 31(1): 365.

Bowers, J. H., B. A. Bailey, et al. (2001). "The impact of plant diseases on world chocolate production." *Plant Health Progress.*

Calli, C. (2009). *Prediction of protein-protein interaction relevance of articles using references.* 24th International Symposium on Computer and Information Sciences (ISCIS 2009), Guzelyurt, IEEE.

Chang, C. and C. Lin (2001). LIBSVM: a library for support vector machines. Software available at http://www.csie.ntu.edu.tw/~cjlin/libsvm.

Cherkasov, A. and S. J. Jones (2004). "An approach to large scale identification of non-obvious structural similarities between proteins." *BMC Bioinformatics* 5: 61.

Cherry, S. and N. Silverman (2006). "Host-pathogen interactions in drosophila: new tricks from an old friend." *Nat Immunol* 7(9): 911-917.

Chiang, J. H. and H. C. Yu (2003). "MeKE: discovering the functions of gene products from biomedical literature via sentence alignment." *Bioinformatics* 19(11): 1417-1422.

Corney, D. P., B. F. Buxton, et al. (2004). "BioRAT: extracting biological information from full-length papers." *Bioinformatics* 20(17): 3206-3213.

Daraselia, N., A. Yuryev, et al. (2004). "Extracting human protein interactions from MEDLINE using a full-sentence parser." *Bioinformatics* 20(5): 604-611.

Daszak, P., A. A. Cunningham, et al. (2000). "Emerging infectious diseases of wildlife--threats to biodiversity and human health." *Science* 287(5452): 443-449.

Davis, F. P., D. T. Barkan, et al. (2007). "Host pathogen protein interactions predicted by comparative modeling." *Protein Sci* 16(12): 2585-2596.

Ding, J., D. Berleant, et al. (2003). "Extracting biochemical interactions from MEDLINE using a link grammar parser."

Donaldson, I., J. Martin, et al. (2003). "PreBIND and Textomy--mining the biomedical literature for protein-protein interactions using a support vector machine." *BMC Bioinformatics* 4: 11.

Doolittle, J. M. and S. M. Gomez (2011). "Mapping protein interactions between Dengue virus and its human and insect hosts." *PLoS neglected tropical diseases* 5(2): e954.

Driscoll, T., M. D. Dyer, et al. (2009). "PIG--the pathogen interaction gateway." *Nucleic Acids Research* 37(Database issue): D647-650.

Driscoll, T., M. D. Dyer, et al. (2009). "PIG--the pathogen interaction gateway." *Nucleic Acids Res* 37(Database issue): D647-650.

Dyer, M. D., T. M. Murali, et al. (2007). "Computational prediction of host-pathogen protein-protein interactions." *Bioinformatics* 23(13): i159-166.

Dyer, M. D., C. Neff, et al. (2010). "The human-bacterial pathogen protein interaction networks of Bacillus anthracis, Francisella tularensis, and Yersinia pestis." *PLoS One* 5(8): e12089.

Espinosa, A. and J. R. Alfano (2004). "Disabling surveillance: bacterial type III secretion system effectors that suppress innate immunity." *Cell Microbiol* 6(11): 1027-1040.

Evans, P., W. Dampier, et al. (2009). "Prediction of HIV-1 virus-host protein interactions using virus and host sequence motifs." *BMC medical genomics* 2: 27.

Fehr, A., P. Thurmann, et al. (2006). "Editorial: drug development for neglected diseases: a public health challenge." *Trop Med Int Health* 11(9): 1335-1338.

Fellbaum, C. (1998). *WordNet : an electronic lexical database*. Cambridge, USA, MIT Press.

Forst, C. V. (2006). "Host-pathogen systems biology." *Drug Discov Today* 11(5-6): 220-227.

Friedman, C., P. Kra, et al. (2001). "GENIES: a natural-language processing system for the extraction of molecular pathways from journal articles." *Bioinformatics* 17(Suppl 1): S74.

Fundel, K., R. Kuffner, et al. (2007). "RelEx--relation extraction using dependency parse trees." *Bioinformatics* 23(3): 365.

Gilbert, D. (2005). "Biomolecular interaction network database." *Brief Bioinform* 6(2): 194-198.

Hao, Y., X. Zhu, et al. (2005). "Discovering patterns to extract protein-protein interactions from the literature: Part II." *Bioinformatics* 21(15): 3294-3300.

Hirschman, L., A. Yeh, et al. (2005). "Overview of BioCreAtIvE: critical assessment of information extraction for biology." *BMC Bioinformatics* 6 Suppl 1: S1.

Hobbs, J. (1978). "Resolving pronoun references." *Lingua* 44(4): 311-338.

Hoffmann, R., M. Krallinger, et al. (2005). "Text mining for metabolic pathways, signaling cascades, and protein networks." *Sci STKE* 2005(283): pe21.

Hoffmann, R. and A. Valencia (2005). "Implementing the iHOP concept for navigation of biomedical literature." *Bioinformatics* 21 Suppl 2: ii252-258.

Hu, Z. Z., M. Narayanaswamy, et al. (2005). "Literature mining and database annotation of protein phosphorylation using a rule-based system." *Bioinformatics* 21(11): 2759-2765.

Huang, M., X. Zhu, et al. (2004). "Discovering patterns to extract protein-protein interactions from full texts." *Bioinformatics* 20(18): 3604-3612.

Jaeger, S., S. Gaudan, et al. (2008). "Integrating protein-protein interactions and text mining for protein function prediction." *BMC Bioinformatics* 9 Suppl 8: S2.

Kahn, R. A., H. Fu, et al. (2002). "Cellular hijacking: a common strategy for microbial infection." *Trends Biochem Sci* 27(6): 308-314.

Kim, S., S. Y. Shin, et al. (2008). "PIE: an online prediction system for protein-protein interactions from text." *Nucleic Acids Research* 36(Web Server issue): W411-415.

Kolchinsky, A., A. Abi-Haidar, et al. (2010). "Classification of protein-protein interaction full-text documents using text and citation network features." *IEEE/ACM transactions on computational biology and bioinformatics / IEEE, ACM* 7(3): 400-411.

Krallinger, M., F. Leitner, et al. (2008). "Overview of the protein-protein interaction annotation extraction task of BioCreative II." *Genome Biology* 9 Suppl 2: S4.

Krallinger, M. and A. Valencia (2005). "Text-mining and information-retrieval services for molecular biology." *Genome Biol* 6(7): 224.

Kumar, R. and B. Nanduri (2010). "HPIDB-a unified resource for host-pathogen interactions." *BMC Bioinformatics* 11(Suppl 6): S16.

Lee, H., G. Yi, et al. (2008). "E3Miner: a text mining tool for ubiquitin-protein ligases." *Nucleic Acids Research* 36(Web Server issue): W416.

Lee, S. A., C. H. Chan, et al. (2008). "Ortholog-based protein-protein interaction prediction and its application to inter-species interactions." *BMC Bioinformatics* 9 Suppl 12: S11.

Lengeling, A., K. Pfeffer, et al. (2001). "The battle of two genomes: genetics of bacterial host/pathogen interactions in mice." *Mamm Genome* 12(4): 261-271.

Leroy, G. and H. Chen (2002). "Filling preposition-based templates to capture information from medical abstracts." *Pac Symp Biocomput*: 350-361.

Leroy, G., H. Chen, et al. (2003). "A shallow parser based on closed-class words to capture relations in biomedical text." *Journal of biomedical informatics* 36(3): 145-158.

Mandell, G. L. and G. C. Townsend (1998). "New and emerging infectious diseases." *Transactions of the American Clinical and Climatological Association* 109: 205-216; discussion 216-207.

Marcotte, E. M., I. Xenarios, et al. (2001). "Mining literature for protein-protein interactions." *Bioinformatics* 17(4): 359-363.

Matthews, L., G. Gopinath, et al. (2009). "Reactome knowledgebase of human biological pathways and processes." *Nucleic Acids Research* 37(Database issue): D619-622.

Maurer, S. M., A. Rai, et al. (2004). "Finding cures for tropical diseases: is open source an answer?" *PLoS Med* 1(3): e56.

Mika, S. and B. Rost (2004). "NLProt: extracting protein names and sequences from papers." *Nucleic Acids Res* 32(Web Server issue): W634-637.

Mika, S. and B. Rost (2004). "Protein names precisely peeled off free text." *Bioinformatics* 20(suppl 1): i241.

Mitchell, J. A., A. R. Aronson, et al. (2003). "Gene indexing: characterization and analysis of NLM's GeneRIFs." *AMIA ... Annual Symposium proceedings / AMIA Symposium. AMIA Symposium*: 460-464.

Miyao, Y., K. Sagae, et al. (2009). "Evaluating contributions of natural language parsers to protein-protein interaction extraction." *Bioinformatics* 25(3): 394-400.

Moran, G. J., D. A. Talan, et al. (2008). "Biological terrorism." *Infect Dis Clin North Am* 22(1): 145-187, vii.

Morse, S. S. (1995). "Factors in the emergence of infectious diseases." *Emerg Infect Dis* 1(1): 7-15.

Munter, S., M. Way, et al. (2006). "Signaling during pathogen infection." *Sci STKE* 2006(335): re5.

Park, J. C., H. S. Kim, et al. (2001). "Bidirectional incremental parsing for automatic pathway identification with combinatory categorial grammar." *Pacific Symposium on Biocomputing. Pacific Symposium on Biocomputing*: 396-407.

Pruitt, K. D., T. Tatusova, et al. (2003). "NCBI Reference Sequence project: update and current status." *Nucleic Acids Research* 31(1): 34-37.

Pyysalo, S., F. Ginter, et al. (2004). *Analysis of link grammar on biomedical dependency corpus targeted at protein-protein interactions.* nternational Workshop on Natural Language

Processing in Biomedicine and its Applications (JNLPBA), Association for Computational Linguistics.
Pyysalo, S., T. Salakoski, et al. (2006). "Lexical adaptation of link grammar to the biomedical sublanguage: a comparative evaluation of three approaches." *BMC Bioinformatics* 7(Suppl 3): S2.
Rodriguez-Esteban, R. (2009). "Biomedical text mining and its applications." *PLoS Comput Biol* 5(12): e1000597.
Rossi, V. and J. Walker (2005). Assessing the Economic Impact and Costs of Flu Pandemics Originating in Asia. Oxford: Abbey House, Oxford Economic Forecasting Group.
Russell, R. B., F. Alber, et al. (2004). "A structural perspective on protein-protein interactions." *Curr Opin Struct Biol* 14(3): 313-324.
Sansonetti, P. (2002). "Host-pathogen interactions: the seduction of molecular cross talk." *Gut* 50 Suppl 3: III2-8.
Santos, C. and D. Eggle (2005). "Wnt pathway curation using automated natural language processing: combining statistical methods with partial and full parse for knowledge extraction." *Bioinformatics* 21(8): 1653.
Seki, K. and J. Mostafa (2005). "A hybrid approach to protein name identification in biomedical texts." *Information Processing & Management* 41(4): 723-743.
Seoud, A., A. Youssef, et al. (2008). *Extraction of protein interaction information from unstructured text using a link grammar parser*, IEEE.
Shapira, S. D., I. Gat-Viks, et al. (2009). "A physical and regulatory map of host-influenza interactions reveals pathways in H1N1 infection." *Cell* 139(7): 1255-1267.
Shatkay, H., A. H^glund, et al. (2007). "SherLoc: high-accuracy prediction of protein subcellular localization by integrating text and protein sequence data." *Bioinformatics* 23(11): 1410.
Shoemaker, B. A. and A. R. Panchenko (2007). "Deciphering protein-protein interactions. Part II. Computational methods to predict protein and domain interaction partners." *PLoS Comput Biol* 3(4): e43.
Sleator, D. and D. Temperley (1995). *Parsing English with a Link Grammar*. Third International Workshop on Parsing Technologies, ACL/SIGPARSE.
Snyder, E. E., N. Kampanya, et al. (2007). "PATRIC: the VBI PathoSystems Resource Integration Center." *Nucleic Acids Research* 35(Database issue): D401-406.
Stebbins, C. E. (2005). "Structural microbiology at the pathogen-host interface." *Cell Microbiol* 7(9): 1227-1236.
Stephens, M., M. Palakal, et al. (2001). "Detecting gene relations from Medline abstracts." *Pac Symp Biocomput*: 483-495.
Sullivan, D. E., J. L. Gabbard, Jr., et al. (2010). "Data integration for dynamic and sustainable systems biology resources: challenges and lessons learned." *Chemistry & biodiversity* 7(5): 1124-1141.
Tanabe, L., N. Xie, et al. (2005). "GENETAG: a tagged corpus for gene/protein named entity recognition." *BMC Bioinformatics* 6(Suppl 1): S3.
Thomas, J., D. Milward, et al. (2000). "Automatic extraction of protein interactions from scientific abstracts." *Pacific Symposium on Biocomputing. Pacific Symposium on Biocomputing*: 541-552.
Thornton, J. M. (2001). "From genome to function." *Science* 292(5524): 2095-2097.

Trouiller, P., P. Olliaro, et al. (2002). "Drug development for neglected diseases: a deficient market and a public-health policy failure." *Lancet* 359(9324): 2188-2194.

Trouiller, P., E. Torreele, et al. (2001). "Drugs for neglected diseases: a failure of the market and a public health failure?" *Trop Med Int Health* 6(11): 945-951.

Tyagi, N., O. Krishnadev, et al. (2009). "Prediction of protein-protein interactions between Helicobacter pylori and a human host." *Molecular bioSystems* 5(12): 1630-1635.

Vapnik, V. N. (1998). *Statistical learning theory*. New York, Wiley.

Wheeler, D., T. Barrett, et al. (2006). "Database resources of the national center for biotechnology information." *Nucleic Acids Research*.

Whitby, S. M. (2001). "The potential use of plant pathogens against crops." *Microbes Infect* 3(1): 73-80.

Williams, E. S., T. Yuill, et al. (2002). "Emerging infectious diseases in wildlife." *Revue scientifique et technique* 21(1): 139-157.

Winnenburg, R., M. Urban, et al. (2008). "PHI-base update: additions to the pathogen host interaction database." *Nucleic Acids Research* 36(Database issue): D572.

Winnenburg, R., M. Urban, et al. (2008). "PHI-base update: additions to the pathogen host interaction database." *Nucleic Acids Research* 36(Database issue): D572-576.

Xenarios, I., E. Fernandez, et al. (2001). "DIP: The Database of Interacting Proteins: 2001 update." *Nucleic Acids Research* 29(1): 239-241.

Yang, Z., H. Lin, et al. (2009). "BioPPIExtractor: A protein-protein interaction extraction system for biomedical literature." *Expert Systems with Applications* 36(2): 2228-2233.

Yin, L., G. Xu, et al. (2010). "Document classification for mining host pathogen protein-protein interactions." *Artificial Intelligence in Medicine* 49(3): 155-160.

Zanzoni, A., L. Montecchi-Palazzi, et al. (2002). "MINT: a Molecular INTeraction database." *FEBS Lett* 513(1): 135-140.

10

Prediction of Novel Pathway Elements and Interactions Using Bayesian Networks

Andrew P. Hodges, Peter Woolf and Yongqun He ξ
University of Michigan, Ann Arbor, MI, USA

1. Introduction

Signalling and regulatory pathways that guide gene expression have only been partially defined for most organisms. Given the increasing number of microarray measurements, it may be possible to reconstruct such pathways and uncover missing connections directly from experimental data. One major question in the area of microarray-based pathway analysis is the prediction of new elements to a particular pathway. Such prediction is possible by independently testing the effects of added genes or variables on the overall scores of the corresponding expanded networks. A general network expansion framework to predict new components of a pathway was suggested in 2001 (Tanay and Shamir, 2001). Many machine learning approaches for identifying hidden or unknown factors have appeared in the literature recently (Gat-Viks and Shamir, 2007; Hashimoto, et al., 2004; Herrgard, et al., 2003; Ihmels, et al., 2002; Needham, et al., 2009; Parikh, et al., 2010; Pena, et al., 2005; Tanay and Shamir, 2001; Yu and Li, 2005).

Compared to existing pathway expansion methods based on correlation, Boolean, or other strategies (Hashimoto, et al., 2004; Herrgard, et al., 2003; Ihmels, et al., 2002; Tanay and Shamir, 2001), Bayesian network-based expansion methods provide distinct advantages. A Bayesian network (BN) is a representation of a joint probability distribution over a set of random variables (Friedman, et al., 2000). Bayesian networks are able to identify causal or apparently causal relationships (Friedman, et al., 2000), and can be used to predict both linear and nonlinear functions. Furthermore, BN analysis is robust to error and noise and easily interpretable by humans. Bayesian network-based expansion has been used for gene expression data analysis (Gat-Viks and Shamir, 2007; Pena, et al.). We have recently developed an algorithm termed "BN+1" which implements Bayesian network expansion to predict new factors and interactions that participate in a specific pathway (Hodges, et al., 2010; Hodges, et al., 2010). This algorithm has been tested using *E. coli* microarray data (Hodges, et al., 2010) and verified with a synthetic network (Hodges, et al., 2010).

This Book Chapter aims to first provide a detailed review on different computational methods for pathway element prediction, introduce how a BN analysis is typically performed, and then describe how this BN+1 algorithm works. We will also introduce our MARIMBA software program (http://marimba.hegroup.org) which can implement the BN+1 algorithm along with many other useful features. So far, the success of BN+1 in new pathway element prediction has been demonstrated in prokaryotic *E. coli* system. This paper will introduce our new study of applying BN+1 to predict new pathway elements for eukaryotic B-cell receptor (BCR) pathway using high throughput microarray data from perturbed B-cells obtained from the Alliance for Cellular Signalling (AfCS) (Zhu, et al., 2004). Finally, we will present current challenges and possible future directions in this field.

2. Overview of different computational methods for prediction of new pathway elements

In this section, we describe several existing methods for pathway expansion. By pathway expansion, we mean the expansion of a known set of variables with some biological role or function to include novel interacting or downstream variables. This definition is highly flexible and can be used for a variety of biological and biomedical situations.

2.1 Correlation methods and pathway expansion

Some of the most prevalent approaches used towards analyzing high-throughput datasets are correlation-based methods. Correlation methods attempt to identify the degree of similarity or dissimilarity between two or more variables (*e.g.*, the expression profiles of two genes) using simple computational distance metrics, such as Manhattan and Pearson metrics (Herrero, et al., 2001). An underlying assumption is that cellular processes often require the participation of multiple gene products which are expected to show correlated expression patterns as well as physical interactions (Meier and Gehring, 2008).

To predict new pathway elements using correlation methods, one or more genes (or other biological entities) are usually selected initially as a target of interest for comparison. A correlation is then determined between each other gene's (or entity's) expression pattern and that of the gene of interest. Those correlations appearing above some established threshold or ranking are then represented as either edges in a network or as a dendrogram in an expression-based heatmap diagram. For example, Herrgard et al defined subset of variables with specific modular behaviors and network structure using correlations and linear multiple regression (Herrgard, et al., 2003). These modules are then expanded to identify other neighboring variables with likely interactions or influences with the module-based sub-networks. Tanay et al (2001) introduced a fitness function-based approach for expanding sets of variables in literature models (Tanay and Shamir, 2001).

One advantage of these correlation-based methods is the ability to compute all pair-wise correlations for genes or features on a gene expression microarray or other high-throughput datasets. However, the correlation networks themselves do not imply any directionality for the interactions, such as which gene activates or represses a correlated gene, or whether those genes are instead co-regulated by another biological entity. The types and sometimes directionality of interactions must be determined using one or more analysis procedures, such as gene enrichment, promoter analysis, and context-dependent (or condition-dependent) analysis (Meier and Gehring, 2008). The correlation-based methods are often sensitive to the underlying distance metrics and assumptions, and are easily misinterpreted when the wrong metrics are employed. In addition, nonlinear (*e.g.* biphasic) interactions cannot usually be detected using correlation-based methods.

2.2 Clustering-based identification of new pathway elements

Various clustering method can be used to group genes based on expression values and identify potential new genes to specific pathways. Unsupervised and supervised clustering methods have been developed (Raychaudhuri, et al., 2001). Unsupervised clusting methods, such as hierarchical clustering (Eisen, et al., 1998), self-organizing maps (Tamayo, et al., 1999), and model-based clustering (*e.g.*, CRCView (Xiang, et al., 2007)), arrange genes and samples in groups/clusters based solely on the similarities in gene expression. Supervised methods, including EASE (Hosack, et al., 2003) and gene set enrichment analysis (GSEA) (Subramanian, et al., 2005), use sample classifiers and gene expression to identify hypothesis-driven correlations. The Gene Ontology program (GO) is frequently used for gene enrichment analysis by many software programs, for example, DAVID (Huang da, et al., 2009) and GOStat (Beissbarth and Speed, 2004). One major disadvantage of such clustering-based methods on identifying new pathway elements is that detailed gene-gene interactions and directionalities cannot be predicted.

2.3 Boolean network-based pathway expansion

In Boolean network modelling, originally introduced by Kauffman (Kauffman, 1969) (Kauffman, 1969) (Kauffman, 1969), gene expression is quantized to only to two levels: ON and OFF. The gene expression level (state) of each gene is functionally related to the expression states of some other genes using logical rules. Probabilistic Boolean Networks (PBN) share the appealing rule-based properties of Boolean networks, but are robust in the face of uncertainty (Shmulevich, et al., 2002). Hashimoto et al. proposed a method to grow genetic regulatory networks from seed genes based on PBN analysis (Hashimoto, et al., 2004). In their study, Boolean functions were implemented towards globally expanding a set of seed genes from known literature-extracted interactions for vascular endothelial growth factor pathway genes using melanoma and glioma data (Hashimoto, et al., 2004). The output of this algorithm depends on the PBN-based objective function. The disadvantage of this approach is that the two-level representation in Boolean network often oversimplifies the complex biological systems.

2.4 Mutual information-based method

Mutual information-based methods have been used for modelling, refining, and expanding biological pathways. In probability theory and information theory, the mutual information of two random variables is a quantity that measures the mutual dependence of the two variables. Recent reports by Luo et al. (Luo, et al., 2008; Luo and Woolf, 2010; Watkinson, et al., 2009) and others have shown the utility and improved modelling of using three-way and higher mutual information influences for a given variable. However, the assembly of these multi-parent interactions into larger global networks is yet a challenging issue.

2.5 Bayesian network pathway refinement and expansion

Bayesian networks have recently been widely used for biological pathway reconstruction and expansion. Since this is the major topic of this book chapter, we will introduce it in more details in the next sections.

3. Bayesian network (BN) analysis

In this section, we introduce Bayesian networks and their uses in biomedical research. Most specifically, models generated for understanding biological pathways and relevant gene regulatory networks are discussed.

3.1 Introduction to Bayesian networks

One exciting development in bioinformatics research was the advent and application of Bayesian networks (BN) in biological research. Basically, BNs are graphical representations of statistical interdependencies amongst sets of nodes. BNs model interactions amongst sets of variables (*e.g.* genes, proteins) as probabilistic dependencies or influences. Judea Pearl introduced the notion of Bayesian networks in 1985 (Pearl, 1985; Pearl, 1988) to emphasize three aspects: (i) Often subjective nature of the input data information; (ii) Reliance on Bayes's conditioning as the basis for information updating; and (iii) Distinction between causal and evidential modes of reasoning. Bayesian networks were later implemented by Heckerman et al, Friedman et al, and various other research labs towards biological research (Cooper and Herskovits, 1992; Friedman, et al., 2000; Heckerman, 1995).

Specifically, a BN for a set of variables $X = \{X_1, X_2, ..., X_n\}$ consists of (1) a network structure S that encodes a set of conditional independence assertions about variables in X, and (2) a set P of conditional probability distributions associated with each variable (Heckerman, 2008). Together, these components denote the joint probability distribution for X. The BN structure S is a directed acyclic graph, meaning that the network is hierarchical and has both top-level and terminal nodes and no directed paths which eventually return to them. We use Pa_i to denote the parents of node Xi in S as well as the variables corresponding to those parents. Given structure S, the joint probability distribution for X is given by

$$p(\mathbf{x}) = \prod_{i=1}^{n} p(x_i|\mathbf{pa}_i) \qquad (1)$$

Different methods have been developed to learn BN structures and will be introduced in detail next.

3.2 Learning Bayesian networks (BNs)

The problem of learning a Bayesian network can be stated as follows: given a training dataset of independent instances, find a network that best matches the dataset. The common approach to this problem is to introduce a statistically sound scoring function that evaluates each network with respect to the training dataset and to search for the optimal network based on this score.

To dissect the processes of learning BNs, we summarize five major steps as follows:

1. Data selection and pre-processing
2. Prior definition (including variables and edges)
3. Selection of network searching strategy (e.g., simulated annealing, greedy)
4. BN execution with a specific scoring method
5. Results output and analysis

These steps will be introduced in detail here for gene expression data analysis:

3.2.1 Data selection and preprocessing

BN is a powerful tool for analyzing high throughput data, e.g., DNA microarray data. Pre-processing is usually required to normalize raw data and possibly filter out those genes that do not show significant changes over all conditions.

3.2.2 Prior definition (including variables and edges)

After selecting appropriate data and variable sets for investigation, settings for the BN simulation must be chosen. Initially, assumptions must be made as to whether structural priors (e.g. the requirement of certain interactions to appear in a model) should be included or not in the BN analysis. It is not necessary to assume any structural priors for the initial set of variables. However, structural priors can be implemented, especially in cases where the biological interactions to be represented are well-established and also fully represented in the underlying biological data used for modelling.

3.2.3 Set up network searching strategy

Once the prior is specified, the BN learning becomes finding a structure that maximizes the BN score according to a BN scoring function. This problem is proven to be NP-complete (Chickering, 1996). Thus heuristic search is needed. The decomposition of the score is crucial for the optimization problem. For example, a local search procedure that changes one edge at a time can efficiently evaluate the gains of a specified score made by adding, removing, or reversing an edge. An example of such a procedure is a greedy random search algorithm with random restarts. Although this procedure does not necessarily achieve a global maximum, it reaches a local maximum and does perform well in practice (Friedman, et al., 2000). Another commonly used method is simulated annealing search algorithm with a temperature schedule that allows for "reannealing" as the temperature is lowered (Heckerman, 1995). Other BN searching strategies include stochastic hill-climbing and genetic algorithm (Friedman, et al., 2000).

3.2.4 Bayesian network scoring approaches

The key part of BN learning is to determine a scoring metric that compares networks and identifies the most likely or 'best supported' networks. Bayesian network scoring is based upon conditional probabilities. One commonly used scoring method is the BDe score (Cooper and Herskovits, 1992; Heckerman, 1995), which is a posterior probability defined as:

$$P(M|D) \propto \prod_{i=1}^{n} \prod_{j=1}^{q_i} \frac{(r_i - 1)!}{\left(N_{ij} + r_i - 1\right)!} \prod_{k=1}^{r_i} N_{ijk}\,! , \tag{2}$$

where n is the number of variables, q_i is the number of parent configurations for given variable i, r_i is the arity of variable i, N_{ij} is the number of observations with selected parent configuration q_i, N_{ijk} is the number of observations of child in state k with parent configuration q_i (Cooper and Herskovits, 1992). The calculation of this score is implemented in many software programs such as BANJO (Smith, et al., 2006).

Another BN scoring method is the Bayesian Information Criterion (BIC), which was specifically designed to compensate for overfitting (Schwarz, 1978).

3.2.5 Bayesian network analysis software

Many BN analysis software programs are available. Dr. Kevin Murphy provides an excellent summary of existing software packages for Bayesian network modelling (http://www.cs.ubc.ca/~murphyk/Bayes/bnsoft.html). Table 1 lists selected BN software programs from Dr. Murphy's website and other resources.

Name	Source	API	GUI	Undir	Exec	Free	Inference	Exp	Reference
Banjo	Java	Y	N	D	W,U, M	Y	N	N	(Bose, et al., 2006)
BayesiaLab	N	N	Y	C,G	W,U, M	N	Jtree, Gibbs	N	(Conrady and Jouffe, 2011)
BNT	Matlab, C	Y	N	D,U	W,U, M	Y	Several options	N	(Murphy, 2001)
BNJ	Java	N	Y	D	C	Y	Jtree, IS	N	http://bnj.sourceforge.net
Causal Explorer	Matlab, C/C++	Y	N	D	W,U, M	Y	N	N	(Aliferis, et al., 2003)
Deal	R	Y	Y	D	I	Y	N	N	(Bøttcher and Dethlefsen, 2003)
Genie	C++	Y	Y	D	C	Y	Jtree	N	(Druzdzel, 1999)
Java Bayes	Java	Y	Y	D	C	Y	Jtree, Varelim	N	http://www.cs.cmu.edu/~javabayes/Home/
LibB	N	Y	N	D	W,L	Y	N	N	http://www.cs.huji.ac.il/labs/compbio/LibB/
MARIMBA	N	N	Y	D	I	Y	N	Y	(Hodges, et al., 2010; Hodges, et al., 2010)
miniTUBA	N	N	Y	D	I	Y	N	N	(Xiang, et al., 2007)
openBUGS	Y	Y	Y	D	W,U, M	Y	Gibbs	N	(Lunn, et al., 2000; McCarthy, 2007)
OpenPNL	C++	Y	Y	D	W,L	Y	Jtree, Gibbs	N	http://sourceforge.net/projects/openpnl/
PEBL	Python	Y	Y	D	W,U, M	Y	N	N	(Shah and Woolf, 2009)
WinMine	N	N	Y	D,U	W	Y	N	N	(Chickering, 2002)

Notes: The catergories listed include: Source, source code; API, application program interface for programmatic access; GUI, graphical user interface; Undir, ability to handle undirected graphes; Exec, the type of execution, including W:Windows, U:Unix, L:Linux, M:Mac, I:OS-independent, or C:any with compiler; Free, the availability of the software as either free (e.g. academic) or commercial; Inference, inferencing ability; Exp, ability for network expansion; and Ref, references.

Table 1. Selected software programs for BN analysis.

3.2.6 BN result output and analysis

To visualize BN results, different methods can be performed. For example, BANJO uses DOT type of BN result output (Reference: http://www.graphviz.org/Documentation/dotguide.pdf). MARIBMA uses DOT and can also export networks as .sif format for use in Cytoscape (http://www.cytoscape.org). Since different BNs are available, it is crucial for a user to select 'best-scoring' networks and/or generate consensus networks. Often methods are also needed to build weighted networks based on computational analysis or from literature and other database queries.

4. Bayesian network expansion methods

Bayesian network (BN) expansion is an approach that is built upon the BN method and aims to identify new pathway elements that participate in a specified network. In this section, we will introduce basic BN expansion methods and then focus on describing our internally developed BN+1 algorithm and its implementation.

4.1 General BN expansion

Compared to the otehr network expansion methods described above, Bayesian network-based expansion methods provide distinct advantages, such as prediction of both linear and nonlinear functions, robustness in noise data analysis, and identification of causal or appearly causal influences representing interactions among genes. In general, Bayesian network expansion can be defined as the addition of new variables to an existing network, followed by rescoring and ranking of those variables.

BN-based expansion has been used for gene expression data analysis (Gat-Viks and Shamir, 2007; Pena, et al.). For example, Pena et al. reported an algorithm AlgorithmGPC that also grows BN models from seed genes (Pena, et al.). This approach starts with one single gene and builds networks around this gene through expansion and pruning with a set number of genes. Gat-Viks et al also generated a Bayesian network-based refinement and expansion method (Gat-Viks and Shamir, 2007). A main limitation of this approach is that it requires high quality of prior knowledge on the signaling pathways. The topology of the biological pathways may not be consistent with networks learned from transcriptional gene expression data obtained via DNA microarray studies. Therefore, a fixed topology as initial seed network may not be appropriate for robust network expansion simulaions.

Other BN expansion methods have also been published (Needham, et al., 2009; Parikh, et al., 2010). These approaches differ from each other but all showed different levels of success in identifying new pathway elements. In the following two sections, we will introduce our BN+1 algorithm (Hodges, et al., 2010; Hodges, et al., 2010), and how it is implemented in the MARIMBA software.

4.2 The BN+1 algorithm

In our recent study, we developed an algorithm termed "BN+1" which implements Bayesian network expansion to predict new factors and interactions that participate in a specific pathway (Hodges, et al., 2010; Hodges, et al., 2010). Broadly, the BN+1 algorithm iteratively tests to see if any single variable added to a given pathway will significantly improve the likelihood of the overall network. This approach is based on the observation that those variables which are hidden and regulate or are regulated by a network are more likely ranked with high posterior probability scores. Using a compendium of microarray gene expression data obtained from *Escherichia coli*, the BN+1 algorithm predicted many novel factors that influence the *E. coli* reactive oxygen species (ROS) pathway. Some of the predicted new ROS and biofilm regulators (*e.g., uspE* and its interaction with *gadX*) were further experimentally verified (Hodges, et al., 2010). In another study, a synthetic network was also designed to further evaluate this algorithm. Based on the synthetic data analysis, the BN+1 method is able to identify both linear and nonlinear relationships and correctly identify variables near the starting network (Hodges, et al., 2010).

The BN+1 algorithm is specified in Figure 1. A few notes are provided here in our BN+1 implementation:

1. The selection of seed (or core) genes is an important step. The seed genes can be selected from an existing pathway database, from literature survey, or from internal experimental results. Since it is computationally expensive to calculate BNs using a large number of variables, it is often necessary to filter out some genes from an initial list using different criteria, for example, filtering out those genes that do not have significant changes among all microarray chips.
2. While we use a top network structure generated from initial core gene simulation as a prior, we prefer not to fix the core network structure for subsequent network expansion. This preference makes our approach differ from a commonly used method of fixing the prior structure. Our argument is that the prior structure is often determined by many layers of studies, including DNA, RNA and protein data analyses. When only RNA transcriptomic data are used, such prior structure may not hold. The fixture of a prior structure would result in obtaining suboptimal networks that do not match the datasets used for BN simulation.

BN+1 Algorithm
Input: *N* variables (e.g., genes) from a dataset (e.g., microarray dataset) with *L* observations each.

Data Preprocessing (Optional)
Filter out *m* variables (e.g., via coefficient of variation (c.v.) <= 1.0)
Number of possible variables for analysis: $N= N\text{-}m$.

BN Core Network Searching
Select *K* variables from the set of *N* variables (e.g. from a pathway database).
Construct matrix data file *D* with $K{*}L$ observations using *K* variables and *L* observations.
Select settings for BN simulation, including data discretization (e.g. q3 quantization), searcher strategy (e.g. simulated annealing), and structural priors.
Execute BN simulation (e.g. using BANJO).
Save top BN network topology *C*

Iterative Core Expansion
Assign the core topology *C* as unfixed structural prior for BN searching
For each variable *a* in the set $\{N\text{-}K\}$, do:
    Generate new data file D^* by concatenating *L* observations for *a* to data file *D*
    Select settings for BN simulation.
    Execute BN simulation.
    Save top network and its posterior probability for *a*.
Rank each variable according to posterior probability.

Output: Rank-ordered BN+1 results.

Fig. 1. BN+1 algorithm.

4.3 Implementation of BN+1 using MARIMBA

MARIMBA is implemented using a three-tiered architecture built on two Dell Poweredge 2580 servers which run the Redhat Linux operating system. Users submit analysis requests and database queries through the web. These queries are then processed using PHP, Perl, Python, JavaScript, and SQL (middle-tier, application server based on Apache) against a MySQL (version 5.0) relational database (back-end, database server). The result of each query is then presented to the user in the web browser.

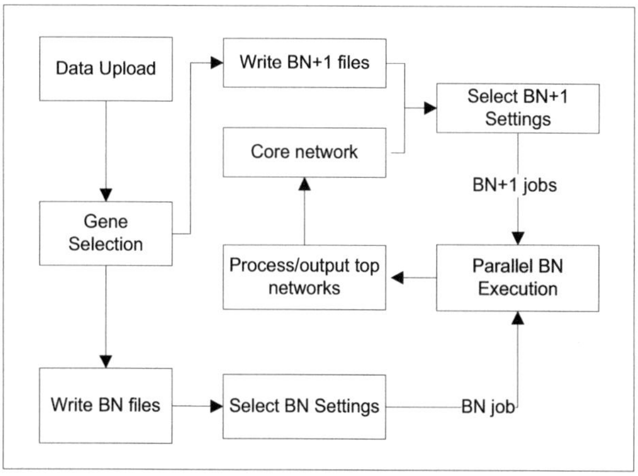

Fig. 2. Workflow of MARIMBA implementation of BN+1 algorithm.

The main MARIMBA system architecture and pipeline for analysis of project data is described in Figure 2 and contains the following steps:

1. *Data Upload:* A user can upload gene expression data using specified format.
2. *Gene Selection:* A gene list as core gene set for BN modeling will be specified by the user.
3. *Write up BN and BN+1 files:* The basic MARIMBA-formatted files are then generated dynamically by MARIMBA for use in Bayesian modeling.
4. *Selection of BN parameters:* BN simulation settings were selected after completing the data and gene selection processes, respectively. A static BN simulation was created to analyze the microarray data. Many settings can be selected by the user. For example, a user can select simulated annealing or greedy method as the network searcher method. We prefer simulated annealing due to its improved performance over greedy searches when no prior knowledge of underlying structure is available (Hartemink, et al., 2002). In our simulated annealing analysis, a relatively low cooling factor is often implemented to allow less restrictive searching of the sample space and potentially identify as many equivalence classes for the top-scoring network as possible. Currently, up to 1,000 networks can be stored in MARIMBA for each run.
5. *Execution of BN and BN+1modeling:* BN files are submitted via the online interface in MARIMBA. Each dataset is transferred to the server prior to simulation by a parallel computer cluster. Each agent runs a unique BANJO simulation. The core BN network is employed as a fixed topology/prior knowledge network in the BN analysis. A BN+1 simulation can also be implemented as defined by a user.
6. *BN Result display and interpretation:* MARIMBA displays top-scoring networks of BN and BN+1 simulations to the user. The images of top-scoring networks are converted directly from their original dot files and are displayed as jpeg images on-the-fly. In addition, MARIBMA is able to calculate conserved edges over a selected number of stored networks. To calculate the conserved edges, MARIMBA determines core BN models by model averaging and equivalence class searching. Here, model averaging is defined as inclusion of an edge between two genes if that edge appeared in more than X percent of the top-scoring networks with identical score. Furthermore, The BN+1 visualization environment in MARIMBA displays a plot for posterior probabilities of saved networks, thus enabling comparison of networks for relevance and likelihood.

Compared to the standalone BANJO system (Bose, et al., 2006), MARIMBA is web-based and allows seamless integration of user project management, analysis construction, BN submission and execuation in a parallel computing environment, and analysis and visualization of results. The user-friendly GUI environment simplifies dataset selection, probeset/gene inclusion, observational file processing, and settings selection for BN execution. Such features are necessitated for efficient querying by biologists who wish to use such BN tools to analyze their data. In addition, the BN+1 algorithm execution and project management is a unique feature in MARIBMA and does not exist in BANJO or any other software program.

5. Use case study: Application of BN+1 to BCR pathway modelling

5.1 Introduction

As an example of the challenge of merging a pathway model and gene expression data, this study focuses on the B-cell receptor pathway (BCR) as described by KEGG (Kanehisa and Goto, 2000; Kanehisa, et al., 2010). The BCR pathway is an integral component of the adaptive immune response mechanism by which B cells respond to foreign antigens (Lucas, et al., 2004). The BCR pathway involves in the activation of specific protein kinase C (PKC) isoforms that induces ultimate activation of the NF-κB transcription factor. Multiple protein species accumulate at the cell membrane in a signalosome complex and are linked to the B cell receptor. Signal propagation from the BCR via kinase-mediated phosphorylation cascades to downstream effectors such as Nfkb, NFAT (nuclear factor of activated T cells), and AP1 is either enhanced or reduced via signalosome interactions with co-stimulatory or co-inhibitory complexes, respectively. BCR signaling guides many important functions such as anergy, B cell ontogeny, and immune response, and is linked to the several imporant pathways: MAPK, coagulation/complement cascades, and actin cytoskeleton (Kanehisa and Goto, 2000; Kanehisa, et al., 2010). NF-κB plays a crucial role in the antigen-induced B lymphocyte proliferation, cytokine production, and B cell survival (Lucas, et al., 2004). While the KEGG pathway database includes a manually curated BCR pathway, this pathway is still considered incomplete (Lucas, et al., 2004).

5.2 Microarray data processing and BN analysis methods

We used gene expression data from perturbed B-cells obtained from the Alliance for Cellular Signaling (AfCS) (Lee, et al., 2006; Zhu, et al., 2004). This dataset is especially attractive because the same tissues were treated with combinations of ligands that perturb different B cell pathways. The AfCS study gathered 424 microarray chips measuring gene expression in B cells from *M. musculus* splenic extracts that are exposed to 33 different ligands (Lee, et al., 2006; Papin and Palsson, 2004; Zhu, et al., 2004). Briefly, B cells purified from splenic preparations from 6- to 8-wk-old male C57BL/6 mice were treated in triplicates or quadruplicates with medium alone, or one of 33 different ligands for 0.5, 1, 2, and 4 h (AfCS protocol PP00000016). RNA was extracted following standard AfCS protocol PP00000009. An Agilent cDNA microarray chip that contains 15,494 cDNA probes printed on 15,832 spots was used. It represents 10,615 unique MGI gene matches (Lee, et al., 2006). Each Agilent array was hybridized with Cy5-labeled cDNA prepared from splenic B cell RNA and Cy3-labeled cDNA prepared from RNA of total splenocytes used as an internal reference (AfCS protocol PP00000019). Hence, each Agilent microarray chip provides one unique observation of relative expression level per selected probe. The arrays were scanned using Agilent Scanner G2505A, and images were processed using the Agilent G2566AA Feature Extraction software version A.6.1.1. The microarray raw data were downloaded from the AfCS repository at ftp://ftp.afcs.org/pub/datacenter/microarray/.

Microarray data were discretized for each variable in the Bayesian networks using quantile normalization with three bins. Though triplicate or quadruplicate microarray experiments were available in most cases per unique treatment and time of drug administration, we assume that each experiment provides an independent source of information. In this analysis, we did not use all BCR pathway genes. We sought to answer here whether expansion of a sub-network from the BCR pathway would preferentially recover other BCR pathway genes. This assumption is advantageous in that the number of variables allows significantly faster simulation searches for the BN and BN+1 simulations. Particularly, those genes most specifically involved in Nfkb-mediated transcriptional regulation were chosen from the KEGG BCR pathway.

A set of 10,000 top-scoring BNs was generated using the eight variables (the core) and 424 observations. Among the eight variables, two variables are Nfkbie probe sets, and two are Ikbkb probe sets. In many cases, one gene has multiple probe sets. We chose to separate them as different variables in our BN analysis since often these probe sets have different values with low correlation (Fig. 3). This BN analysis was accomplished by running 100 independent simulations and saving the top 100 simulations for each of those runs.

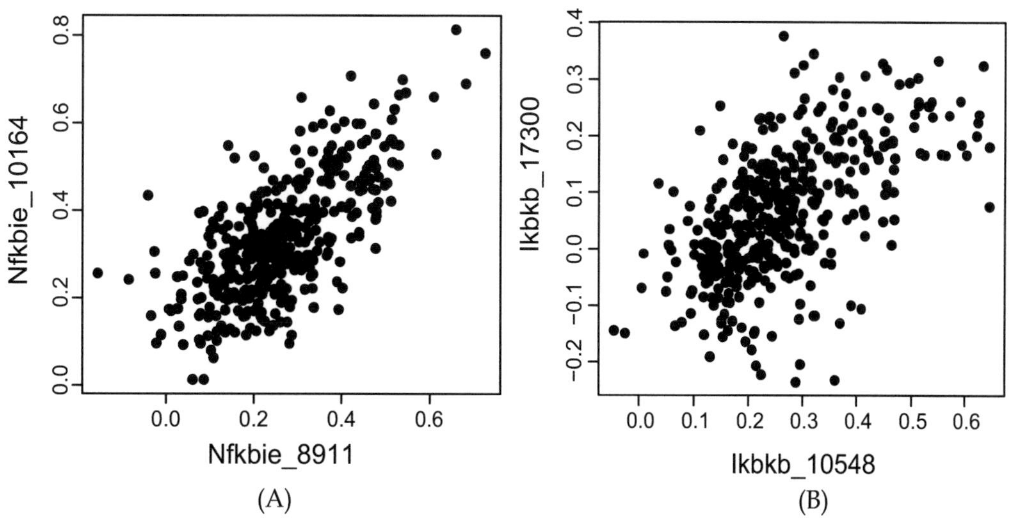

Fig. 3. Scatter plots for Nfkbie and Ikbkb probes from AfCS study. Agilent probe identifiers are listed next to each respective gene. This figure indicates that the probe sets Nfkbie_10164 and Nfkbie_8911 correlate relatively well with a Pearson correlation coefficient of 0.69 (A). However, the correlation between Ikbkb_17300 and Ikbkb_10548 is low (Pearson correlation coefficient: 0.58) (B).

5.3 Results

5.3.1 Defining the core network

Fig. 4 depicts the shared set of interactions appearing in all of the top networks sharing the same best score.

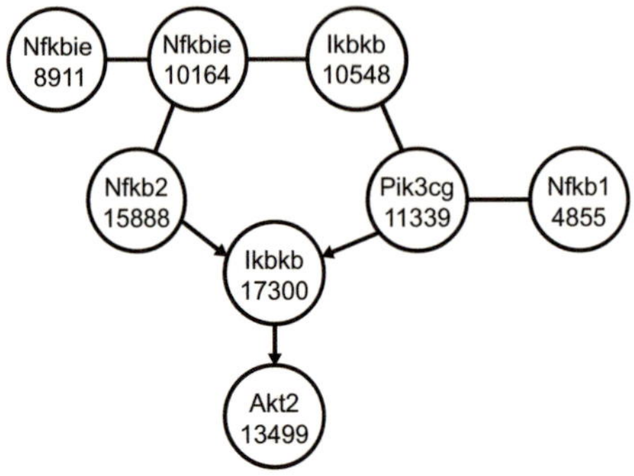

Fig. 4. Consensus of top scoring Bayesian networks for eight probes representing BCR receptor signaling pathway genes. Gene symbols and corresponding Agilent probe identifiers are represented in nodes in the network. Directed edges represent those influences appearing in the same direction in all top-scoring Bayesian networks, while undirected edges appear at least once in the opposite direction though appearing cumulatively with 100% frequency in all of the top networks.

Compared with the KEGG BCR pathway, the consensus network found in our BN analysis (Fig. 2) has a 75% overlap with known interactions (3 out of 4 were correctly predicted), with only one interaction missing (Fig. 5).

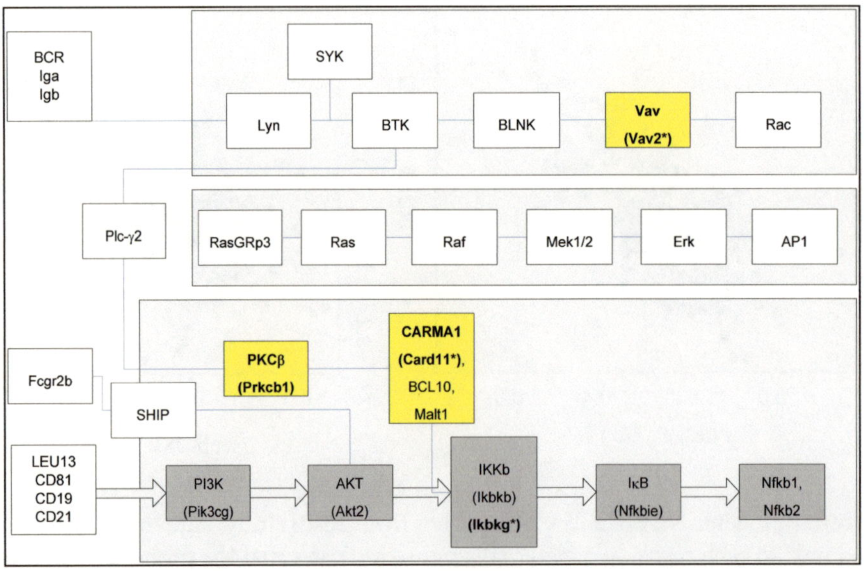

Fig. 5. Schematic representation of the BN+1 analysis results in the content of KEGG BCR pathway. The three blue boxes represent three major sub-networks within the BCR pathway with distinct regulatory and functional roles. The BN core network was defined using members from the third sub-network (dark grey boxes) which reflect major components of Nfkb signalling. Bolded gene names are those genes which were not included in the core network, yet were recovered during BN+1 analysis in the top 100 results. Note that not all members of the listed Nfkb signalling pathway were included in the core network (e.g. Ikbkg), and in some cases were not available on the microarray platform.

5.3.2 Defining BN+1 genes

One of the top-scoring networks used to generate the consensus shown in Fig. 4 was used as a core network for subsequent BN+1 expansion. BN+1 searching was executed for 14,353 individual probes with 50 million networks searched per probe. If only those genes in close neighbourhood in the KEGG BCR pathway are considered, out of 19 selected genes, nine genes were found to be connected to the core network in our analysis. Furthermore, four of these nine genes are in close proximity (within top 10% of top-scoring BN+1 genes with at least one connection to the core network) with these core genes in the KEGG protein signalling pathway: Card11, Prkcb1, Ikbkg, and Vav2. These results suggest that the neighbourhood of transcriptional regulation around the core network as well as distance between the elements in the protein signalling pathway are related to each other.
Analysis of the top BN+1 variables recovered during simulation revealed several interesting results. First, the top set of BN+1 variables is listed in Table 1.

Rank	Agi_ID	GeneID	Symbol	BN1_score	Neighbors
1	11062	77619	Prelid2	-3402.0	Nfkb2
2	9502	20744	Strbp	-3517.0	Nfkbie
3	14138	20823	Ssb	-3545.2	Nfkb2
4	6276	12530	Cdc25a	-3569.2	Nfkb2
5	11361	108829	Jmjd1c	-3586.8	Ikbkb(both), Pik3cg
6	14614	75964	Trappc8	-3587.8	Ikbkb, Pik3cg
7	15876	108786	Cxcl13*	-3593.1	Nfkb2
8	10759	73132	Slc25a16	-3594.8	Ikbkb, Pik3cg
9	5275	67887	Tmem66	-3596.0	Nfkb1, Pik3cg
10	9036	109339	2700018L05Rik	-3599.1	Pik3cg

Notes: Identifier information for each ranked gene is provided, including Agilent probe ID (Agi_ID), Entrez gene ID (GENEID), and gene symbol. Probe variables from the core network which directly connect to the BN+1 variables in the top-scoring networks are listed in the "Neighbors" column.

Table 2. Top ten predicted BN+1 genes.

Many interesting findings were observed from this analysis. Many genes, for example, the Sjorgen syndrome antigen B gene (Ssb) (Brenet, et al., 2009), have been shown to be associated with the Nf-kB and BCR pathways. Ssb plays an important role in polysome translation (Brenet 2009), and is an early DNA-damage responder in apoptotic cells and those treated with cytotoxic chemicals (Al-Ejeh, et al., 2007). Interestingly, we identified Jmjd1c, a member of the jumonji family proteins, as a top predicted gene in our BN+1 simulation. Jmjd1c is conserved in several mammalian species and has documented roles in metal ion binding, oxidoreductase activity, and transcriptional regulation (Katoh, 2007). The murine Jmjd1c mRNA is expressed in multiple tissues, including hematopoietic and undifferentiated ES stem cells, fertilized egg, pancreatic islet, etc (Katoh, 2007). Jmjd1c has a promoter region orthologous to humans with binding sites for the AP-1 transcription factor, which is considered a member of the BCR signalling pathway and is included in the KEGG representation as AP1 (downstream of the Raf/MEK sub-network in Figure 5 though not in our core network). Fig. 6 illustrates the strongly-correlated relationships uncovered between the Jmjd1c genes and connected core network members. As another example, the Cxcl13 is a chemokine ligand in B cells with a C-X-C motif. It has already been established that Cxcl13 induction requires activation of canonical and non-canonical Nf-kB pathways (Suto, et al., 2009), which confirms the prediction of this gene in our network. These data strongly support the predictions generated by our analysis.

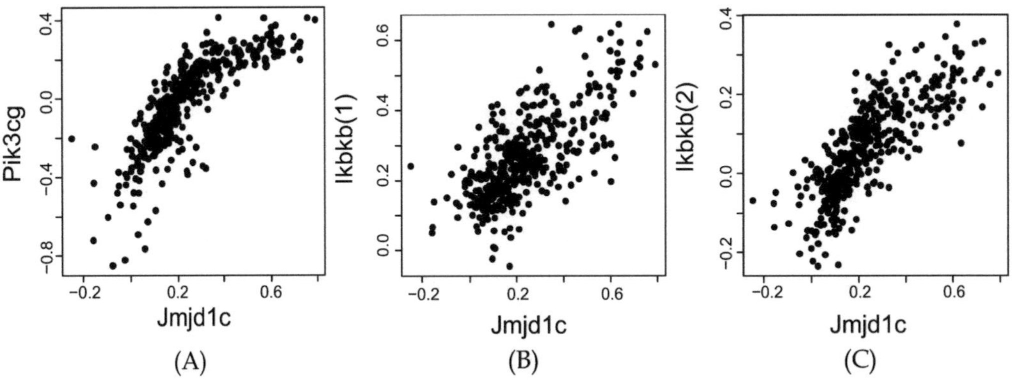

Fig. 6. Scatter plot of expression values for core genes Pik3cg and Ikbkb (both probes) versus BN+1 gene Jmjd1c. A non-linear association between Pik3cg and Jmjd1c is observed (A). A roughly linear relation is observed between Jmjd1c and Ikbkb(1) (Pearson correlation coefficient: 0.71) (B) and between Jmjd1c and Ikbkb(2) (Pearson correlation coefficient: 0.79) (C).

One property of interest, as shown in the table, is that the core genes which recruit the top BN+1 genes are not always the same. From this analysis and previous studies, we have observed that BN+1 variables which show high correlations to at least one core network variable often appear as top BN+1 results. However, in some cases, the BN+1 variable may connect to multiple variables in the core network, and yet show moderate to low correlations with each of them. It is observed that many BN+1 variables have multiple core network variables as parent nodes in the predicted top network. Multi-parent relationships are less common, though statistically more meaningful due to the nature of the implemented conditional probability tables in BDe scoring.
Different methods, such as clustering and GO gene enrichment, can be used to further analyze BN+1 genes.

5.3.3 Clustering analysis of core genes and BN+1 genes

A clustering method provides a way to group BN+1 genes based on gene expression values. A heapmap clustering analysis was performed using 8 probe sets in the core network and 10 probe sets from the BN+1 analysis (Fig. 7). As shown in this heatmap, all NF-κB genes (core genes in our BN simulation) are clustered together, indicating their close association. Our analysis also found that Jmjd1c is closely associated with these NF-κB genes. This further strengthens our BN+1 prediction of the important role of this gene in the NF-κB pathway in B cell signalling.

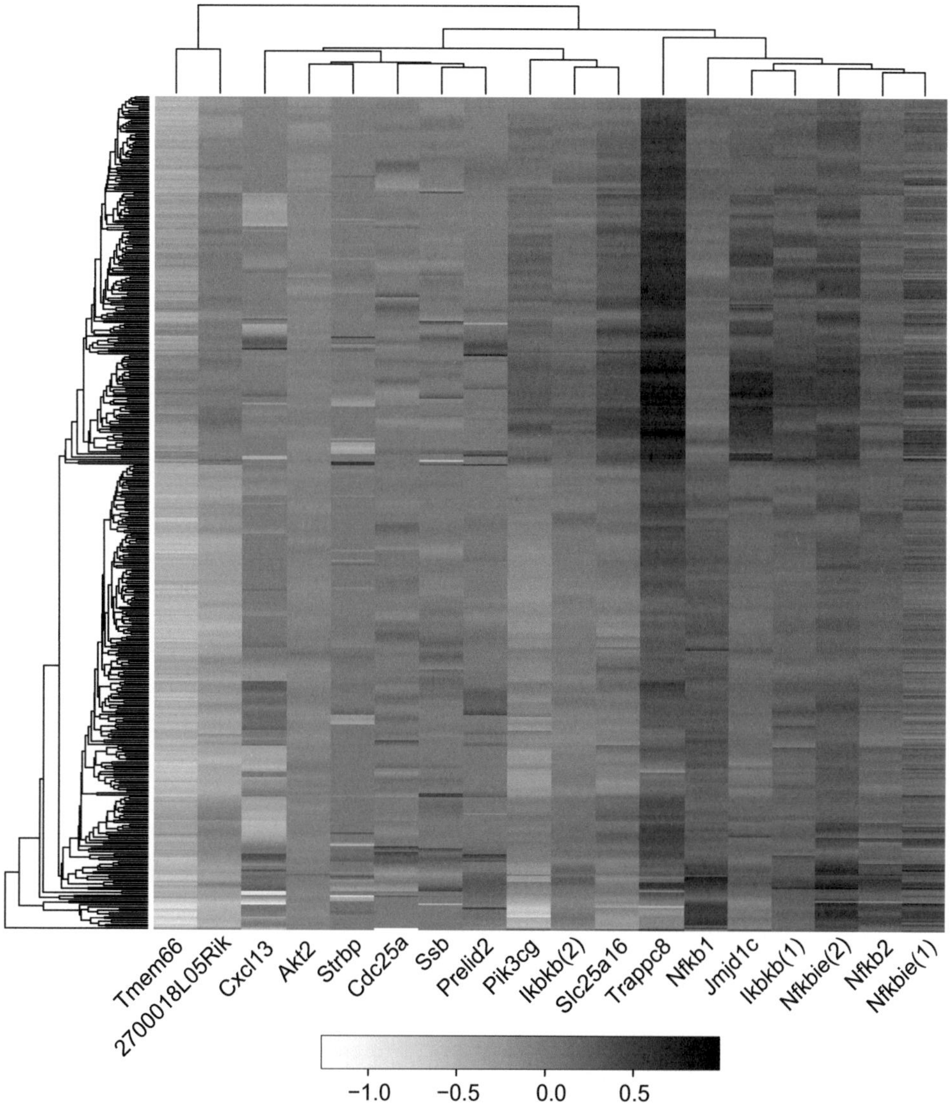

Fig. 7. Heatmap of expression data for top BN+1 and core variables. Parentheses indicate specific probe identities.

5.3.4 GO enrichment of predicted BN+1 genes

Our previous studies indicate that the top few hundred BN+1 genes (i.e. those genes predicted by the BN+1 algorithm) often interact with the seed gene network and biologically active relevant to the pathway of interest (Hodges, et al., 2010; Hodges, et al., 2010). A GO gene enrichment analysis was performed using 250 top BN+1 genes (Table 3). Given the nature of the Nfkb-selected core network and their roles in nuclear localization and transcriptional initiation, it was not surprising that many of the recovered genes show some nuclear compartmentalization. Interestingly, many apoptotic and death-related genes were enriched (Table 3).

Term	*Count*	*P-Value*	*Benjamini P-value*
Biological Process			
GO:0009987~cellular process	106	8.29E-06	0.00981
GO:0070227~lymphocyte apoptosis	4	1.44E-04	0.0823
GO:0008219~cell death	15	1.73E-04	0.0663
GO:0016265~death	15	2.20E-04	0.0634
GO:0048569~post-embryonic organ development	4	2.36E-04	0.0546
GO:0006915~apoptosis	14	2.65E-04	0.0512
GO:0012501~programmed cell death	14	3.12E-04	0.0517
Cellular Compartment			
GO:0005622~intracellular	125	2.93E-08	5.68E-06
GO:0044424~intracellular part	119	4.32E-07	4.19E-05
GO:0043229~intracellular organelle	105	2.76E-06	1.78E-04
GO:0043226~organelle	105	2.84E-06	1.38E-04
GO:0043231~intracellular membrane-bounded organelle	93	4.08E-05	0.00158
GO:0043227~membrane-bounded organelle	93	4.24E-05	0.00137
GO:0005634~nucleus	58	0.001749	0.0474

Notes: Entrez gene identifiers were input for the top 250 BN+1 results into the DAVID tool for GO analysis. The 250 results mapped to 188 unique *Mus musculus* and seven unknown species genes, revealing that some of the top genes were represented by multiple Agilent probes in the top results. Benjamini-derived p-values of 0.01 were used as cutoffs here.

Table 3. GO enrichment results for top 100 predicted variables in the BN+1 analysis.

6. Conclusion

In this paper, different bioinformatics methods for network expansion and detection of new pathway elements are surveyed. Bayesian network-based expansion methods are specifically introduced. Particularly, we outline our BN+1 Bayesian network method that can be used to iteratively compare BDe scores and rank those genes that are likely critical to a specific pathway or network. BN+1 has been successfully demonstrated in *E. coli* system and synthetic data simulation. In this paper, we first demonstrate its use in BCR pathway, a eukaryotic signalling pathway. Our study shows that BN+1 can also be used to predict pathway elements and gene interactions in important eukaryotic pathways. Therefore, the BN+1 algorithm appears to be a generic BN expansion system that can be used to study other prokaryotic and eukaryotic pathways.

Many future directions are envisioned. For example, we can extend the BN+1 algorithm to BN+2, BN+3, or BN+n algorithm by iteratively adding more than one variable to the seed gene network. The principle used in the development of the BN+1 algorithm can also be used for dynamic BN analysis. We are currently in the process of developing a DBN+1 algorithm and using it for temporal data analysis.

7. Acknowledgment

This research was supported in part by NIH Grant U54-DA-021519, NIH Training Grant (5 T32 GM070449-04), 2008 Rackham Spring/Summer Research Grant at the University of Michigan, and the University of Michigan Bioinformatics Program.

8. References

Al-Ejeh, F., *et al.* (2007) In vivo targeting of dead tumor cells in a murine tumor model using a monoclonal antibody specific for the La autoantigen, *Clin Cancer Res*, 13, 5519s-5527s.

Aliferis, C.F., *et al.* (2003) Causal explorer: A causal probabilistic network learning toolkit for biomedical discovery. Citeseer, pp. 371-376.

Beissbarth, T. and Speed, T.P. (2004) GOstat: find statistically overrepresented Gene Ontologies within a group of genes, *Bioinformatics*, 20, 1464-1465.

Bose, R., *et al.* (2006) Phosphoproteomic analysis of Her2/neu signaling and inhibition, *Proc Natl Acad Sci U S A*, 103, 9773-9778.

Bøttcher, S.G. and Dethlefsen, C. (2003) *deal: A package for learning Bayesian networks.* Department of Mathematical Sciences, Aalborg University.

Brenet, F., *et al.* (2009) Akt phosphorylation of La regulates specific mRNA translation in glial progenitors, *Oncogene*, 28, 128-139.

Chickering, D.M. (1996) Learning Bayesian networks is NP-complete, *Learning from data: Artificial intelligence and statistics v*, 112, 121-130.

Chickering, D.M. (2002) The winmine toolkit, *Microsoft Research MSR-TR-2002-103*.

Conrady, S. and Jouffe, L. (2011) Breast Cancer Diagnostics with Bayesian Networks.

Cooper, G.F. and Herskovits, E. (1992) A Bayesian method for the induction of probabilistic networks from data, *Machine Learning*, 9, 309-347.

Druzdzel, M.J. (1999) SMILE: Structural Modeling, Inference, and Learning Engine and GeNIe: a development environment for graphical decision-theoretic models. JOHN WILEY & SONS LTD, pp. 902-903.

Eisen, M.B., *et al.* (1998) Cluster analysis and display of genome-wide expression patterns, *Proc Natl Acad Sci U S A*, 95, 14863-14868.

Friedman, N., *et al.* (2000) Using Bayesian networks to analyze expression data, *J Comput Biol*, 7, 601-620.

Gat-Viks, I. and Shamir, R. (2007) Refinement and expansion of signaling pathways: the osmotic response network in yeast, *Genome Res*, 17, 358-367.

Hartemink, A.J., *et al.* (2002) Combining location and expression data for principled discovery of genetic regulatory network models, *Pac Symp Biocomput*, 437-449.

Hashimoto, R.F., *et al.* (2004) Growing genetic regulatory networks from seed genes, *Bioinformatics*, 20, 1241-1247.

Heckerman, D. (2008) A tutorial on learning with Bayesian networks, *Innovations in Bayesian Networks*, 33-82.

Heckerman, D., Geiger, D. (1995) Learning Bayesian networks: the combination of knowledge and statistical data, *Machine Learning*, 20, 197-243.

Herrero, J., Valencia, A. and Dopazo, J. (2001) A hierarchical unsupervised growing neural network for clustering gene expression patterns, *Bioinformatics*, 17, 126-136.

Herrgard, M.J., Covert, M.W. and Palsson, B.O. (2003) Reconciling gene expression data with known genome-scale regulatory network structures, *Genome Res*, 13, 2423-2434.

Hodges, A., Woolf, P. and He, Y. (2010) BN+ 1 Bayesian network expansion for identifying molecular pathway elements, *Communicative & Integrative Biology*, 3, 59-64.

Hodges, A.P., *et al.* (2010) Bayesian network expansion identifies new ROS and biofilm regulators, *PLoS One*, 5, e9513.

Hosack, D.A., *et al.* (2003) Identifying biological themes within lists of genes with EASE, *Genome Biol*, 4, R70.

Huang da, W., Sherman, B.T. and Lempicki, R.A. (2009) Systematic and integrative analysis of large gene lists using DAVID bioinformatics resources, *Nat Protoc*, 4, 44-57.

Ihmels, J., *et al.* (2002) Revealing modular organization in the yeast transcriptional network, *Nat Genet*, 31, 370-377.

Kanehisa, M. and Goto, S. (2000) KEGG: kyoto encyclopedia of genes and genomes, *Nucleic Acids Res*, 28, 27-30.

Kanehisa, M., *et al.* (2010) KEGG for representation and analysis of molecular networks involving diseases and drugs, *Nucleic Acids Res*, 38, D355-360.

Katoh, M. (2007) Comparative integromics on JMJD1C gene encoding histone demethylase: conserved POU5F1 binding site elucidating mechanism of JMJD1C expression in undifferentiated ES cells and diffuse-type gastric cancer, *Int J Oncol*, 31, 219-223.

Kauffman, S.A. (1969) Metabolic stability and epigenesis in randomly constructed genetic nets, *J Theor Biol*, 22, 437-467.

Lee, J.A., *et al.* (2006) Components of the antigen processing and presentation pathway revealed by gene expression microarray analysis following B cell antigen receptor (BCR) stimulation, *BMC Bioinformatics*, 7, 237.

Lucas, P.C., McAllister-Lucas, L.M. and Nunez, G. (2004) NF-kappaB signaling in lymphocytes: a new cast of characters, *J Cell Sci*, 117, 31-39.

Lunn, D.J., *et al.* (2000) WinBUGS-a Bayesian modelling framework: concepts, structure, and extensibility, *Statistics and Computing*, 10, 325-337.

Luo, W., Hankenson, K.D. and Woolf, P.J. (2008) Learning transcriptional regulatory networks from high throughput gene expression data using continuous three-way mutual information, *BMC Bioinformatics*, 9, 467.

Luo, W. and Woolf, P.J. (2010) Reconstructing transcriptional regulatory networks using three-way mutual information and Bayesian networks, *Methods Mol Biol*, 674, 401-418.

McCarthy, M.A. (2007) *Bayesian methods for ecology*. Cambridge Univ Pr.

Meier, S. and Gehring, C. (2008) A guide to the integrated application of on-line data mining tools for the inference of gene functions at the systems level, *Biotechnol J*, 3, 1375-1387.

Murphy, K. (2001) The bayes net toolbox for matlab, *Computing science and statistics*, 33, 1024-1034.

Needham, C.J., *et al.* (2009) From gene expression to gene regulatory networks in Arabidopsis thaliana, *BMC Syst Biol*, 3, 85.

Papin, J.A. and Palsson, B.O. (2004) The JAK-STAT signaling network in the human B-cell: an extreme signaling pathway analysis, *Biophys J*, 87, 37-46.

Parikh, A., *et al.* (2010) New components of the Dictyostelium PKA pathway revealed by Bayesian analysis of expression data, *BMC Bioinformatics*, 11, 163.

Pearl, J. (1985) *Bayesian networks: A model of self-activated memory for evidential reasoning.* Computer Science Department, University of California.

Pearl, J. (1988) *Probabilistic reasoning in intelligent systems: networks of plausible inference.* Morgan Kaufmann.

Pena, J.M., Bjorkegren, J. and Tegner, J. (2005) Growing Bayesian network models of gene networks from seed genes, *Bioinformatics*, 21 Suppl 2, ii224-229.

Raychaudhuri, S., *et al.* (2001) Basic microarray analysis: grouping and feature reduction, *Trends Biotechnol*, 19, 189-193.

Schwarz, G. (1978) Estimating the dimension of a model, *The annals of statistics*, 461-464.

Shah, A. and Woolf, P. (2009) Python environment for Bayesian learning: inferring the structure of Bayesian Networks from knowledge and data, *The Journal of Machine Learning Research*, 10, 159-162.

Shmulevich, I., *et al.* (2002) Probabilistic Boolean Networks: a rule-based uncertainty model for gene regulatory networks, *Bioinformatics*, 18, 261-274.

Smith, V.A., *et al.* (2006) Computational inference of neural information flow networks, *PLoS Comput Biol*, 2, e161.

Subramanian, A., *et al.* (2005) Gene set enrichment analysis: a knowledge-based approach for interpreting genome-wide expression profiles, *Proc Natl Acad Sci U S A*, 102, 15545-15550.

Suto, H., *et al.* (2009) CXCL13 production by an established lymph node stromal cell line via lymphotoxin-beta receptor engagement involves the cooperation of multiple signaling pathways, *Int Immunol*, 21, 467-476.

Tamayo, P., *et al.* (1999) Interpreting patterns of gene expression with self-organizing maps: methods and application to hematopoietic differentiation, *Proc Natl Acad Sci U S A*, 96, 2907-2912.

Tanay, A. and Shamir, R. (2001) Computational expansion of genetic networks, *Bioinformatics*, 17 Suppl 1, S270-278.

Watkinson, J., *et al.* (2009) Inference of Regulatory Gene Interactions from Expression Data Using Three Way Mutual Information, *Annals of the New York Academy of Sciences*, 1158, 302-313.

Xiang, Z., *et al.* (2007) miniTUBA: medical inference by network integration of temporal data using Bayesian analysis, *Bioinformatics*, 23, 2423-2432.

Xiang, Z., Qin, Z. and He, Y. (2007) CRCView: A web server for analyzing and visualizing microarray gene expression data using model-based clustering, *Bioinformatics*.

Yu, T. and Li, K.C. (2005) Inference of transcriptional regulatory network by two-stage constrained space factor analysis, *Bioinformatics*, 21, 4033-4038.

Zhu, X., *et al.* (2004) Analysis of the major patterns of B cell gene expression changes in response to short-term stimulation with 33 single ligands, *J Immunol*, 173, 7141-7149.

11

MicroRNA Identification Based on Bioinformatics Approaches

Malik Yousef[1], Naim Najami[1,2] and Walid Khaleifa[1]
[1]*The Galilee Society Institute of Applied Research,*
[2]*Department of Biology, The Academic Arab College of Education, Haifa, Israel*

1. Introduction

One of the most fascinating aspects of RNA interference (RNAi) is the non-cell-autonomous nature of silencing. Seminal studies on RNAi focused on the ability of transgene silencing to propagate systemically throughout an organism, such as from a single Agrobacterium infiltrated leaf to other parts of the plant, or from a grafted silenced stock into a non-silenced scion[1, 2]

The discovery of RNAi was preceded first by observations of transcriptional inhibition by antisense RNA expressed in transgenic plants[3] and more directly by reports of unexpected outcomes in experiments performed by plant scientists in the U.S. and The Netherlands in the early 1990s[4] In an attempt to alter flower colors in petunias, researchers introduced additional copies of a gene encoding chalcone synthase, a key enzyme for flower pigmentation into petunia plants of normally pink or violet flower color. Soon after, a related event termed *quelling* was noted in the fungus *Neurospora crassa* [5], although it was not immediately recognized as related. Further investigation of the phenomenon in plants indicated that the downregulation was due to post-transcriptional inhibition of gene expression via an increased rate of mRNA degradation[6]. This phenomenon was called *co-suppression of gene expression*, but the molecular mechanism remained unknown.

Not long after, plant virologists working on improving plant resistance to viral diseases observed a similar unexpected phenomenon. While it was known that plants expressing virus-specific proteins showed enhanced tolerance or resistance to viral infection, it was not expected that plants carrying only short, non-coding regions of viral RNA sequences would show similar levels of protection. Researchers believed that viral RNA produced by transgenes could also inhibit viral replication[7]. The reverse experiment, in which short sequences of plant genes were introduced into viruses, showed that the targeted gene was suppressed in an infected plant. This phenomenon was labeled "virus-induced gene silencing" (VIGS), and the set of such phenomena were collectively called post transcriptional gene silencing [8][15].

The spread of RNA silencing is not limited to plants or viruses: the first reported experiments of RNAi in *Caenorhabditis elegans* (*C. elegans*) demonstrated a systemic silencing response induced by locally injected or ingested double-stranded RNA (dsRNA) molecules[9, 10]. In plants, as in *C. elegans*, the systemic silencing signal acts in a sequence-specific manner, invoking the involvement of an RNA component. Sequence-specific RNA silencing that acts non-cellautonomously has tremendous implications, not only practically as an experimental tool but in biological processes as well. The long-distance movement of RNA silencing through the vasculature forms a crucial component of the antiviral defence system and has been implicated in microRNA (miRNA)-regulated stress responses[11] [12, 13] RNA dependent gene silencing can also move from cell to cell to elicit short-range signaling responses, such as in the patterning of leaves and roots[14, 15].

After these initial observations in plants, many laboratories around the world searched for the occurrence of this phenomenon in other organisms[16] [16]. Craig C. Mello and Andrew Fire's 1998 *Nature* paper reported a potent gene silencing effect after injecting double stranded RNA into *C. elegans* [9]. In investigating the regulation of muscle protein production, they observed that neither mRNA nor antisense RNA injections had an effect on protein production, but double-stranded RNA successfully silenced the targeted gene. Fire and Mello's discovery was particularly notable because it represented the first identification of the causative agent of a previously inexplicable phenomenon. Fire and Mello were awarded the Nobel Prize in Physiology or Medicine in 2006 for their work.

MicroRNAs are the most thoroughly characterized. These single-stranded RNAs are typically 19 to 25 nucleotides in length and are thought to regulate gene expression post-transcriptionally by binding to the 3′ untranslated regions (UTRs) of target mRNAs, inhibiting their translation[17]. Recent experimental evidence suggests that the number of unique miRNAs in humans could exceed 800 [18], though several groups have hypothesized that there may be up to 20,000[19] [20] noncoding RNAs that contribute to eukaryotic complexity.

RNA polymerase II transcribes miRNA genes, generating long primary transcripts (pri-miRNAs) that are processed by the RNase III–type enzyme Drosha, yielding hairpin structures (pre-miRNAs). Pre-miRNA hairpins are exported to the cytoplasm where they are further processed into unstable miRNA duplexes by the RNase III protein Dicer. The less stable of the two strands in the duplex is incorporated into a multiple-protein nuclease complex, the RNA-induced silencing complex (RISC), which regulates protein expression. In mammalian cells, these RISCs, guided by the miRNA, interact with the 3′ UTR of target mRNAs at regions exhibiting imperfect sequence homology, inhibiting protein synthesis by a mechanism that has yet to be fully elucidated.

Although hundreds of miRNAs have been discovered in a variety of organisms, little is known about their cellular function. Several unique physical attributes of miRNAs, including their small size, lack of polyadenylated tails, and tendency to bind their mRNA targets with imperfect sequence homology, have made them elusive and challenging to study.

Endogenously expressed miRNAs, including both intronic and intergenic miRNAs, are most important in translational repression and in the regulation of development, especially the timing of morphogenesis and the maintenance of undifferentiated or incompletely differentiated cell types such as stem cells[21]. The role of endogenously expressed miRNA in downregulating gene expression was first described in *C. elegans* in 1993 [25]. In plants this function was discovered when the "JAW microRNA" of *Arabidopsis* was shown to be involved in the regulation of several genes that control plant shape[22]. In plants, the majority of genes regulated by miRNAs are transcription factors [23]; thus miRNA activity is particularly wide-ranging and regulated entire gene networks during development by modulating the expression of key regulatory genes, including transcription factors as well as F-box_proteins[24]. In many organisms, including humans, miRNAs disruption have also been linked to the formation of tumors and dysregulation of the cell cycle. Here, miRNAs

can function as both oncogenes and tumor suppressors[25]. Another example, miRNAs are aberrantly expressed in: liver, pancreatic, oesophageal, stomach, colon, haematopoietic, ovarian, breast, pituitary, prostate, thyroid, testicular and brain cancers[26] [27] [28] [29] [30]; central nervous system disorders (e.g. schizophrenia and Alzheimer's disease) [31]; and cardiovascular disease[32] [33][36,37].
It is becoming clear that a comprehensive understanding of human biology must include both small and large non-coding RNAs, and that it is perhaps only through inclusion of these elements in the biomedical research agenda, including studies to determine the mechanistic basis of the causative variations identified by genome-wide association studies, that complex human diseases will be completely deciphered.

2. Computational methods

The discovery that microRNAs are synthesized as hairpin-containing precursors with many shared features has stimulated the development of several computational approaches to the discovery of new microRNA genes in various animal species. Many of these approaches rely heavily on conservation of sequence within and between species, while others emphasize machine learning methods to screen hairpin candidates for structural features shared by known microRNA precursors. The identification of animal microRNA targets is a particularly difficult problem because an exact match to the target sequence is not required. We discuss the most recently devised algorithms for microRNA and target discovery.

2.1 Machine learning approaches to miRNA discovery

Methods derived from the machine learning field have recently been applied to miRNA discovery with good success. Machine mearning depends on the development of algorithms and methods that allow a specific computer program to *learn* from data already collected on verified miRNAs. These algorithms require a training set for the learning process that consists of positive examples (that define the miRNA characteristics) and negative examples (the control set of non-miRNA sequences). The known microRNAs used as positive examples can be downloaded from the database miRBase [34, 35] and random sequences can be one choice of negative set. One of the most important tasks associated with the learning process is the identification of characteristics and the definition of the rules that define the positive class. This is especially important in this case as these characteristics are not always explicitly defined, Readers who wish to pursue machine learning in greater detail may consult a recent review [36].
Examples of supervised machine learning algorithms include, naïve Bayes, support vector machines (SVM), hidden Markov models (HMM), neural networks and the k-nearest neighbor algorithm. Naïve Bayes is a classification model obtained by applying a relatively simple method to a training dataset [37] A Naïve Bayes classifier calculates the probability that a given instance (example) belongs to a certain class. Support Vector Machines (SVMs) are widely used machine learning algorithms developed by Vapnik [38]. In this technique, the numbers describing each feature of a microRNA are combined into a single vector in an n-dimensional space. The algorithm compares the vectors from the positive class with those from the negative class, and finds a "hyperplane" which produces the best separation (margin) between the two classes. The "support vectors" are the samples from the two classes which are closest together but still separable--they "support" the separating

hyperplane, (See Figure 1). The performance of this algorithm, as compared to other algorithms, has proven to be particularly useful for the analysis of various classification problems, particularly when the two classes are closely related or non-uniform, and has recently been widely used in the bioinformatics field [39, 40].

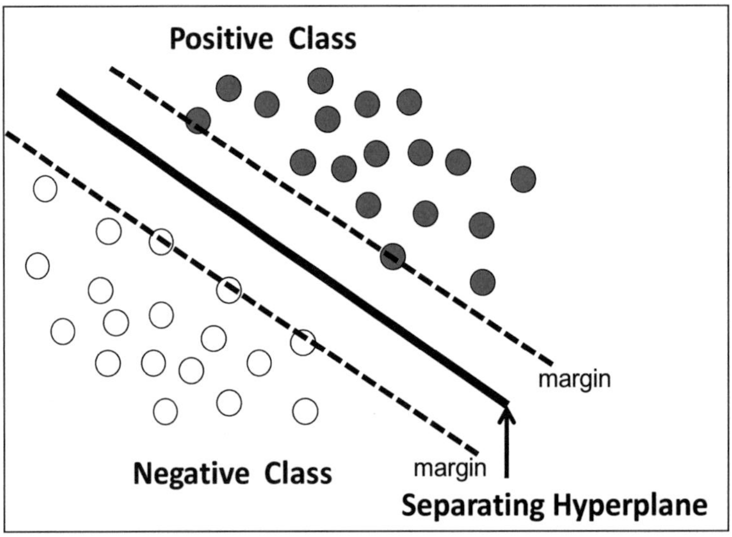

Fig. 1. The solid line is the Separating Hyperplane and the dashed lines are the margins for a SVM trained with samples from two classes. Samples (point) on the margin are called the support vectors

2.2 MicroRNA discovery tools

Numerous computational approaches (in addition to machine learning) have been implemented for miRNA gene prediction using methods based on sequence conservation and/or structural similarity[41]; [42],[43] ; [44]; [45]. Some of these tools are listed in Table 1. Lim and others [41] developed a program for identification of miRNAs, called MiRscan, with 70% specificity at a sensitivity of 50%. MiRscan uses seven miRNA features with associated weights to build a computational tool, which assigns scores to hairpin candidates. The weights are estimated using statistics based on the previously known miRNAs from *C.elegans*. Grad, et al., (2003), developed a computational method using sequence conservation and structural similarity to predict miRNAs in the *C.elegans* genome. Lai, et al., (2003) used similar ideas to develop a different computational tool for the Drosophila genome, called miRseeker. These efforts were previously reviewed by Bartel [46]. Others have used homology searches for revealing paralog and ortholog miRNAs ([42]; [47]; [48]; [49]; [50]). Additionally, Wang and others[51] developed a method based on sequence and structure alignment for miRNA identification.

ProMiR [52] is based on machine learning for miRNA discovery. ProMiR uses a highly specific probabilistic model (HMM) whose topology and states are handcrafted based on prior knowledge and assumptions, and whose exact probabilities are derived from the accumulated data. Pfeffer, et al., (2005) used support vector machines (SVMs) for predicting conserved miRNAs in herpesviruses. The features that defined the positive class were extracted from the sequence and structure features in the stem loop to form the positive class. The negative class was generated from mRNAs, rRNAs, or tRNAs from human and viral genomes which should not include any miRNA sequences. The same approach was also applied to analysis of clustered miRNAs [53] using a tool named mir-abela, while Xue, et al.,(2005) developed a SVM classifier as a 2-class tool that does not rely on comparative genomic approaches. They defined a negative class called pseudo pre-miRNAs. The criteria for this negative class included a minimum of 18 paired bases, a maximum of -15 kcal/mol folding free energy and no multiple loops. The tool is called triplet-SVM. BayesMiRNAfind [54] is a machine learning approach based on the Naïve Bayes classifier for predicting miRNA genes. This method differs from previous efforts in two ways: 1) they generate the model automatically and identify rules based on the miRNA gene structure and sequence, allowing prediction of non-conserved miRNAs and 2) they use a comparative analysis over multiple species to reduce the false positive rate. This allows for a trade-off between sensitivity and specificity. The resulting algorithm demonstrates higher specificity and similar sensitivity to algorithms that use conserved genomic regions to reduce false positives [41, 43-45]. Grundhoff, et al.,(2006) have developed an approach to identify miRNAs that is based on bioinformatics and array-based technologies. The bioinformatics tool, VMir [55], does not rely on evolutionary sequence conservation. RNAmicro [56] is another miRNA prediction tool developed by Hertel and Stadler that relies mainly on comparative sequence analysis rather then structural features using two-class SVM.

Sheng, et al.,(2007) describe a computational method, mirCoS [57], that applies three support vector machine models, based on sequence, secondary structure, and conservation, sequentially to discover new conserved miRNA candidates in mammalian genomes.

Defining the negative class is a major challenge in developing machine learning algorithms for miRNA discovery. Two machine learning approaches have recently appeared for identifying microRNAs without the necessity of defining a negative class. Yousef, et al., (2008) presented a study using one-class machine learning for microRNA using only positive data to build the classifier (One-ClassMirnaFind [58]). Several different classifiers, including two classes SVM were used to compare the one-class approach to the corresponding two-class methods. Although the two-class procedure was generally found to be superior, it was more complex to implement.

Xu, et al., (2008) recently developed a tool called miRank. MiRank [59] is a novel ranking algorithm based on a random walk through a graph consisting of known miRNA examples and unknown candidate sequences. Each miRNA is a vertex connected to its neighbor by an edge which is weighted by its similarity of the miRNA features. The score or *relevance* of a vertex increases with the number of its connections. The vertices are then ranked by relevance score, and an arbitrary cutoff of the ranked list includes both the positive examples and the most similar of the predicted unknowns. The strength of miRank is its ability to identify novel miRNAs in newly sequenced genomes where there are few annotated miRNAs (positive examples). The authors found miRank to be superior to SVM classifiers, and attribute its success to the fact that it structures the list and ranks the candidate examples as well as the query sequences during the training and classification steps.

We should note in passing that high-throughput methods for sequencing isolated small RNAs provide a new tool for discovering new microRNA species [60] and a new method for amplifying low-concentration microRNAs allows easier testing of predictions [61] .

Algorithm	Web link	References
MiRseeker		Lai et al., 2003
MiRscan	http://genes.mit.edu/mirscan/	Lim et al., 2003a,b
miRank	*MiRank is programmed in Matlab*	Xu, et al.,2008
ProMiR II	http://cbit.snu.ac.kr/~ProMiR2/	Nam et al., 2005
PalGrade		Bentwich et al., 2005
mir-abela	http://www.mirz.unibas.ch/cgi/pred_miRNA_genes.cgi	Sewer et al., 2005
triplet-SVM	http://bioinfo.au.tsinghua.edu.cn/mirnasvm/	Xue, et al., 2005
Vmir	*http://www.hpi-hamburg.de/fileadmin/downloads/VMir.zip*	Grundhoff et al., 2006
RNA micro	http://www.bioinf.uni-leipzig.de/~jana/software/index.html	Hertel and Stadler 2006
mirCoS	Based on LIBSVM library package [62]	Sheng et al., 2007
BayesMiRNAfind	https://bioinfo.wistar.upenn.edu/miRNA/miRNA/login.php	Yousef et al., 2006,
One-ClassMirnaFind	http://wotan.wistar.upenn.edu/OneClassmiRNA/	Yousef et al., 2008

Table 1. Summary information about computational tools for miRNA predictions.

3. Target identification

Although recent findings [63] suggest MicroRNAs may affect gene expression by binding to either 5′ or 3′ untranslated regions of messenger RNA, most studies have found that microRNA mark their target mRNAs for degradation or suppress their translation by binding to the 3′-untranslated region (3′UTR) and most target programs search there . These studies have suggested that the microRNA seed segment which includes 6-8 nucleotides at the 5′ end of the mature miRNA sequence is very important in the selection of the target site (see Figure 2). Thus, most of the computational tools developed to identify mRNA target sequences depend heavily on complementarity between the miRNA seed sequence and the target sequence. Diana-microT [64] was one of the first computational tools for target prediction that identified specific interaction rules based on bioinformatics and experimental approaches. The tool successfully recovered all validated *C. elegans* miRNA targets

Several additional methods for the prediction of miRNA targets have been subsequently developed. These methods mainly use sequence complementarities, thermodynamic stability calculations, and evolutionary conservation among species to determine the likelihood of a productive miRNA:mRNA duplex formation [46, 65]. John et al., (2004) developed the miRanda [66] algorithm for miRNA target prediction. MiRanda uses dynamic programming to search for optimal sequence complementarities between a set of mature microRNAs and a given mRNA MicroRNA.org (http://www.microrna.org) [67] is a comprehensive resource of microRNA target predictions and miRNA expression profiles. Target predictions are based on the miRanda algorithm while miRNA expression profiles are derived from a comprehensive sequencing project of a large set of mammalian tissues and cell lines of normal and disease origin. Another algorithm RNAhybrid [68] [69] is similar to a RNA secondary structure prediction algorithm like the Mfold program [70] but it determines the most favorable hybridization site between two sequences.

Bennecke and others [71] have recently suggested that the 3′ out-seed segment of the miRNA:mRNA duplex can compensate for imperfect base pairing of the target with the seed segment and a recent computational approach [72] has considered the contributions of both seed and the out-seed miRNA segments in target identification. Using sequence conservation reduces false positive predictions but as a result some less-conserved target-sites may be missed. This presents a dilemma, which is how to avoid rejection of these less highly conserved target sites while still reducing the very large numbers of predictions that are found when seed region conservation in the target is not required. In order to reduce the false positive predictions inherent in methods that heavily weight specific target sequence conservation, Lewis, et al.,(2005) developed TargetScanS [73]. TargetScanS scores target sites based on the conservation of the target sequences between five genomes (human, mouse, rat, dog and chicken) as evolutionarily conserved target sequences are more likely to be true targets. In testing, TargetScanS was able to recover targets for all 5300 human genes known at the time to be targeted by miRNAs.

PicTar [74] is a computational method to detect common miRNA targets in vertebrates, *C. elegans*, and *Drosophila*. PicTar is based on a statistical method applied to eight vertebrate genome-wide alignments (multiple alignments of orthologous nucleotide sequences (3' UTRs)). PicTar was able to recover validated miRNA targets at an estimated 30% false-positive rate. In a separate study PicTar was applied to target identification in *Drosophila melanogaster* [75] . These studies suggest that one miRNA can target 54 genes on average and that known microRNAs are projected to regulate a large fraction of all *D. melanogaster* genes (15%). This is likely to be a conservative estimate due to the incomplete input data.

TargetBoost [76] is a machine learning algorithm for miRNA target prediction using only sequence information to create weighted sequence motifs that capture the binding characteristics between microRNAs and their targets. The authors suggest that TargetBoost is stable and identifies more of the already verified true targets than do other existing algorithms.

Sung-Kyu, et al., (2005), also reported the development of a machine learning algorithm using SVM. The best reported results [77] were 0.921 sensitivity and 0.833 specificity. More recent Yan and others, used a machine learning approach that employs features extracted from both the seed and out-seed segments [72]. The best result obtained was an accuracy of 82.95% but it was generated using only 48 positive human and 16 negative examples, a relatively small training set to assess the algorithm.

In 2006, Thadani and Tammi [78] launched MicroTar, a novel statistical computational tool for prediction of miRNA targets from RNA duplexes which does not use sequence homology for prediction. MicroTar mainly relies on a quite novel approach to estimate the duplex energy. However, the reported sensitivity (60%) is significantly lower than that achieved using other published algorithms. At the same time, a microRNA pattern discovery method, RNA22 [79] was proposed to scan UTR sequences for targets . RNA22 does not rely upon cross-species conservation but was able to recover most of the known target sites with validation of some of its new predictions.

More recently, Yousef, et al.,(2007) described a target prediction method, (NBmiRTar [80]) using instead machine learning by a Naïve Bayes classifier. NBmiRTar does not require sequence conservation but generates a model from sequence and miRNA:mRNA duplex information derived from validated target sequences and artificially generated negative examples. In this case, both the seed and "out-seed" segments of the miRNA:mRNA duplex are used for target identification. NBmiRTar technique produces fewer false positive predictions and fewer target candidates to be tested than miRanda [66]. It exhibits higher sensitivity and specificity than algorithms that rely only on conserved genomic regions to decrease false positive predictions.

Algorithm	Web link	References
TargetScanS	http://genes.mit.edu/targetscan	Lewis, et al., 2005
miRanda	http://www.microma.org	John, et al., 2004
PicTar	http://pictar.bio.nyu.edu	Krek, et al. 2005
RNAhybrid	http://bibiserv.techfak.uni-bielefeld.de/rnahybrid	Rehmsmeier, et al., 2004
Diana-microT	http://www.diana.pcbi.upenn.edu/cgi-bin/micro_t.cgi	Kiriakidou, et al. 2004
Target Boost	https://demo1.interagon.com/demo	SaeTrom, et al. 2005
Rna22	http://cbcsrv.watson.ibm.com/rna22_targets.html	Miranda, et al. 2006
MicroTar	http://tiger.dbs.nus.edu.sg/microtar/	Thadani and Tammi 2006
NBmiRTar	http://wotan.wistar.upenn.edu/NBmiRTar	Yousef, et al. 2007
miRecords	http://mirecords.umn.edu/miRecords/	Xiao, et al., 2009

Table 2. MicroRNA Target prediction tools

In a 2004 review Lai [65] noted that there is almost no overlap among the predicted targets identified by the various methods and suggested that each tool captures a subset of the entire target class as a function of the specific features they have incorporated into their prediction models. More recently, Sethupathy, et al., (2006) conducted a comparison of the 5 most used tools for mammalian target prediction. This study indicated that 30% of the experimentally validated target sites are nonconserved, supporting the need for the development of different or complementary computational approaches to capture new target sites. Furthermore, the large number of predictions that each of these tools is producing suggests that the heavy reliance on homology or comparative sequence analysis is not sufficient to generate accurate predictions with a high sensitivity and there are yet to be identified recognition parameters that must be considered.

4. Databases for microRNA and targets

There is a variety of very useful databases that provide a significant amount of information on miRNA and Target predictions,(Table 3). The most extensive database for both miRNA and target sequences is miRBase[34]. MiRBase contains both miRNA mature sequences, hairpin sequences of precursors and associated annotation. Release 12.0 of the database contains 8619 entries representing hairpin precursor miRNAs, expressing 8273 mature miRNA products, in primates, rodents, birds, fish, worms, flies, plants and viruses. MiRBase also contains predicted miRNA target genes in miRBase Targets, and provides a gene naming and nomenclature function in the miRBase Registry. The miRNA target genes are predicted by the miRanda tool [66] and not necessarily experimentally validated.

TarBase [81] contains a set of experimentally supported targets in different species that are collected manually from the literature. TarBase version 5 has more than 1300 experimentally supported miRNA target interactions. The database has information about the target site described by the duplex of miRNA and gene. It also includes information on the experiments that were conducted to test the target, the sufficiency of the site to induce translational repression and/or cleavage, and a reference to the paper used to extract the information.

Argonaute [82] is a compilation of comprehensive information on mammalian miRNAs, their origin and regulated target genes in an exhaustively curated database. The source information of Argonaute is from both literature and other databases.

The most recently released database, miRecords [83], is an integrated resource for animal miRNA-target interactions. miRecords stores predicted miRNA targets produced by 11 established miRNA target prediction programs.

DataBase	Web Link
MiRBase	http://microrna.sanger.ac.uk/
TarBase	http://diana.cslab.ece.ntua.gr/tarbase/
Argonaute	http://www.ma.uni-heidelberg.de/apps/zmf/argonaute/
miRecords	http://mirecords.umn.edu/miRecords/

Table 3. Databases for microRNA and Targets

5. References

[1] Voinnet, O. and D.C. Baulcombe, *Systemic signalling in gene silencing.* Nature, 1997. 389(6651): p. 553-553.

[2] Palauqui, J.-C., et al., *Systemic acquired silencing: transgene-specific post-transcriptional silencing is transmitted by grafting from silenced stocks to non-silenced scions.* EMBO J, 1997. 16(15): p. 4738-4745.

[3] Napoli, C., C. Lemieux, and R. Jorgensen, *Introduction of a Chimeric Chalcone Synthase Gene into Petunia Results in Reversible Co-Suppression of Homologous Genes in trans.* The Plant Cell Online, 1990. 2(4): p. 279-289.

[4] Romano, N. and G. Macino, *Quelling: transient inactivation of gene expression in Neurospora crassa by transformation with homologous sequences.* Molecular Microbiology, 1992. 6(22): p. 3343-3353.

[5] Van Blokland, R., et al., *Transgene-mediated suppression of chalcone synthase expression in Petunia hybrida results from an increase in RNA turnover.* The Plant Journal, 1994. 6(6): p. 861-877.

[6] Covey, S.N., et al., *Plants combat infection by gene silencing.* Nature, 1997. 385(6619): p. 781-782.

[7] Ratcliff, F., B.D. Harrison, and D.C. Baulcombe, *A Similarity Between Viral Defense and Gene Silencing in Plants.* Science, 1997. 276(5318): p. 1558-1560.

[8] Guo, S. and K.J. Kemphues, *par-1, a gene required for establishing polarity in C. elegans embryos, encodes a putative Ser/Thr kinase that is asymmetrically distributed.* Cell, 1995. 81(4): p. 611-620.

[9] Fire, A., et al., *Potent and specific genetic interference by double-stranded RNA in Caenorhabditis elegans.* Nature, 1998. 391(6669): p. 806-811.

[10] Timmons, L. and A. Fire, *Specific interference by ingested dsRNA.* Nature, 1998. 395(6705): p. 854-854.

[11] Schwach, F., et al., *An RNA-Dependent RNA Polymerase Prevents Meristem Invasion by Potato Virus X and Is Required for the Activity But Not the Production of a Systemic Silencing Signal.* Plant Physiology.

[12] Pant, B.D., et al., *MicroRNA399 is a long-distance signal for the regulation of plant phosphate homeostasis.* The Plant Journal, 2008. 53(5): p. 731-738.

[13] Buhtz, A., et al., *Identification and characterization of small RNAs from the phloem of Brassica napus.* The Plant Journal, 2008. 53(5): p. 739-749.

[14] Chitwood, D.H., et al., *Pattern formation via small RNA mobility.* Genes & Development, 2009. 23(5): p. 549-554.

[15] Ecker, J.R. and R.W. Davis, *Inhibition of gene expression in plant cells by expression of antisense RNA.* Proceedings of the National Academy of Sciences, 1986. 83(15): p. 5372-5376.

[16] Pal-Bhadra, M., U. Bhadra, and J.A. Birchler, *Cosuppression in Drosophila: Gene Silencing of Alcohol dehydrogenase by white-Adh Transgenes Is Polycomb Dependent.* Cell, 1997. 90(3): p. 479-490.

[17] Ambros, V., *The functions of animal microRNAs.* Nature, 2004. 431(7006): p. 350-355.

[18] Bentwich, I., et al., *Identification of hundreds of conserved and nonconserved human microRNAs.* Nat Genet, 2005. 37(7): p. 766-770.

[19] Okazaki, Y.e.a., *Analysis of the mouse transcriptome based on functional annotation of 60,770 full-length cDNAs.* Nature, 2002. 420(6915): p. 563-573.

[20] Imanishi, T., et al., *Integrative Annotation of 21,037 Human Genes Validated by Full-Length cDNA Clones.* PLoS Biol, 2004. 2(6): p. e162.

[21] Carrington, J.C. and V. Ambros, *Role of MicroRNAs in Plant and Animal Development.* Science, 2003. 301(5631): p. 336-338.

[22] Lee, R.C., R.L. Feinbaum, and V. Ambros, *The C. elegans heterochronic gene lin-4 encodes small RNAs with antisense complementarity to lin-14.* Cell, 1993. 75(5): p. 843-854.

[23] Zhang, B., et al., *Plant microRNA: A small regulatory molecule with big impact.* Developmental Biology, 2006. 289(1): p. 3-16.

[24] Jones-Rhoades, M.W., D.P. Bartel, and B. Bartel, *MicroRNAs AND THEIR REGULATORY ROLES IN PLANTS.* Annual Review of Plant Biology, 2006. 57(1): p. 19-53.

[25] Zhang, B., et al., *microRNAs as oncogenes and tumor suppressors.* Developmental Biology, 2007. 302(1): p. 1-12.

[26] Visone, R. and C.M. Croce, *MiRNAs and Cancer.* Am J Pathol, 2009. 174(4): p. 1131-1138.

[27] Pang, J., et al., *Oncogenic role of microRNAs in brain tumors.* Acta Neuropathologica, 2009. 117(6): p. 599-611.

[28] Voorhoeve, P.M., et al., *A Genetic Screen Implicates miRNA-372 and miRNA-373 As Oncogenes in Testicular Germ Cell Tumors.* Cell, 2006. 124(6): p. 1169-1181.

[29] Khoshnaw, S.M., et al., *MicroRNA involvement in the pathogenesis and management of breast cancer.* Journal of Clinical Pathology, 2009. 62(5): p. 422-428.

[30] Novakova, J., et al., *MicroRNA involvement in glioblastoma pathogenesis.* Biochemical and Biophysical Research Communications, 2009. 386(1): p. 1-5.

[31] Kocerha, J., S. Kauppinen, and C. Wahlestedt, *microRNAs in CNS Disorders.* NeuroMolecular Medicine, 2009. 11(3): p. 162-172.

[32] Barringhaus, K.G. and P.D. Zamore, *MicroRNAs: Regulating a Change of Heart.* Circulation, 2009. 119(16): p. 2217-2224.

[33] Sen, C.K., et al., *Micromanaging Vascular Biology: Tiny MicroRNAs Play Big Band.* Journal of Vascular Research, 2009. 46(6): p. 527-540.

[34] Griffiths-Jones, S., et al., *miRBase: tools for microRNA genomics.* Nucl. Acids Res., 2008. 36(suppl_1): p. D154-158.

[35] Griffiths-Jones, S., *The microRNA Registry.* Nucleic Acids Res, 2004. 32(90001): p. D109-111.

[36] Larranaga, P., et al., *Machine learning in bioinformatics.* Brief Bioinform, 2006. 7(1): p. 86-112.

[37] Mitchell, T., *Machine Learning*1997: McGraw Hill.

[38] Vapnik, V., *The Nature of Statistical Learning Theory*1995: Springer.

[39] Haussler, D., *Convolution kernels on discrete structures*, 1999, Baskin School of Engineering, University of California: Santa Cruz. p. Technical Report UCSCCRL - 99-10.

[40] Pavlidis, P., et al. *Gene functional classification from heterogeneous data* in *Proceedings of the fifth annual international conference on Computational biology* 2001 Montreal, Quebec, Canada ACM Press.

[41] Lim, L.P., et al., *Vertebrate MicroRNA Genes.* Science, 2003. 299(5612): p. 1540.

[42] Weber, M.J., *New human and mouse microRNA genes found by homology search.* FEBS Journal, 2005. 272(1): p. 59-73.

[43] Lim, L.P., et al., *The microRNAs of Caenorhabditis elegans.* Genes Dev., 2003. 17(8): p. 991-1008.

[44] Lai, E., et al., *Computational identification of Drosophila microRNA genes.* Genome Biology, 2003. 4(7): p. R42.

[45] Grad, Y., et al., *Computational and Experimental Identification of C. elegans microRNAs.* Molecular Cell, 2003. 11(5): p. 1253-1263.

[46] Bartel, D.P., *MicroRNAs: Genomics, Biogenesis, Mechanism, and Function.* Cell, 2004. 116(2): p. 281-297.

[47] Lagos-Quintana, M., et al., *Identification of Novel Genes Coding for Small Expressed RNAs.* Science, 2001. 294(5543): p. 853-858.

[48] Lau, N.C., et al., *An Abundant Class of Tiny RNAs with Probable Regulatory Roles in Caenorhabditis elegans.* Science, 2001. 294(5543): p. 858-862.

[49] Lee, R.C. and V. Ambros, *An Extensive Class of Small RNAs in Caenorhabditis elegans.* Science, 2001. 294(5543): p. 862-864.

[50] Pasquinelli, A.E., et al., *Conservation of the sequence and temporal expression of let-7 heterochronic regulatory RNA.* Nature, 2000. 408(6808): p. 86.

[51] Wang, X., et al., *MicroRNA identification based on sequence and structure alignment.* Bioinformatics, 2005. 21(18): p. 3610-3614.

[52] Nam, J.-W., et al., *Human microRNA prediction through a probabilistic co-learning model of sequence and structure.* Nucleic Acids Res, 2005. 33(11): p. 3570-3581.

[53] Sewer, A., et al., *Identification of clustered microRNAs using an ab initio prediction method.* BMC Bioinformatics, 2005. 6(1): p. 267.

[54] Yousef, M., et al., *Combining multi-species genomic data for microRNA identification using a Naive Bayes classifier.* Bioinformatics, 2006. 22(11): p. 1325-1334.

[55] Grundhoff, A., C.S. Sullivan, and D. Ganem, *A combined computational and microarray-based approach identifies novel microRNAs encoded by human gamma-herpesviruses.* RNA, 2006. 12(5): p. 733-750.

[56] Hertel, J. and P.F. Stadler, *Hairpins in a Haystack: recognizing microRNA precursors in comparative genomics data.* Bioinformatics, 2006. 22(14): p. e197-202.

[57] Sheng, Y., P. Engstrom, G., and B. Lenhard, *Mammalian MicroRNA Prediction through a Support Vector Machine Model of Sequence and Structure.* PLoS ONE, 2007. 2(9): p. e946.

[58] Yousef, M., et al., *Learning from positive examples when the negative class is undetermined-microRNA gene identification*, 2008. p. 2.

[59] Xue, C., et al., *Classification of real and pseudo microRNA precursors using local structure-sequence features and support vector machine.* BMC Bioinformatics, 2005. 6(1): p. 310.

[60] Glazov, E.A., et al., *A microRNA catalog of the developing chicken embryo identified by a deep sequencing approach*, 2008. p. gr.074740.107.

[61] Berezikov, E., et al., *Many novel mammalian microRNA candidates identified by extensive cloning and RAKE analysis*, 2006. p. 1289-1298.

[62] Chang, C.-C. and C.-J. Lin, *LIBSVM: a library for support vector machines*, 2001.
[63] Lytle, J.R., T.A. Yario, and J.A. Steitz, *Target mRNAs are repressed as efficiently by microRNA-binding sites in the 5²€ UTR as in the 3²€ UTR*. Proceedings of the National Academy of Sciences, 2007. 104(23): p. 9667-9672.
[64] Kiriakidou, M., et al., *A combined computational-experimental approach predicts human microRNA targets*. Genes & Development, 2004. 18(10): p. 1165-1178.
[65] Lai, E., *Predicting and validating microRNA targets*. Genome Biology, 2004. 5(9): p. 115.
[66] John, B., et al., *Human MicroRNA Targets*. PLoS Biology, 2004. 2(11): p. e363.
[67] Betel, D., et al., *The microRNA.org resource: targets and expression*. Nucl. Acids Res., 2008. 36(suppl_1): p. D149-153.
[68] Rehmsmeier, M., et al., *Fast and effective prediction of microRNA/target duplexes*. RNA, 2004. 10(10): p. 1507-1517.
[69] Kruger, J. and M. Rehmsmeier, *RNAhybrid: microRNA target prediction easy, fast and flexible*. Nucl. Acids Res., 2006. 34(suppl_2): p. W451-454.
[70] Zuker, M., *Mfold web server for nucleic acid folding and hybridization prediction*. Nucleic Acids Res, 2003. 31 (13): p. 3406-3415.
[71] Brennecke, J., et al., *Principles of microRNA-target recognition*. PLoS Biol., 2005. 3: p. e85.
[72] Yan, X., et al., *Improving the prediction of human microRNA target genes by using ensemble algorithm*. FEBS Letters, 2007. 581(8): p. 1587.
[73] Lewis, B.P., et al., *Prediction of mammalian microRNA targets*. Cell, 2003. 115: p. 787.
[74] Krek, A., et al., *Combinatorial microRNA target predictions*. Nat Genet, 2005. 37(5): p. 495-500.
[75] Grun, D., et al., *microRNA Target Predictions across Seven Drosophila Species and Comparison to Mammalian Targets*. PLoS Computational Biology, 2005. 1(1): p. e13.
[76] SaeTrom, O.L.A., O.J. Snove, and P.A.L. SaeTrom, *Weighted sequence motifs as an improved seeding step in microRNA target prediction algorithms*. RNA, 2005. 11(7): p. 995-1003.
[77] Sung-Kyu, K., et al. *A Kernel Method for MicroRNA Target Prediction Using Sensible Data and Position-Based Features*. in *Computational Intelligence in Bioinformatics and Computational Biology*. 2005. Proceedings of the 2005 IEEE Symposium
[78] Thadani, R. and M. Tammi, *MicroTar: predicting microRNA targets from RNA duplexes*. BMC Bioinformatics, 2006. 7(Suppl 5): p. S20.
[79] Miranda, K.C., et al., *A Pattern-Based Method for the Identification of MicroRNA Binding Sites and Their Corresponding Heteroduplexes*. 2006. 126(6): p. 1203-1217.
[80] Yousef, M., et al., *Naive Bayes for microRNA target predictions machine learning for microRNA targets*, 2007. p. 2987-2992.
[81] Sethupathy, P., B. Corda, and A.G. Hatzigeorgiou, *TarBase: A comprehensive database of experimentally supported animal microRNA targets*. RNA, 2006. 12(2): p. 192-197.
[82] Shahi, P., et al., *Argonaute--a database for gene regulation by mammalian microRNAs*. Nucl. Acids Res., 2006. 34(suppl_1): p. D115-118.
[83] Xiao, F., et al., *miRecords: an integrated resource for microRNA-target interactions*. Nucl. Acids Res., 2009. 37(suppl_1): p. D105-110.

12

Motif Discovery with Compact Approaches - Design and Applications

Cinzia Pizzi
University of Padova
Italy

1. Introduction

In the post-genomic era, the ability to predict the behavior, the function, or the structure of biological entities (such as genes and proteins), as well as interactions among them, plays a fundamental role in the discovery of information to help biologists to explain biological mechanisms.

In this context, appropriate characterization of the structures under analysis, and the exploitation of combinatorial properties of sequences, are crucial steps towards the development of efficient algorithms and data structures to be able to perform the analysis of biological sequences.

Several functional and structural properties, and also evolutionary mechanisms, can be predicted either by the comparison of new elements with already classified elements, or by the comparison elements with a similar structure of function to infer the common mechanism that is at the basis of the observed similar behavior. Such elements are commonly called *motifs*. Comparison-based methods for sequence analysis find their application in several biological contexts, such as extraction of transcription factor binding sites, identification of structural and functional similarities in proteins, and phylogeny reconstruction. Therefore, the development of adequate methodologies for motif discovery is of undoubt interests for several different fields in computational biology.

In motif discovery in biosequences, it is common to assume that statistically significant candidates are those that are likely to hide some biologically significant property. For this purpose all the possible candidates are ranked according to some statistics on words. Then they are presented in output for further inspection that need to be carried out by a biologist, who identifies the most promising patterns. These, in turn, are tested in laboratory to confirm their biological significance. Therefore, when designing algorithms for motif discovery, besides obviously aim at time and space efficiency, particular attention should be devoted to the output representation. In fact, even considering fixed length strings, the size of the candidate set becomes exponential if exhaustive enumeration is applied. This is already true when only exact matches are considered as candidate occurrences, and worsen when the intrinsic variability of biological sequences is taken into account.

Alternatively to methods based on exhaustive enumeration, heuristics could be used. However, heuristics cannot guarantee to find the optimal solution. Therefore some degree

of uncertainty remains whether motifs, that are statistically as significant as those reported in output, have been left out.
Computational power of nowadays computers can partially reduce the effects of exhaustive enumeration approaches, in particular for short length candidates. However, if the size of the output is too big to be analyzed by human inspection the risk is to provide biologists with very fast tools that produce mostly useless output.
A possible solution to these problems relies on compact approaches. Compact approaches are based on the partition of the search space into classes.
The final user can then be presented with an output that has the size of the partition, rather than the size of the candidate space, with obvious advantages for the human-based analysis that follows the computer-based filtering of the pattern discovery algorithms.
Compact approaches find applications both in searching and discovery frameworks. They can also be applied to several motif models: exact patterns, approximate patterns, position matrices, etc. And under both independent and identically distribution (i.i.d.) and Markov distributions.
The purpose of this chapter is to describe the basis of compact approaches, to provide the readers with the conceptual tools for applying compact approaches to the design of their algorithm for biosequence analysis. This will be achieved by overwieving examples of compact approaches that have been successfully developed for several motif models that will be illustrated with the sustain of examples and experiments to discuss their power.

2. Background

The methodologies to study the Science of Life dramatically changed during the past years. The advent of the web made it possible for the scientific community to share the massive quantity of data produced by high throughput techniques, thus accelerating the analysis of the available data and the discovery of related properties and associations. The development of high throughput technologies has as a consequence not only an increase in the amount of data, but also a diversification of the type of data available, opening new perspectives of investigations. Disciplines such as Bioinformatics and Computational Biology try to combine the efforts and competency of the communities of biologists and computer scientists in a single more powerful combination of human knowledge and efficiency, thanks to automatic approaches to data analysis.

2.1 Definition of the problem

One of the key aspects in the analysis of biological sequences is the identification of interesting patterns. "Interestingness" is a wide concept that may embrace very different definitions depending on the contexts in which the analysis is carried out. At an higher level we can define an interesting pattern as a pattern that shows an unusual behavior from what it is expected in terms of presence within the sequence under analysis.
More in details, searching for shared or over-represented patterns is motivated by a simple commonly accepted principle: if two or more sequences perform the same functions or have the same structure, then the common elements among the sequences might be somehow responsible for the observed similarity.
The problem of finding biologically significant patterns is then moved to the problem of finding statistically significant patterns.

solid word
wildcards-mismatches
insertion-deletion
generalized patterns
alignments
position weight matrices
hidden Markov models

Table 1. Various choices to model motifs in biological sequences. Starting from solid words the level of sensibility increases (allowing for variations), but the level of specificity consequently decreases, thus making more difficult the process of detection of the signal.

The problem of searching for similar regions among biological sequences faces several issues. First of all, the genetic code must be fault-tolerant to deal with errors that may occur during transcription or are due to random mutations, so that an intrinsic variability characterizes biological motifs. This variability has as a consequence the possible explosion of the size of the search space under study, due to the rich underlying combinatorics. Searching the whole pattern space then is feasible only for very short patterns. Note that this is also true for exact words, because their number increases exponentially with the length. On the other hand, heuristics are not guaranteed to find a globally optimal solution.

A critical step of the process is the choice of an appropriate structure to model the motifs. In some cases, deterministic patterns do not have enough expressive power to describe the specificity of the contributions of each symbol in any position of the site. Statistical matrices or graph-based models might offer a better framework in these cases. Several options have been considered during the past decades to model signals in biosequences, and to take into account for this intrinsic variability. Some of these models of choice are listed in Table 1, sorted in increasing order of expressive power (and consequent increase of difficulty in the design of related algorithm, and of their intrinsic complexity).

The choice of an appropriate model to describe motifs is a trade-off between the expressiveness of the model to describe particular biological properties, and the efficiency of the algorithms that can be applied when that model is chosen.

The scoring function chosen to evaluate the output also plays an important role in the identification of the searched sites. However, simple statistics are often unable to discriminate interesting motifs from motifs that are likely to occur by chance, so that different measures of statistical significance need to be considered. In summary, for a given choice of a statistical measure S of a motif m, one could ask three questions:

- What is the value of $S(m)$?
- How surprising is to measure $S(m)$ with respect to the value that was expected according to some background distribution?
- How likely is it for the recorded values to occur by chance?

These three questions can be answered by the means of different computations. The first one, for example, can be answered by exact counts or estimates. To answer the second, we need some score that measures *over-representation*, such as the *z-score*. Finally the third one is solved by resort to so called *p-value* of a statistic.

Once the model and the scoring function have been fixed, the next step is the development of efficient approaches to extract the searched information. There is a vast literature about algorithms for motif discovery. However, most of them either have been developed to solve very specific instances of the problem, such as the *Motif Challenge Problem* (Pevzner & Sze, 2000), or are based on exhaustive search of the pattern space (among which (Queen et al., 1982; Staden, 1989; Tompa, 1999; van Helden et al., 1998; Waterman et al., 1984)), or rely on heuristics (for example (Hertz & Stormo, 1999; Stormo & III, 1989)). Comparison of several techniques in fact showed how performances substantially depend on the underlying model of the motif to discover (Tompa et al., 2005).

3. Compact scoring

Despite the intrinsic possible explosion of the size of the search space, it is possible to conceive a compact representation of the patterns such that only *representative* patterns are scored, and no critical information is lost in the process.

The classes must be designed in such a way that the score used to rank the candidates has a monotone behavior within each class. This allows the identification of a representative of each class, which is the element with the highest score. Consequently, it suffices to compute, and to report in output, only the score for the representatives. In fact, we are guaranteed that for each element that has not been ranked there is another one (the representative of the class it belongs to) that is at least as significant.

In such a framework, the output size would depend upon the number of classes in which patterns can be grouped, rather than on all the existing patterns that belong to the search space. This approach can also be used as a filter to detect unusual classes of strings that need to be scrutinized further.

The compact scoring needs two important steps to be carried out with critical attention:

1. definition of the search space
2. efficient partitioning

The definition of the search space clearly depends implicitly on the motif model, as discussed above. Nevertheless there are also two working frameworks in which one can pose his search (Brazma et al., 1998).

In the pattern-driven framework the search space consists of all possible patterns (of a given size) that can be generated over a given alphabet Σ. The input sequence is then tested for occurrences of each and every motif in a family of *a priori* generated, abstract *models* (for example (Keich & Pevzner, 2002)). Although more correct in principle, this method may pose severe computational issues.

In the sequence-driven framework the search space consists of strings that actually occur at least once in the given input, or to some more or less controlled neighborhood of those substrings (for example (Lawrence et al., 1993)). This may be less firm methodologically, but leads to time and space savings.

The choice of techniques and data structures that can be used to perform the partition are strictly related to the definition of the model and search space. In turn they affect the space complexity reduction that can be achieved by the partitioning, as it will be clear in the following discussion.

4. Solid words

In this section we describe a methodology for compact representation and scoring of single solid words, and pairs of solid words. We will outline the basic concepts used in the approach. Details of formulas and proofs of theorems can be found in (Apostolico et al., 2003).

4.1 Compact indexes for single words

Let us consider the problem of extracting frequent substrings from a string x of length n. The number of substrings in x is equal to the number of possible choices of starting and ending indexes that univocally identify the substring. These are obviously $O(n^2)$. There are several observations that can be done with respect to this output size:

- for very long strings (consider genome wide analysis) the computation of all the occurrences of all these substrings might become prohibitive in terms of both time and space needed;
- the list of candidates might be too long to allow deep inspection by the final user;
- some substrings, of increasing length, occur the same number of times and in the same exact position: the output set might hide some notable redundancy.

In order to overcome these issues a compact approach might be applied. The compact representation and scoring of solid words can be achieved by exploiting the characteristics of appropriate data structures, such as the *suffix tree*, and by an in-depth analysis of the properties of monotonicity for some measure of over- or under-representation.

4.1.1 Suffix trees

A suffix tree is a data structure used to efficiently store and retrieve information about all the substrings of a given text. Given a text x of length n defined over an alphabet Σ, and a symbol $\$ \notin \Sigma$, the suffix tree associated with x is the digital search trie of all the suffixes of x. There are two versions of suffix tree: the expandend suffix tree (also called suffix trie), and the compact suffix tree. In the expanded version each arc is labelled with a symbol, except for the leaves that are labelled with the corresponding suffix. The space required to store an expanded suffix tree is $O(n^2)$ in the worst case (Aho et al., 1974). On the other hand, in the compact representation chains of nodes with one outgoing edge are collapsed together in a single arc. The symbol \$ guarantees that each node in the compact suffix tree (except the leaf nodes) is branching. This property, together with the observation that there are n leaves, corresponding to the n suffixes, implies that there are $O(n)$ nodes in the suffix tree. Each arc is labeled with two indexes, the start and the end positions of the corresponding substring in the text (or, equivalently, the start position and the length of the path from the root node to the node at which the arc ends). With such a representation, the overall space needed to store a compact suffix tree is $O(n)$. The brute force construction of a suffix tree requires $O(n^2)$ time. However, more clever algorithms allow for linear time construction of the tree (McCreight, 1976; Ukkonen, 1995; Weiner, 1973). The word spelled by a path from the root to a node α is indicated with $w(\alpha)$, and α is called the *proper locus* of $w(\alpha)$. The *locus* of a word w is the unique node β of the suffix tree such that w is a proper prefix of β, and $father(\beta)$ is a proper prefix of w. The frequency of a word w can be obtained in time proportional to the length of w and to the number of its occurrences. For this purpose we reach the locus β of

w and then visit the subtree rooted at β to count the number of its leaves. Alternatively, one could annotate the nodes of the tree in a bottom-up fashion with the count of their children. Then, to know the frequency of any query word w, only the time to reach its locus is needed. Alternatively to the suffix tree, one could use a suffix array (Manber & Myers, 1993). As it is shown in (Abouelhoda et al., 2004) these two data structures are equivalent. The advantage of the suffix array is that it usually takes less space. This comes at expenses of an increase of the difficulty in the designing of related algorithms.

4.1.2 Example

For a string $x = AGCTAGCTAAA$ of length 11, the number of substrings is $\frac{n(n+1)}{2} = 66$ (number of choices of starting and ending position for a substring). In the specific case of the string in our example, if we remove duplications there will remain 52 different substrings. By building the suffix tree (see Fig.1), and considering only the strings corresponding to nodes the number of candidates reduces to 15.

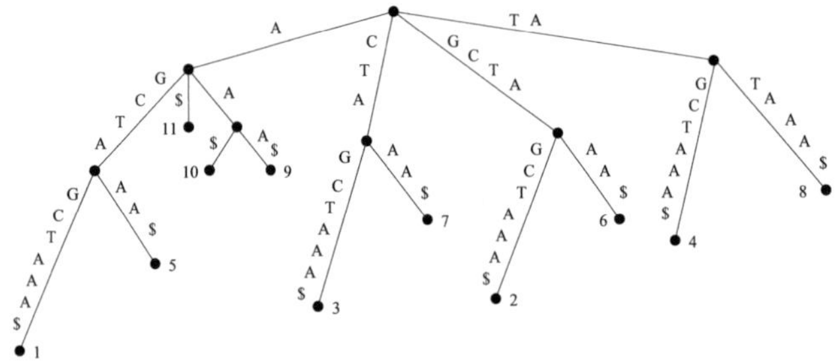

Fig. 1. The suffix tree for the string $x = AGCTAGCTAAA$. The numbering of the leaves indicates the corresponding suffix in x, i.e. their starting position.

4.1.3 Compact representation for the frequency of single words

An interesting property of the suffix tree is that all the words that end within an arc have the same frequency, equal to the frequency of the word corresponding to their locus. Thus a suffix tree partitions the $\Theta(n^2)$ subwords of a text in $O(n)$ classes, such that the words that belong to the same class have the same locus. Take, for example, the strings "C", "CT", and "CTA" in the suffix tree of Fig.1. They all occur at positions 3 and 7 in x, as it can be easily verified by eye inspection of the string. Therefore they start at the same positions in the text, and have the same frequency.

The word $w(\beta)$, corresponding to the locus β of the class C_β can be considered as *maximal in length*, since any extension of $w(\beta)$ will have a different frequency, and thus would not belong to the class C_β. These words are called *representatives*.

In Table 2 we enumerate all the classes identified by the partition induced by the suffix tree of Fig.1.

The compact output related to the frequency analysis of the substrings of x is represented by the following Table 3. It is easy to note by eye inspection that this output gives a much clearer and immediate representation of the frequency distribution of substrings of x with respect to a full enumeration of all 66 possible substrings in it.

$\{\mathbf{A}\} = [1,5,9,10,11]$
$\{AG, AGC, AGCT, \mathbf{AGCTA}\} = [1,5]$
$\{AGTCAG, AGTCAGC, AGTCAGCT, AGTCAGCTA, AGTCAGCTAA, \mathbf{AGTCAGCTAAA}\} = [1]$
$\{AGTCAA, \mathbf{AGTCAAA}\} = [5]$
$\{\mathbf{AA}\} = [9,10]$
$\{\mathbf{AAA}\} = [9]$
$\{C, CT, \mathbf{CTA}\} = [3,7]$
$\{CTAG, CTAGC, CTAGCT, CTAGCTA, CTAGCTAA, \mathbf{CTAGCTAAA}\} = [3]$
$\{CTAA, \mathbf{CTAAA}\} = [7]$
$\{G, GC, GCT, \mathbf{GCTA}\} = [2,6]$
$\{GCTAG, GCTAGC, GCTAGCT, GCTAGCTA, GCTAGCTAA, \mathbf{GCTAGCTAAA}\} = [2]$
$\{GCTAA, \mathbf{GCTAAA}\} = [6]$
$\{T, \mathbf{TA}\} = [4,8]$
$\{TAG, TAGC, TAGCT, TAGCTA, \mathbf{TAGCTAAA}\} = [4]$
$\{TAT, TATA, TATAA, \mathbf{TATAAA}\} = [8]$

Table 2. The maximal partition that is obtained with a suffix tree for the string $x = AGCTAGCTAAA$. Each row enumerates the strings of each class, and their occurring positions. The representative of each class is in bold. The set of starting positions for all the strings in the class are listed within square brackets.

A	5
AGCTA	2
CTA	2
AA	2
GCTA	2
TA	2
AGTCAGCTAAA	1
AGTCAAA	1
AAA	1
CTAGCTAAA	1
CTAAA	1
GCTAGCTAAA	1
GCTAAA	1
TAGCTAAA	1
TATAAA	1

Table 3. The compact output for the frequency count of the substrings in x.

4.1.4 Compact representation for single words statistics

In some kind of analysis the frequency count might be insufficient to extract "interesting" patterns, since some of them might occur often simply by chance. In these cases an evaluation of over- or under- representation can give a better solution to the problem of the extraction of interesting patterns. It has been shown in (Apostolico et al., 2003) that the compact approach can be extended to over-representation scores, such as z-scores. Z-scores have the characteristic to compare the counted frequency with the frequency expected assuming a given background distribution. Some examples are $\frac{f}{e}$, $f - e$, $\frac{f-e}{e}$, where f and e are the counted and expected frequency respectively.

The partition induced by the suffix tree is such that the words that belong to a class are prefixes of increasing length of the representative substring. The expected frequency of words decreases with word length. We show this with a simple example. Consider the case of i.i.d. hypothesis, with an uniform distribution, i.e. all the symbols in the alphabet have the same

probability p to occur. The probability of a word of length l is then p^l. Since $0 < p < 1$, we have that the term p^l decreases with l. The reasoning can be extended and proved for a general distribution and under both i.i.d. and Markov chain hypothesis.
Within each class two conditions hold:

1. the counted frequency is constant;
2. the expected frequency is monotonically decreasing and reach its minimum at the representative.

Hence z-scores with a constant frequency f and decreasing frequency e will have a monotonically increasing behavior within each class, and reach their maximum in correspondence of the representative of the class. Using over-representation to extract interesting patterns allows to use a threshold to filtering the results further. The threshold can be arbitrarily fixed (for example we are interested in patterns that occur at least the double of what we expect), or by fixing a p-value (that it is a probability of seeing a pattern by chance) and computing the corresponding absolute threshold for the score (Staden, 1989).
In Table 4 we can see an example of compact output of the z-score $\frac{f}{e}$, for a uniform distribution and for a general distribution. We still refer to the string x of the previous examples. It can be seen that the two background distribution produce a slightly different order in the output, and a quite different score associated to the strings. By setting a threshold $T_h = 1000$ we obtain a further reduction of the output size that in the case of the uniform distribution maintains 9 strings, while in the case of the considered general distribution it maintains 12 strings.

Substring	Score (uniform)	Substring	Score (general)
AGTCAGCTAAA	4194304	*AGTCAGCTAAA*	694444444
GCTAGCTAAA	1048574	*GCTAGCTAAA*	69444444
CTAGCTAAA	262144	*CTAGCTAAA*	41666667
TAGCTAAA	65536	*TAGCTAAA*	4166667
AGTCAAA	16384	*AGTCAAA*	833333
GCTAAA	4096	*TATAAA*	250000
TATAAA	4096	*GCTAAA*	83333
AGCTA	2048	*CTAAA*	50000
CTAAA	2048	*AGCTA*	16667
GCTA	512	*GCTA*	1667
CTA	128	*CTA*	1000
AAA	64	*AAA*	1000
AA	32	*AA*	200
TA	32	*TA*	100
A	20	*A*	50

Table 4. The compact output for the over-representation score of the substrings in x. In the first two columns the substrings are scored according to a uniform distribution. In the last two columns the substrings are scored according to the following distribution:
$p_a = 0.1, p_c = 0.1, p_g = 0.6, p_t = 0.2$

4.2 Compact indexes for co-occurrences

The suffix tree property of partitioning the set of substrings of a string in $O(n)$ classes can be exploited also for the computation of co-occurrences between substrings of an input string.
In (Apostolico et al., 2004) the compact approach was extended to the efficient computation of a table to hold the number of co-occurrences of substrings within a text. In this context

the problem is: given two words y and z, and a distance d, compute the number of times that z follows y at a distance at most d. In (Apostolico et al., 2004) further restrictions were set so that z must follow an occurrence of y but it must occur before the next occurrences of y. Moreover, if many occurrences of z are found at the right side of the same occurrence of y, only the closest one is counted. In Fig.2 it can be seen how the co-occurrence count can vary. A simple count, without restriction would give a value of 4: $\{(p_i, q_k), (p_i, q_{k+1}), (p_i, q_{k+2}), (p_{i+1}, q_{k+2})\}$. If no interleaving occurrences of y are allowed the count would reduce to 3: $\{(p_i, q_k), (p_i, q_{k+1}), (p_{i+1}, q_{k+2})\}$. If also no interleaving occurrences of z are allowed the count would drop to 2: $\{(p_i, q_k), (p_{i+1}, q_{k+2})\}$.

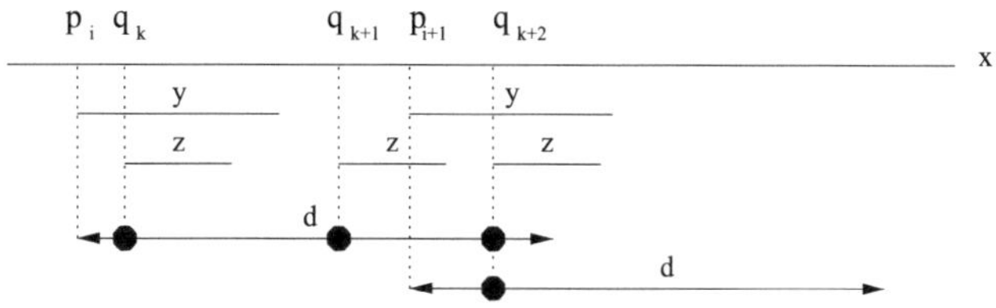

Fig. 2. Illustration of co-occurrences between two substrings y and z.

A naive approach would count the number of co-occurrences between all $O(n^2)$ substrings of x, thus requiring $O(n^4)$ time and space. In contexts other than bioinformatics some algorithms have been developed, on related, even though more general problems (Arimura et al., 2000; Wang et al., 1994). However, they can also solve the given problem in time $O(d^2n^3 \log n)$ and $O(n^3)$ respectively. Here we focus on the description of the algorithm in (Apostolico et al., 2004) that solves the problem of computing the simple frequency. Eliminating multiple occurrences in fact comes at no extra cost but also does not change the space complexity, that is the focus of compact approaches.
The algorithm exploits the suffix tree property described in Sec. 4.1.3 to partition the $O(n^2)$ substrings in an $O(n)$ classes corresponding to the nodes, and computes the co-occurrence count only for words corresponding to node pairs. The key observation is that if a pair (y', z') is left out, then the following conditions hold:

1. there exists already a corresponding pair (y, z) such that y' and z' are prefixes of y and z respectively;
2. the score of (y, z) is at least equal to the score of (y', z').

These facts follow from the property of the suffix tree that all the substrings ending within an arc have the same starting positions of the string corresponding to their locus. Since the distance d is measured from the beginning of the first component, this implies that classes of substrings that share the same set of starting positions will occur, within distance d, the same number of times. Again, the strings corresponding to the proper loci of the suffix tree can be selected as representative of the classes, and indexed. In Fig.3 the pair (ACG, T) co-occurs the same number of times of the pair $(ACGTA, TA)$.
There are $O(n)$ nodes in the suffix tree, and each of them can be chosen as first component. A second suffix tree is used to store the number of co-occurrences between the fixed first component, and all other nodes in the suffix tree. The annotation is made as follows:

1. assign a null weight to all nodes of the suffix tree used for the counting of co-occurrences
2. let y be the first fixed component, and $\{p_1 \dots p_k\}$ its set of occurrences;

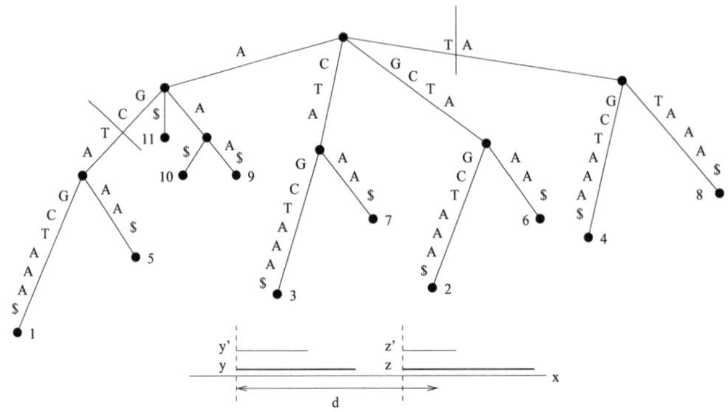

Fig. 3. Illustration of co-occurrences classes.

3. add 1 to all the leaves that correspond to positions $\{p_1, p_1 + 1, \ldots p_1 + d\}$, $\{p_2, p_2 + 1, \ldots p_2 + d\}, \ldots \{p_k, p_k + 1, \ldots p_k + d\}$;
4. annotate the tree bottom-up, so that internal nodes have a weight corresponding to the sum of the weights of their children.

After the annotation the suffix tree will hold the number of co-occurrences between the fixed y and any string z with a proper locus in the tree. The annotation of the tree is performed in linear time. Since this must be repeated for every possible choice of the first component y, the resulting complexity is $O(n^2)$.

4.2.1 Example

Turning back at the example in the previous section, the co-occurrences will be counted only for $15 \times 15 = 225$ pairs of substrings of x. If an exhaustive index would have been build instead, the number of entries of the output would have been 4356, or 2704 if duplication was removed. In the former case we achieved a 95% reduction of the table size, in the latter a 92% reduction.

5. Compact approaches for words with mismatches

Co-occurrence counts are useful for the detection of motifs that are particularly conserved at the sides and allow for high variability in the middle, i.e. so called dyads. However, other distributions of variability are possible and compact approaches to deal with them have also been developed. However, dealing directly with variability implies an higher order of complexity in the solution of the problem. It is possible to identify in literature two main frameworks in which compact approaches have been developed:

1. the candidate motifs occur at least once exactly in the input string; the errors (mismatches) can occur in any position of the motifs;
2. the candidate motifs might never occur in the input string exactly; the position of the errors (wildcards) are fixed in the motif template.

5.1 Compact approaches for motif with mismatches

In (Apostolico & Pizzi, 2007) a compact approach for the extraction of motifs with mismatches, with an *i.i.d.* background, was proposed. The approach was next extended to deal with a

markovian background distribution (Pizzi & Bianco, 2009). The assumptions are that: the motif must occur at least once exactly in the sequences; its instances can appear with exactly, or at most, k mismatches; the mismatches can occur in any position of the motif.
We first observe that, because of the introduction of mismatches, it is not possible to use a suffix tree to partition the search space. In fact if we consider a string w with exactly k mismatches, and its extension wa, with $a \in \Sigma$, with exactly k mismatches we are not guaranteed that the frequency of wa is the same of w, even if both have the same locus. In fact, in the count of wa with k mismatches we would add the number of occurrences of w with k mismatches followed by an a, and the number of occurrences of w with $k-1$ mismatches followed by a symbol $b \neq a$, and finally subtract the number of occurrences of w with k mismatches followed by a symbol $b \neq a$. Depending on how the symbols occur in the text the number of occurrences of wa might be less, equal, or greater than those of w.
Hence, to compact the output, one has to isolate intervals of words of increasing length with the same frequency (with mismatches. However, this is not sufficient. The next step is to verify that within those intervals the expectation with mismatches has indeed a monotone behavior.
In (Apostolico & Pizzi, 2007; 2008), besides the proposal of polynomial algorithms for the computation of the expected frequency, a full study about the score behavior of motifs with mismatches has been presented. We redirect the readers to the original papers for proofs and report here only the main results.

1. the word length is increased, and the number of mismatches is fixed: the expected frequency with mismatches decreases;
2. the number of mismatches is increased, keeping the word length fixed, counting an exact number of mismatches: the expected frequency with mismatches increases (at least until a number of mismatches equal to half of the length of the motif, then it might decrease);
3. the number of mismatches is increased, keeping the word length fixed, counting at most a given number of mismatches: the expected frequency with mismatches increases.

The case 1) is that of interest for the intervals of increasing length and constant counted frequency with mismatches. In fact, in these intervals we have that the z-scores are monotone, and we can take as a representative of the interval the corresponding string.
Experiments to measure the effectiveness of the interval definition can be carried out in the following way. Let us consider words of length $m \pm \delta$, for a given m and δ and compute the score for runs of words with the same frequency (with mismatches) only once.
Tables 5 and 6 report the percentage of entry savings, with respect to a full enumeration, when the frequency is counted for exactly or at most k mismatches respectively. The input string was a sample 10k bases from the yeast genome.

5.2 Compact approaches for words with wildcards

In case of regular patterns the words are defined over an alphabet $\Sigma' = \Sigma \cup \{\text{"."}\}$ that includes also the wildcard "." symbol. The classes of equivalence are identified by the subwords that are generated by the corresponding grammars. Moreover, the property of maximality must hold both in length and in composition. A pattern is maximal in composition if the substitution of any of its wildcards with a given symbol of the alphabet Σ alters the number of occurrences of the pattern. Maximal regular patterns have been used to devise combinatorial algorithms

k		$m = 12$	$m = 13$	$m = 14$	$m = 15$	$m = 16$	$m = 17$	$m = 18$
0	avg length	6.74	6.91	6.96	6.98	6.98	6.98	6.99
	% saving	84.21	85.17	85.49	85.58	85.63	85.65	85.67
1	avg length	2.33	2.35	2.38	2.54	2.74	2.92	2.97
	% saving	17.71	23.04	26.51	29.41	36.41	45.07	50.37
2	avg length	2.30	2.32	2.35	2.37	2.40	2.48	2.61
	% saving	9.46	13.14	18.36	23.43	26.29	28.26	31.67
3	avg length	2.14	2.21	2.28	2.31	2.36	2.40	2.44
	% saving	4.85	7.29	9.98	13.75	18.79	23.53	27.34
4	avg length	2.02	2.07	2.15	2.23	2.28	2.31	2.34
	% saving	1.01	2.91	5.39	7.84	10.69	14.70	19.50

Table 5. Average run length and table size reductions for frequency with exactly k mismatches ($\delta = 3$).

k		$m = 12$	$m = 13$	$m = 14$	$m = 15$	$m = 16$	$m = 17$	$m = 18$
0	avg length	6.74	6.91	6.96	6.98	6.98	6.99	6.99
	% saving	84.20	85.17	85.49	85.59	85.63	85.65	85.67
1	avg length	5.02	5.86	6.50	6.82	6.93	6.97	6.98
	% saving	68.24	77.92	82.75	84.65	85.33	85.53	85.60
2	avg length	3.29	3.96	4.72	5.55	6.27	6.70	6.88
	% saving	37.84	51.70	65.09	75.47	81.32	83.98	85.04
3	avg length	2.32	2.68	3.21	3.84	4.57	5.39	6.19
	% saving	12.43	23.74	36.76	50.53	63.84	74.31	80.65
4	avg length	2.03	2.14	2.34	2.69	3.20	3.83	4.55
	% saving	1.05	4.80	12.61	23.75	36.73	50.44	63.46

Table 6. Average run length and table size reductions for frequency with at most k mismatches ($\delta = 3$).

for the detection of frequent patterns in biosequences (Califano, 2000; Rigoutsos & Floratos, 1998). In both works, the extracted patterns can be said to be a compact representation for motifs in which variability is allowed, but only at specific positions.

Although the number of maximal motifs with wildcard can be exponential, in (Parida et al., 2000) the authors present a way of extracting the inner structure that characterize such type of motif so that the output size could be further reduced. This can be obtained by defining a basis of motifs. A basis B is a subset of all the motifs from which it is possible to recover all the other motifs. More in details a motif is characterized by its list of occurrences, and all motifs not in the basis can be obtained by a combination of the location lists of some of the motifs in B. The motifs that belong to the basis are called *irredundant*. Several works have been done on such motif representation, among which (Apostolico & Parida, 2004; Pelfrêne et al., 2003; Ukkonen, 2009). In (Apostolico, Parida & Rombo, 2008) the concept of irredundant basis was further extended to 2D patterns.

In the context of patterns with wildcards, a pattern is taken into consideration if it occurs for a number of time that is defined by a quorum $q \geq 2$. In (Pisanti et al., 2005) an in dept study about the complexity of the number of irredundant motif showed that there exists a family of motifs for which the number of motifs in the basis is $\Omega(n^2)$ for $q = 2$. In the same paper the authors propose a new definition for a basis, that is stronger than that in (Parida et al., 2000). Therefore their basis is smaller and included in the previous one. Motifs that belong to the

new definition are called *tiling*. For basis on tiling motifs the number of elements is linear in the size of the input string for $q = 2$. However, it has been proved that for all basis there is an exponential dependency in the quorum when $q > 2$, so that no polynomial algorithm exists to extract a basis in this case.
Basis are clearly a compact representation of the motifs space, hence a powerful tool for the analysis of biological sequences when variability is taken into account in the form of wildcards.

6. Position weight matrices

Positions weight matrices are one of the most widely used models for biological signals, being used both in genomic and proteomic studies.
A position weight matrix is a scoring matrix M, where each row represents a position and each column a symbol from the alphabet Σ (in literature it is common to find also the vice versa). The score of the matrix against a segment $x_j x_{j+1} \dots x_{j+m-1}$ of a sequence $x = x_1 x_2 \dots x_n$, is given by:

$$S(M,j) = \sum_{i=0}^{m-1} M[i][x_{j+i}]$$

Given a threshold T, the matrix M is said to have an hit at position j of x if $S(M,j) \geq T$.
The threshold can be given as an absolute value or as a p-value or MSS score T', and then converted in terms of absolute score with respect to the matrix M (Staden, 1989).
The hits of a matrix M in the sequence x can be found naively by an $O(mn)$ algorithm that for each position j check if $S(M,j)$ is above the threshold. The computation of the score takes $O(m)$ steps, and need to be computed for $n - m + 1$ possible starting positions, hence the claimed complexity.

6.1 The look-ahead technique

The look-ahead technique (Wu et al., 2000) can be used to stop the computation of the score whenever one is sure that the threshold will never be reached. Let $S_k(M,j) = \sum_{i=k+1}^{m-1} M[i][x_{j+i}]$ be the score of the segment $x_j \dots x_{j+k-1}$ with respect to the matrix M, the condition to stop the comparison is:

$$S_k(M,j) + \sum_{i=k+1}^{m-1} \max_{s \in \Sigma} M[i][s] < T$$

i.e. even if in the rest of the comparison we will add the maximum score of the matrix for those positions, the final score will be below the threshold.
It is easy to compute an array t_{la} of size m that holds the value of the partial thresholds that need to be reached in order to have the chance to reach the threshold. The entries of the array are given by:

$$t_{la}[i] = T - S_k(M, i+1)$$

These can be computed in linear time in m, by setting $t_{la}[m-1] = T$, and recursively computing the other entries by the formula:

$$t_{la}[i] = t_{la}[i+1] - \max_{s \in \Sigma} M[i+1][s], \text{ for } i = m-2 \ldots 0$$

6.2 The Minimum Gain

A dual concept to that of look-ahead score is given by the *Minimum Gain* (Pizzi et al., 2011). The minimum gain relative to a position j in the matrix is the minimum score that can be obtained by summing the scores from positions $j+1$ to m.

The minimum gain can be used similarly to look-ahead to stop the comparison of the matrix against a segment earlier. In this case the condition to be verified is:

$$S_k(M, j) + \sum_{i=k+1}^{m-1} \min_{s \in \Sigma} M[i][s] \geq T$$

If this is the case, then no matter what the following $m-k$ symbols are, there will be a hit to report at position j.

Similarly to look-ahead, we can build a minimum gain array t_{mg} corresponding to the partial thresholds that need to be reached to ensure that the hit. Starting from $t_{mg}[m-1] = 0$, we have:

$$t_{mg}[i] = t_{mg}[i+1] - \min_{s \in \Sigma} M[i+1][s], \text{ for } i = m-1 \ldots 0$$

6.3 Compact approach to automaton construction

In (Pizzi et al., 2011) an algorithm with optimal $O(n)$ searching time was proposed to solve the problem of profile matching (i.e. given a string x, a threshold T and a matrix M, find all the hits of M in x).

This algorithm is based on the classic multi-pattern matching algorithm based on the Aho-Corasick automaton (Aho et al., 1974). In summary, an automaton is built that contains all the words that are a match for the given threshold and matrix.

The look-ahead technique can be used in this context to generate all and only the words of length m that are hits. The words are generated directly in a trie, that will later be annotated with failure links to build the Aho-Corasick automaton. Starting from the empty trie, only symbols which score is above $t_{la}[0]$ are expanded. Each node will take track of the partial score of the path from the root to the node itself, and will recursively expand further levels in a similar way with comparison with the look-ahead partial thresholds.

Depending on the given threshold, and on the distribution of the score within the matrix, the size of the trie is very variable, but can become prohibitive. In fact by decreasing the threshold (e.g. increasing the p-value), the number of words that are hits for the matrix increases, and can possibly reach an exponential number (the worst case given by $|\Sigma|^m$).

The minimum gain can then be used to substantially reduce the size of the automaton, implementing the compact approach philosophy. The idea is to limit the length of the words that are hits for the matrix whenever a prefix of the word is enough to establish that there will be a hit. While building the trie this means that we do not need to build the subtree of this prefix. The cost of the further comparison with t_{mg} at construction time is irrelevant compared to the time saved by avoiding to build the subtree. Moreover, there could be considerable space savings. See, for example what happens when we consider matrix M00003 from the

Jaspar database (Sandelin et al., 2004), and the threshold score , corresponding to MSS=0,85 (see Table 7). The entries of the matrix have been already multiplied by the factors needed to compute the score relative to MSS (Quandt et al., 1995).
The matrix entries have been sorted in ascending order, so each table entry on the first four columns contains a pair (symbol,score), and the last two columns hold the look-ahead and the minimum gain score, respectively.

0	1	2	3	LA	MG
A,0	C,0	T,0	G,18500	8799	53914
A,0	T,0	G,0	C,18500	27279	53914
A,0	T,0	G,0	C,18500	45779	53914
A,110	T,230	G,230	C,355	46134	54024
A,95	T,240	C,285	G,305	46439	54119
T,114	C,264	A,330	G,402	46841	54233
T,176	C,480	A,848	G,1456	48297	54409
T,494	C,608	A,722	G,5266	53503	54903
A,180	T,380	C,1560	C,1580	55083	55083

Table 7. Computation of scores for the matrix M00003 of the Jaspar database.

When we build the trie using the look-ahead technique, we have that at the first level only the symbol G has a score that is above the partial threshold $t_{la}[0]$. Hence we will have only one child node for the root, with partial score 18500. At the second level we have again that only the score of one symbol, (C,18500) summed up with the current partial threshold 18500, will give a score (37000) above the partial threshold 27279. Again we add a single child to the previous node, and label the edge with C. At the third step only C have a score that summed up with the path partial score (37000) will give a value (55500) that is above $t_{la}[2] = 45779$. The trie is then extended a further level using only symbol C. From now on we can notice that the partial threshold of this path is already above any following look-ahead partial thresholds. This means that all symbols will be considered at each level, thus obtaining a full subtree of high 6. This implicity means that the prefix GCC is enough to establish whether there is an hit or not. If when building the trie we compare the value of the path partial threshold with the minimum gain partial threshold we can stop early the computation, and set as the final state of the Aho-Corasick automaton the current node.
The minimum gain basically defines classes of equivalence within the space of hits of the matrix. Hits that share the same prefix do not need to be totally inserted in the trie. It suffices to insert their common representative prefix.
In case of matrix M00003, we will save the construction of $\sum_{i=1}^{6} 4^i$ nodes, that is the number of nodes of the full subtrees rooted at the common prefix. This means that instead of having 5464 nodes we just have 4 (including the root).

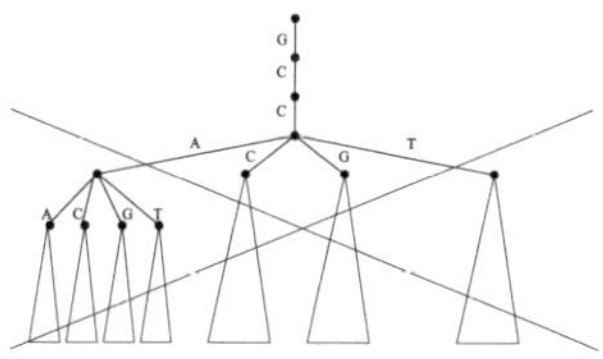

Fig. 4. Space saving with the compact approach for the matrix M00003 of the Jaspar database.

In terms of a compact approach, the minimum gain partitions the set of words that are a hit for a given matrix and threshold in classes in which the representative is the shortest prefix with a score that, summed up with the minimum sum of score that can follow, is above the threshold.

7. Conclusion

In this chapter we discussed the compact approach for the discovery of significant signals (motifs) in biological sequences. Compact approaches are characterized by a significant reduction of the size of the output, without loss of crucial information. Compact approaches that have been developed in these years for several different motif models have been illustrated and discussed, also with the help of practical examples.

8. References

Abouelhoda, M. I., Kurtz, S. & Ohlebusch, E. (2004). Replacing suffix trees with enhanced suffix arrays, *J. Discrete Algorithms* 2(1): 53–86.

Aho, A., Hopcroft, J. & Ullman, J. (1974). *The design and analysis of computer algorithms.*, Addison-Wesley, Reading Mass.

Apostolico, A., Bock, M. E. & Lonardi, S. (2003). Monotony of surprise and large-scale quest for unusual words, *Journal of Computational Biology* 10(2/3): 283–311.

Apostolico, A. & Parida, L. (2004). Incremental paradigms of motif discovery, *Journal of Computational Biology* 11(1): 15–25.

Apostolico, A., Parida, L. & Rombo,S.E. (2008). Motif patterns in 2D, *Theoretical Computer Science* 390(1): 40–55.

Apostolico, A. & Pizzi, C. (2007). Motif discovery by monotone scores, *Discrete Applied Mathematics* 155(6-7): 695–706.

Apostolico, A. & Pizzi, C. (2008). Scoring unusual words with varying mismatch errors, *Mathematics in Computer Science* 1(4): 639–653.

Apostolico, A., Pizzi, C. & Satta, G. (2004). Optimal discovery of subword association in strings, *Proceedings of the seventh International Conference on Discovery Science (DS04)*, Vol. 3245 of *LNCS*, Springer-Verlag, pp. 270–277.

Arimura, H., Arikawa, S. & Shimozono, S. (2000). Efficient discovery of optimal word-association patterns in large text databases, *New Generation Computing* 18: 49–60.

Brazma, A., Jonassen, I., Eidhammer, I. & Gilbert, D. (1998). Approaches to the automatic discovery of patterns in biosequences, *Journal of Computational Biology* 5(2): 277–304.

Califano, A. (2000). Splash: structural pattern localization analysis by sequential histograms, *Bioinformatics* 16(4): 341–357.

Hertz, G. Z. & Stormo, G. D. (1999). Identifying dna and protein patterns with statistically significant alignments of multiple sequences, *Bioinformatics* 15: 563–577.

Keich, U. & Pevzner, P. A. (2002). Finding motifs in the twilight zone, *RECOMB*, pp. 195–204.

Lawrence, C., Altschul, S., Bugoski, M., Liu, J., Neuwald, A. & Wootton, J. (1993). Detecting subtle sequence signals: a gibbs sampling strategy for multiple alignment, *Science* 262: 208–214.

Manber, U. & Myers, E. W. (1993). Suffix arrays: A new method for on-line string searches, *SIAM J. Comput.* 22(5): 935–948.

McCreight, E. M. (1976). A space-economical suffix tree construction algorithm, *Journal of the ACM* 23(2): 262–272.

Parida, L., Rigoutsos, I., Floratos, A., Platt, D. E. & Gao, Y. (2000). Pattern discovery on character sets and real-valued data: linear bound on irredundant motifs and an efficient polynomial time algorithm, *SODA*, pp. 297–308.

Pelfrêne, J., Abdeddaïm, S. & Alexandre, J. (2003). Extracting approximate patterns, *Combinatorial Pattern Matching*, pp. 328–347.

Pevzner, P. A. & Sze, S.-H. (2000). Combinatorial approaches to finding subtle signals in DNA sequences, *Proceedings of the Eight International Conference on Intelligent Systems for Molecular Biology (ISMB00)*, AAAI Press, pp. 269–278.

Pisanti, N., Crochemore, M., Grossi, R. & Sagot, M.-F. (2005). Bases of motifs for generating repeated patterns with wild cards, *IEEE/ACM Trans. Comput. Biology Bioinform.* 2(1): 40–50.

Pizzi, C. & Bianco, M. (2009). Expectation of Strings with Mismatches under Markov Chain Distribution, *Proceedings of the sixteenth International Symposium on String Algoeirthms and Information Retrieval (SPIRE09)*, Vol. 5721 of *LNCS*, Springer-Verlag, pp. 222–233.

Pizzi, C., Rastas, P. & Ukkonen, E. (2011). Finding significant matches of position weight matrices in linear time, *IEEE/ACM Trans. Comput. Biology Bioinform.* 8(1): 69–79.

Quandt, K., Frech, K., Karas, H., Wingender, E. & Werner, T. (1995). Matind and matinspector: new fast and versatile tools for detection of consensus matches in nucleotide sequence data, *Nucleic Acids Research* .

Queen, C., Wegman, M. & Korn, L. (1982). Improvements to a program for dna analysis: a procedure to find homologies among many sequences, *Nucleic Acid Research* 10: 449–456.

Rigoutsos, I. & Floratos, A. (1998). Combinatorial pattern discovery in biological sequences, *Bioinformatics* 14: 55–67.

Sandelin, A., Alkema, W., Engstrom, P., Wasserman, W. & Lanhard, B. (2004). Jaspar: an open-access database for eukaryotic transcription factor binding profiles, *Nucleic Acids Research* 32.

Staden, R. (1989). Methods for calculating the probabilities of finding patterns in sequences, *Computer Applications in the Biosciences* 5(2): 89–96.

Stormo, G. & III, G. H. (1989). Identifying protein-binding sites from unaligned dna fragments., *Proceedings of the National Academy of Science USA*, Vol. 86, pp. 1183–1187.

Tompa, M. (1999). An exact method for finding short motifs in sequences, with application to the ribosome binding site problem, *ISMB*, pp. 262–271.

Tompa, M., Li, N., Bailey, T. L., Church, G. M., Moor, B. D., Eskin, E., Favorov, A. V., Frith, M. C., Fu, Y., Kent, W. J., Makeev, V. J., Mironov, A. A., Noble, W. S., Pavesi, G., Pesole, G., Regnier, M., Simonis, N., Sinha, S., Thijs, G., van Helden, J., Vandenbogaert, M., Weng, Z., Workman, C., Ye, C. & Zhu, Z. (2005). Assessing computational tools for the discovery of transcription factor binding sites, *Nature Biotech.* 23(1): 137 – 144.

Ukkonen, E. (1995). On-line construction of suffix trees, *Algorithmica* 14(3): 249–260.

Ukkonen, E. (2009). Maximal and minimal representations of gapped and non-gapped motifs of a string, *Theor. Comput. Sci.* 410(43): 4341–4349.

van Helden, J., André, B. & Collado-Vides, J. (1998). Extracting regulatory sites from the upstream region of yeast genes by computational analysis of oligonucleotide frequencies, *J. Mol. Biol.* 281: 827–842.

Wang, J. T.-L., Chirn, G.-W., Marr, T. G., Shapiro, B. A., Shasha, D. & Zhang, K. (1994). Combinatorial pattern discovery for scientific data: Some preliminary results, *in* R. T. Snodgrass & M. Winslett (eds), *Proceedings of the 1994 ACM SIGMOD International Conference on Management of Data, Minneapolis, Minnesota, May 24-27, 1994,* ACM Press, pp. 115–125.

Waterman, M., Arratia, R. & Galas, D. (1984). Pattern recognition in several sequences: consensus and alignment, 46: 515–527.

Weiner, P. (1973). Linear pattern matching algorithms.

Wu, T. D., Nevill-Manning, C. G. & Brutlag, D. L. (2000). Fast probabilistic analysis of sequence function using scoring matrices, *Bioinformatics* 16(3): 233–244.

Part 3

Omics-Based
Molecular and Cellular Experimental Systems - Examples and Applications

13

Data Mining Pubmed Using Natural Language Processing to Generate the β-Catenin Biological Association Network

Fengming Lan, Xiao Yue, Lei Han, Peiyu Pu and Chunsheng Kang
Department of Neurosurgery, Laboratory of Neuro-Oncology, Tianjin Medical University General Hospital, China

1. Introduction

β-catenin, originally identified in Drosophila as the segment polarity protein armadillo, is a multifunctional protein that is encoded in humans by the *CTNNB1* gene. β-catenin is found in at least three cellular pools: (i) at the adherens junctions, where β-catenin binds to the cytoplasmic domain of type I cadherins and modulates cadherin-dependent cell-cell adhesion by linking the cadherin/catenin complex to the cortical actin cytoskeleton through the binding of a-catenin; (ii) the cytoplasm, where β-catenin plays a critical role in the canonical Wnt signaling cascade by interacting with APC and GSK3β linked destruction complex, leading to its ubiquitination and subsequent degradation by the proteasome; and (iii) the nucleus, in association with other transcription factors. The crucial event in the canonical Wnt signalling cascade is the cytoplasmic stabilization of β–catenin, leading to its subsequent nuclear localization and gene transcription activity (Liu & Millar 2010). To date, numerous β-catenin target genes have been identified in diverse biological systems (http://www.stanford.edu/%7ernusse/wntwindow.html), yet little is understood about β–catenin outside of its roles in Wnt and cadherin signaling.

Biomedical literature is growing at a double-exponential pace, with more than 19,000,000 publications in MEDLINE of which more than three million were published in the last 5 years alone (Abrams et al., 1998). Over the last 10 years, the total size of MEDLINE (the database searched by PubMed) has grown at a ~4.1% compounded annual growth rate, and the number of new entries in MEDLINE each year has grown at a compounded annual growth rate of ~3.1% (Albert, 1999, 2002, 2005). Thus, a massive wealth of information is embedded in the literature and waiting to be discovered and extracted. Literature mining is a promising strategy to utilize this untapped information for knowledge discovery. Most mining is performed on the abstracts of biomedical articles, which represent a readily available resource of highly concentrated information and result in high quality extracted relations (Apte & Weiss, 1997; Card et al., 1996; Chien et al., 2007) . Text mining of biomedical literature has been applied successfully to various biological problems including the discovery and characterization of molecular interactions (protein-protein , gene-protein, gene-drug, protein sorting and molecular binding, consolidating information into a more accessible form (Babu et al., 2004; Balaji et al., 2006; Giot et al., 2003; Lee et al., 2006).

Recently, considerable interest and effort has been focused on the construction and analysis of genome-wide gene networks (McCraith et al., 2000). The task is complicated for heavily investigated transcription factors such as β-catenin due to the large volume of manuscripts published. As no searchable records are available to efficiently retrieve information relevant to the β-catenin gene network, we extracted gene/protein interactions by text mining Pubmed abstracts and constructed the β-catenin biologic association network. Our text-mining by natural language processing established an association of the β-catenin network with survival signaling, clarifying the fragmentary data that was previously available describing this relationship and confirming the crucial role of β-catenin in growth and development.

2. Methods

2.1 Natural Language Processing (NLP)

Medline/PubMed was used as the information source for bioinformatics text mining. Medline abstracts were retrieved using National Center for Biotechnology Information (NCBI) PubMed portal. We queried Pubmed with: (catenin OR CTNNB OR CTNNB1) AND ("1980/01/01"[PDAT]: "2009/05/24"[PDAT]). All abstracts were downloaded as HTML text without images and converted into XML documents. Sentence tokenization was performed with Lingpipe tools. Subsequent analysis was based on the sentence as the basic unit. Gene mentions (including the β-catenin gene) were tagged using ABNER (Egghe & Rousseau, 1990). To solve the matter of the plethora of gene aliases, all gene mentions were normalized to Entrez gene (http://www.ncbi.nlm.nih.gov/Entrez/) official gene symbols. A genetic interaction of the verb dictionary was established from BioNLP item (http://bionlp.sourceforge.net/), containing verbs such as repress, regulate, inhibit, interact, phosphorylate, downregulate, upregulate and all other verbs and their variants. Verbs in abstracts were tagged using Lingpipe and the interaction verb dictionary (Ghannad-Rezaie et al., 2006). Only sentences with the β-catenin gene, a proper interaction verb and another gene were selected. In order to test the null hypothesis 'the relationship between β-catenin and another gene is random', the hypergeometric distribution test was employed (Kim et al., 1997).

N represents the total number of PubMed abstracts and m and n represent the number gene mentions in PubMed for β-catenin and a related gene, respectively.

$$p=1-\sum_{i=0}^{k-1} p(i \mid n,m,N)$$

Where:

$$p(i \mid n,m,N)=\frac{n!(N-n)!m!(N-m)!}{(n-i)!i!(n-M)!(N-n-M+i)!N!}$$

The β-catenin-gene' relations with p-value<0.05 were then summarized and subjected to a relational database for further analysis. The flowchart of our NLP pipeline is shown as Figure 1.

2.2 Gene ontology analysis

Gene ontology analysis was performed using the GSEABase package of BioConductor (http://www.bioconductor.org/). A gene set enrichment analysis was performed on the 543 β-catenin-related genes based on the gene ontology (GO) categories (Rual et al., 2005).

2.3 Pathway and gene network analysis

Expression Analysis Systematic Explorer (EASE) was used to analyze KEGG pathways. Over representation of genes in a KEGG pathway was present if a larger fraction of genes within that pathway was differentially expressed compared to all other genes in the genome. The 'β-catenin-verb-gene' relationships retrieved by our NLP system were filtered by pathway enrichment analysis. The links between β-catenin and related genes were visualized using Cytoscape software (Lopez & Blobel, 2008) (http://www.cytoscape.org/). Genes were grouped according to pathway. Genes that are involved in multiple pathways were assigned to a single pathway with the smallest enrichment p-value. Integrating PubMed text mining, homology prediction, gene neighbor, protein-protein interaction, gene fusion and other data sources through the Search Tool for the Retrieval of Interacting Genes/Proteins(STRING) , we created the β-catenin related genes knowledge-driven network (Sousa et al., 2004; Uetz et al., 2000).

3. Experimental results

3.1 Identification of β-catenin interaction genes

Query of β-catenin on Pubmed with: (catenin OR CTNNB OR CTNNB1) AND ("1980/01/01"[PDAT]: "2009/05/24"[PDAT]) led to the identification of 10018 articles describing putative interactions between β-catenin and other genes. Titles containing the β-catenin gene along with a proper interaction verb and another gene were selected for further analysis. A total of 543 genes with published interaction with β-catenin were identified (scheme describing the NLP process, Fig. 1; visualization of the 543 genes Fig. 2).

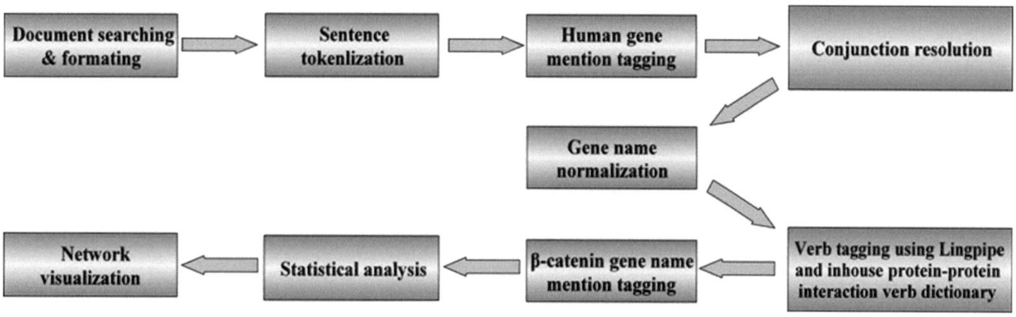

Fig. 1. Schematic representation of the literature-based gene network analysis. Literature mining by natural language processing was performed on all Pubmed abstracts available from 1/1/1980-5/24/2009. Mining identified 10018 articles describing putative interactions between β-catenin and another gene product, revealing 543 distinct β-catenin interacting proteins.

Hepatocyte nuclear factor 4 alpha (HNF4A) was the most prevalent gene identified, commonly referred to as a β-catenin "target" in the literature. Table 1 provides a list of the 10 most frequently published interactions with β-catenin, and their putative relationship.

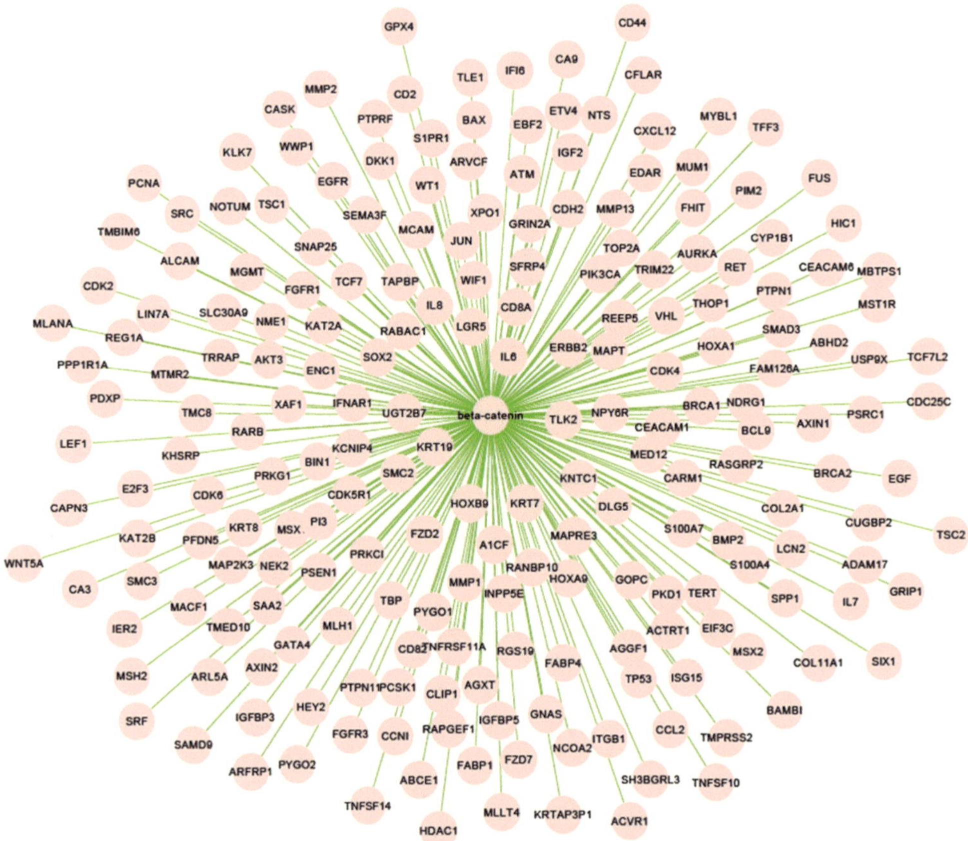

Fig. 2. Visualization of β-catenin interacting proteins. Data mining revealed 543 β-catenin interacting proteins, visualized using the Cytoscape software as described in Methods (http://www.cytoscape.org/).

Gene sym	Pubmed count	Putative interaction
HNF4A	121	target
APC	62	regulate
LEF1	39	activate
EGFR	30	associate
MAPK8	28	activate
LRP6	25	phosphorylate
MUC1	25	interact
IQGAP1	24	interact
CTNNBIP1	21	inhibit
DKK1	21	associate
CD44	20	associate

Table 1. Description of the 10 highest published interacting partners with β-catenin. The gene symbol, number of hits on Pubmed and putative interaction of the 10 highest frequency hits among the 543 interacting proteins identified by literature mining, as described in Methods.

Among the 543 gene products, 18 distinct putative protein-protein relationships were identified involving β-catenin, with the distribution of the seven most frequent relationships
included in Fig. 3. The most common relationship, complexing with β-catenin, was identified for 213 (39.2%) of the gene products (Fig. 3), including TCF4/TCF7L2 and LEF/LEF1. 54 (9.5%) gene products were identified as β-catenin targets, including cell cycle regulating proteins cyclinD1 and CDC42, proteins that influence cellular migratory behavior including uPA, Timp3, and CD44 and proteins that play a role in differentiation such as BMP-7, FGF8 and PPAR-d.

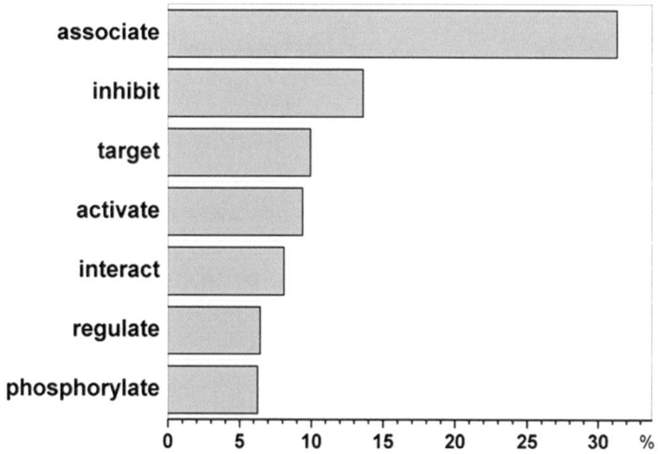

Fig. 3. Frequency distribution of protein 'relationships' with β-catenin. Literature mining revealed 18 types of relationships exhibited by β-catenin with the 543 identified gene products. Relationship type is expressed as the percent of total proteins analyzed. Analysis revealed that the most frequent relationship is "associate" (31.31%), while the least frequent relationship, occurring for only one associated protein, is "dephosphorylate".

3.2 β-catenin-related gene function

To better understand the biological role of the 543 β-catenin-related genes identified, and demonstrate the complexity of the β-catenin-related genes interaction network, we performed a Gene Ontology (GO) enrichment analysis.

GO provides structured, controlled ontologies for describing gene products in terms of their associated molecular function, biological process, or cellular compartment. Enrichment for molecular function revealed that most β-catenin-associated genes, including KIT, HSF1, XPO1, HSPA5, FGF8 and ALCAM, encode proteins that bind to β-catenin. Genes such as CAPN3, EPHA7, PTGER4, SRC, BMP4 composed the second largest category, encoding for proteins that act as signal transducers (Fig. 4, left panel). Enrichment for biological process revealed that the most common functions of gene products associated with β-catenin include developmental processes, cell communication and signaling transduction (Fig. 4, middle panel). Finally, enrichment for the cellular compartments where β-catenin associated gene products can be found primarily included the cytoplasm, nucleus and plasma membrane (Fig. 4, right panel).

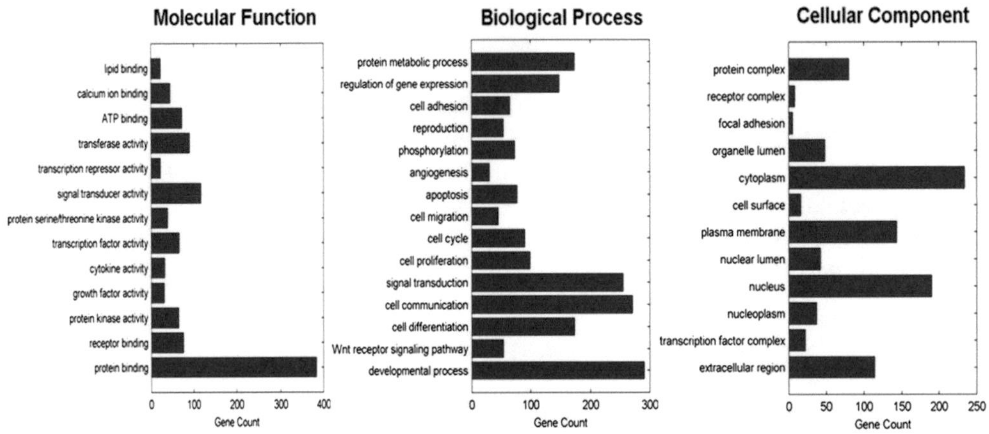

Fig. 4. Gene Ontology (GO) analysis of the β-catenin molecular network. GO enrichment sorted all data by Molecular Function, Biological Process and Cellular Compartment. GO analysis revealed that protein binding is the most prevalent molecular function of β-catenin interacting proteins, development, cell communication and signal transduction are the most common biological processes involved and the gene products are active primarily in cytoplasm, nucleus and plasma membrane.

3.3 Pathway and gene network analysis

Pathway information is required for understanding of gene function.For each of the 543 genes identified, we searched the Kyoto Encyclopedia of Genes and Genomes (KEGG)

Term	Count	%	P-Value
Wnt signaling pathway	54	10.06%	6.85E-26
Focal adhesion	36	6.70%	8.79E-08
MAPK signaling pathway	40	7.45%	1.07E-06
Adherens junction	21	3.91%	4.61E-08
p53 signaling pathway	18	3.35%	1.63E-06
ErbB signaling pathway	19	3.54%	1.03E-05
Apoptosis	17	3.17%	1.25E-04
Insulin signaling pathway	22	4.10%	2.18E-04
VEGF signaling pathway	15	2.79%	2.27E-04
Toll-like receptor signaling pathway	18	3.35%	4.18E-04
GnRH signaling pathway	17	3.17%	4.83E-04
Cell cycle	19	3.54%	5.61E-04
mTOR signaling pathway	11	2.05%	0.001869395
TGF-ß signaling pathway	14	2.61%	0.007074483

Table 2. Pathway analysis of the 543 β-catenin interacting proteins. β-catenin interacting proteins are involved in 14 different cell signaling pathways, identified using the Cytoscape software following NLP analysis.

database to identify their pathway information. 14 signaling pathways where identified whose corrected P-value was less than 0.01 (Table 2), and 321 of the genes belonged to these 14 pathways. Pathway analysis was visualized using the Cytoscape software (Fig. 5). The Using the Search Tool for the Retrieval of Interacting Genes/Proteins (STRING), we created a β-catenin related genes network (Fig. 6). STRING incorporates known and predicted protein interaction information from HPRD, BioGrid, MINT, BIND, DIP, and imports known reactions from Reactome and KEGG pathways to generate a generalized source of protein interaction information. STRING analysis of β-catenin related genes revealed several hub genes, or genes in which high connection exists giving these genes an influential role in network stability. Hub genes identified include AKT1, CCND1, CTNNB1, JUN, TP53 and VEGFA (Fig. 7).

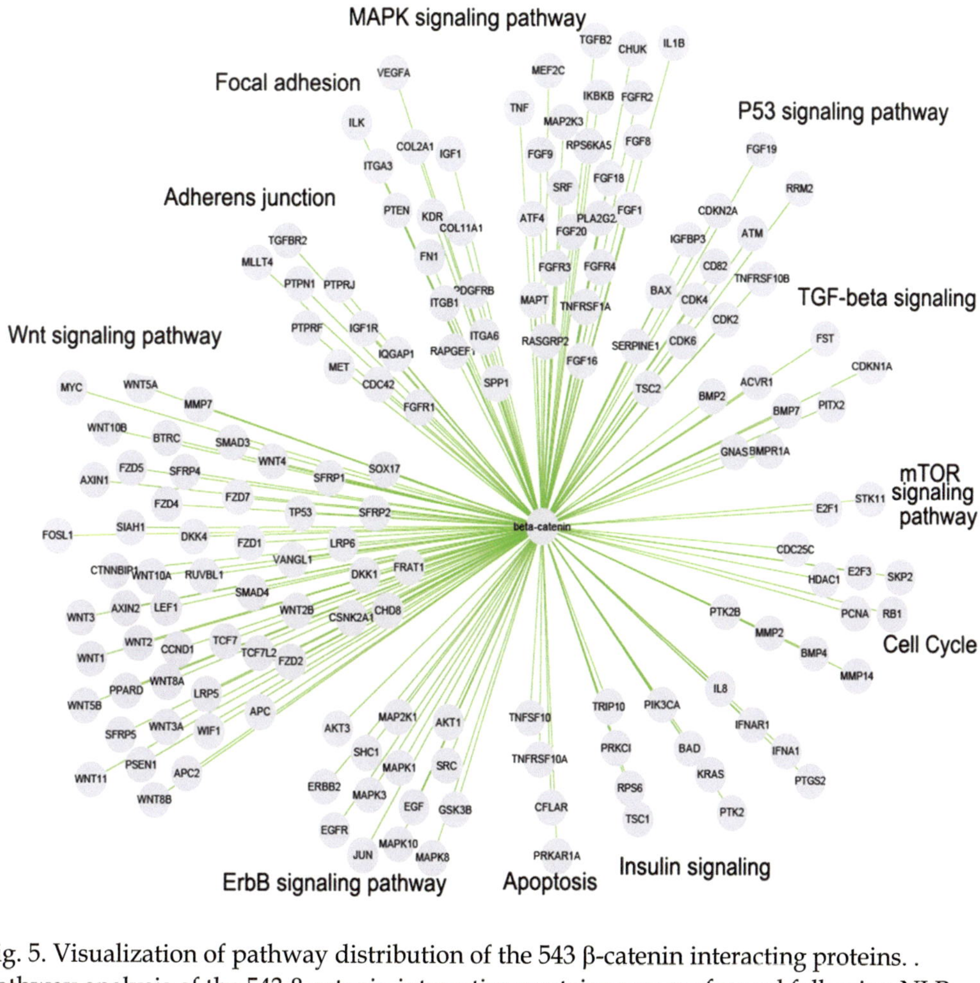

Fig. 5. Visualization of pathway distribution of the 543 β-catenin interacting proteins. . Pathway analysis of the 543 β-catenin interacting proteins was performed following NLP analysis and visualized using the Cytoscape software, as described in Methods. Interacting proteins fit into 14 different cell signaling pathways, including several pathways involved in cell survival signaling.

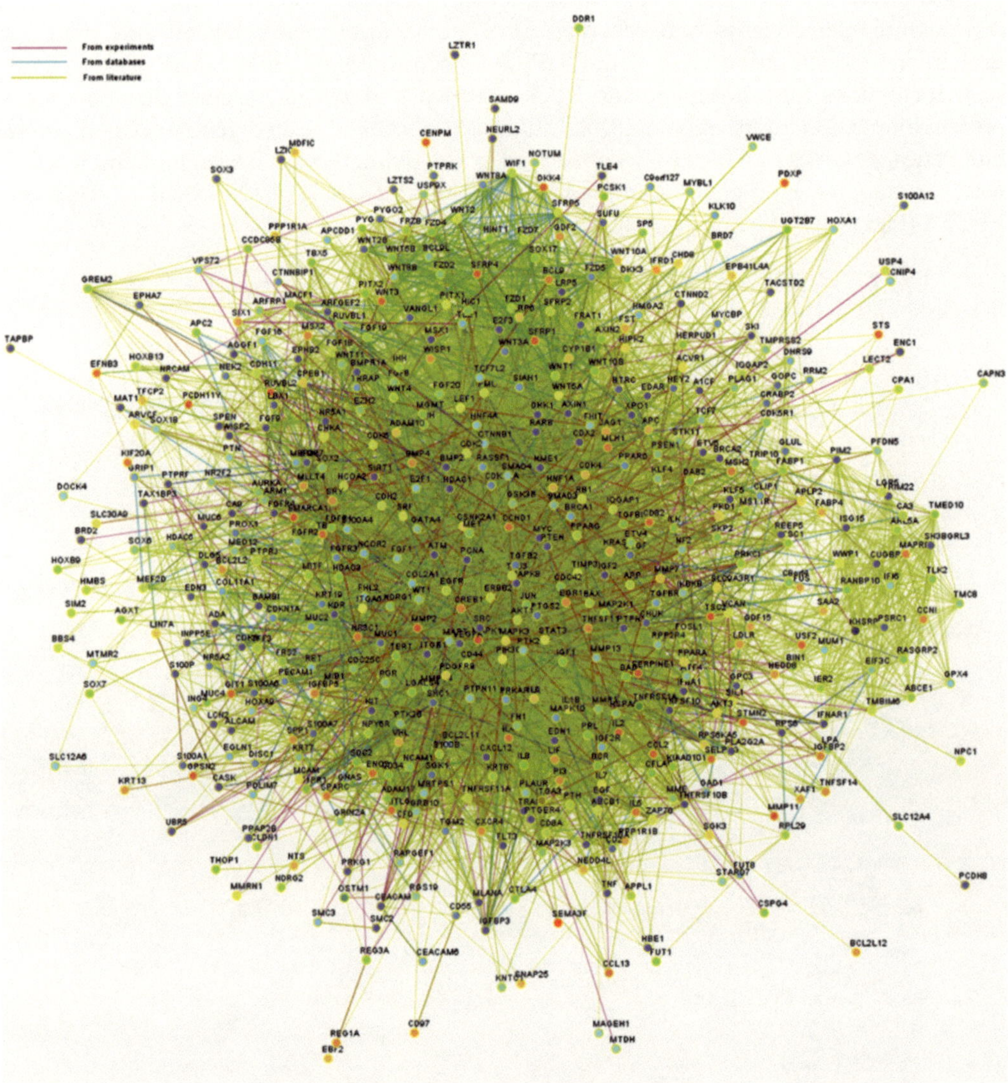

Fig. 6. Visualization of the β-catenin related genes network. Network analysis was performed using the Expression Analysis Systematic Explorer (EASE) to analyze KEGG pathways and Search Tool for the Retrieval of Interacting Genes/Proteins(STRING). Visualization was performed using Cytoscape. Pink lines indicate connections experimentally confirmed by other researches, Cyan lines indicate connections derived from databases (including the KEGG pathway and MIPS) and Green lines indicate connections compiled from co-citation data from literature mining PubMed abstracts.

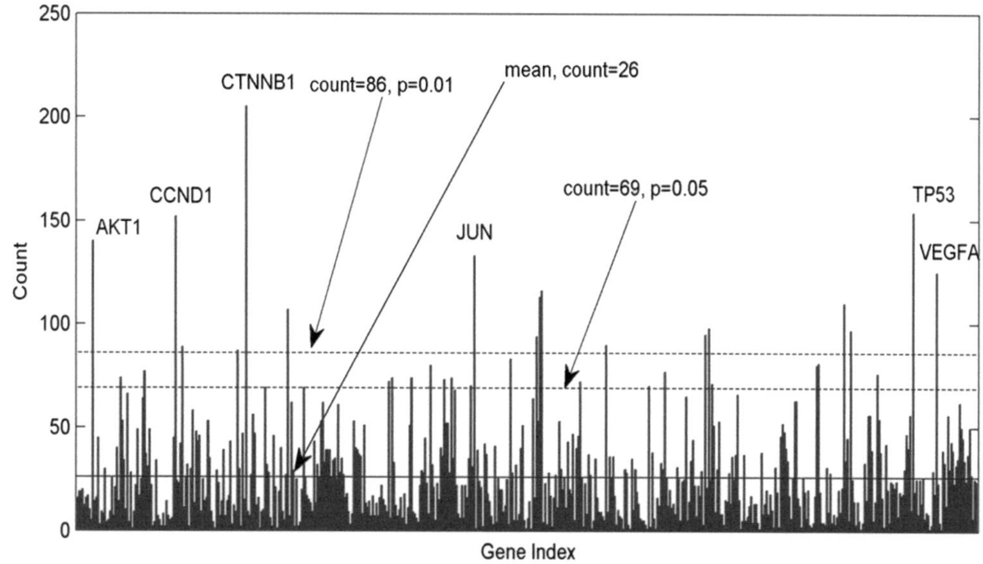

Fig. 7. Connectivity analysis of the β-catenin related genes network. Connectivity analysis was performed using the Search Tool for the Retrieval of Interacting Genes/Proteins(STRING) to generate the β-catenin related genes knowledge-driven network, as described in Methods. Analysis revealed AKT1, CCND1, CTNNB1, JUN, TP53 and VEGFA are important hub genes in the β-catenin network, with mean frequency counts >86 ($p < 0.001$).

4. Discussion

Gene/protein interaction networks provide critical information for a thorough understanding of cellular processes. Thus, detailed characterization of interactions between individual genes or proteins has become a primary focuses of biological research (Barabasi & Oltvai, 2004) . The complete biomedical literature database, containing a massive amount of information attained over a long period of time, is a largely untapped repository of information for study of gene/protein interaction networks (Chalmers et al., 1998; Gónzalez & Ochoa ,2008). Here, we generated a molecular network of β-catenin associated proteins. Analysis revealed that these proteins interacted with β-catenin via 18 different relationships, perform 13 biologic functions, take part in 15 cellular processes, localize to 12 cellular compartments, and signal in 14 different pathways. Significantly, this analysis identified a vast and mostly uncharacterized role for β-catenin in signal transduction pathways distinct from the Wnt and cadherin pathways. In particular, β-catenin may have a significant role in cell survival signaling.

Our analysis was performed using only abstracts instead of full text manuscripts. Our studies suggest that full-text articles contain too much detail for high-throughput analysis of biomedical research and development, while abstracts usually have higher information density and result in better quality relations extracted by text mining techniques. Further, abstracts are freely available through most public databases. Due to the large number of abstracts (10018) analyzed, we believe that the network data generated is both comprehensive and significant (Mackinnon et al., 2008; Ouzounis & Karp, 2000; Ptacek et al., 2005; Thieffry et al., 1998) .

Wnt/β-catenin signaling is involved in almost every aspect of embryonic development and controls homeostatic self-renewal in lots of adult tissues. Gene Otology identified three biological process (developmental process, cell communication and signaling transduction) that are significantly overrepresented within the β-catenin associated network. Validating these observations, several severe phenotypes in multiple tissues and organs can be observed in flies, frogs, fish, and mice following loss of Wnt/β-catenin signaling components. In adults, Wnt/β-catenin signaling remains essential throughout life for driving tissue renewal in rapidly self-renewing organs, including the intestine and skin. In addition, deregulation of Wnt/β-catenin signaling upsets the homeostatic balance in self-renewing tissues and leads to a variety of abnormalities and disease including bone defects and cancer. Pathway analysis of the β-catenin associated network revealed a close relationship between β-catenin and survival signaling (i.e. the Wnt, MAPK, insulin, and adhesion junction pathways), supporting an important role for β-catenin pathway in growth and development. Integration of PubMed text mining, homology prediction, gene neighbor, protein-protein interaction, gene fusion and other data sources identified AKT1, CCND1, CTNNB1, JUN, TP53 and VEGFA as hub genes in the β-catenin signaling network. Network analysis reveals an extremely high connectivity of these genes with other β-catenin associated genes. The involvement of these six genes in survival signaling, including anti-apoptosis, cell cycle and cell migration, provides further surport for a vital role for β-catenin in growth and development.

5. Conclusion

We performed natural language processing, a literature mining tool that can cluster a list of genes with keywords that are auto-extracted from their up-to-date related literature and then manually curated by the user, to establish the β-catenin biologic association network. Our results establish a significant association of this network with survival signaling. These data demonstrate the power of data-mining strategies as tools for biological discovery, suggesting that the use of similar strategies to consolidate all existing data for specific disease states, specifically cancer, may yield yield important discoveries in disease pathogenesis and identification of novel therapeutic targets. Further analysis of the β-catenin biologic association network may provide a deeper understanding of β-catenin signaling, particularly in relation to cell survival signaling.

6. Acknowledgements

This work was financially supported by the China National Natural Scientific Fund (81001128 and 30971136), the Tianjin Science and Technology Committee (09JCZDJC17600), and a Program for New Century Excellent Talents in University (NCET-07-0615), and Tianjin City High School Science & Technology Fund (2009CD01). We thank two anonymous referees for numerous excellent suggestions that led to substantial improvement of the manuscript.

7. References

Abrams, D., Baecker, R., and Chignell, M. Information Archiving with Bookmarks: PersonalWeb Space Construction and Organization, in *Proceedingsof CHI'98* (Los Angeles CA, April 1998), ACM Press, 41-48.

Albert, R.; Yeong H. & Barabasi, A. L. (1999). Diameter of the world-wide web. *Nature* 401 (September 1999) 130.

Albert, R. & Barabasi, A. L. (2002). Statistical mechanics of complex networks. *Rev. Modern Phys,* 74, (January 2002) 47-97

Albert, R. (2005). Scale-free networks in cell biology. *Jcell Sci,* 118, (Pt 21) (October 2005)4947-57.

Apte C.; Weiss S. (1997) Data mining with decision trees and decision rules, *Future Generation Computer Systems,* Vol. 13, No 2-3, pp 197-210

Babu, M.M.; Luscombe N.M.; Aravind, L.; Gerstein, M. & Teichmann, S.A. (2004). Structure and evolution of transcriptional regulatory networks. *Curr Opin Struct Biol* 14, (2004 Jun) 283-291.

Balaji, S.; Babu, M.M.; Iyer, L.M.; Luscombe, N.M. & Aravind, L (2006). Comprehensive analysis of combinatorial regulation using the transcriptional regulatory network of yeast. *J Mol Biol,* 360, (June 2006) 213-27.

Barabasi, A.L. & Oltvai, Z.N. (2004). Network biology: Understanding the cell's functional organization. *Nat. Rev. Genet,* 5, (February 2004) 101-13.

Card, S.K., Robertson, G.C., and York, W. The WebBook and the Web Forager: An Information Workspace for the World-Wide Web, in *Proceedings of CHI'96*(Vancouver BC, April 1996), ACM Press, 111-117

Chalmers, M., Rodden, K., and Brodbeck, D. The Order of Things: Activity- Centred Information Access, In Proceedings of 7th International Conference on the WWW, 1998. (Brisbane Australia, April 1998), 359-367.

Chien C; Wanga W; Chenga J (2007) Data mining for yield enhancement in semiconductor manufacturing and an empirical study, Expert Systems with Applications, Vol 33, No 1,pp 192-198

Egghe, L and Rousseau, R. *Introduction to Informetrics: Quantitative methods in library, documentation, and information science.* Elsevier, New York, NY, 1990.

Ghannad-Rezaie M.; Soltanian-Zadeh H.; Siadat M.-R.; K.V. Elisevich (2006) Medical Data Mining using Particle Swarm Optimization for Temporal Lobe Epilepsy, *Proceedings of the IEEE World Congress on Computational Intelligence,* Vancouver, Canada, July 15-21

Giot, L.; Bader, J.S & Brouwer, C. et al. (2003). A protien interaction map of drosophila melanogaster. *Science,* 302, 5651, (December 2003) 1727-1736

Górnzalez S. & Ochoa A. Resolution of Japanese puzzles using Data Mining and Cultural Algorithms. (Accepted paper) In Proceedings of COMCEV'2008, México; 2008.

Kim, S. H.; Hyun Ju Noh (1997) Predictability of Interest Rates Using Data Mining Tools, EXPERT SYSTEMS WITH APPLICATIONS, Vol 13; No 2, pp 85-96

Lee, T. I.; Rinaldi, N. J & Robert, F. et al. (2002). Transcriptional regulatory networks in saccharomyces cerevisiae. *Sciencie,* 298, 5594, (October 2002) 799-804

Liu F. and Millar S.E. (2010).Wnt/beta-catenin signaling in oral tissue development and disease, In: Journal of Dental Research, Vol. 89, NO. 4, (January 8, 2010), PP. (318-330), ISSN (on-line) 1569-1551 (print) 1748-3050

Lopez DM, Blobel BG. (2008). A development framework for semantically interoperable health information systems. Int J Med Inform. 2008 Jul 11. [Epub ahead of print].

Mackinnon AD, Billington RA, Adam EJ, Dundas DD, Patel U. (2008). Picture archiving and communication systems lead to sustained improvements in reporting times and productivity: results of a 5-year audit. Clin Radiol. 2008 Jul;63(7):796-804. Epub 2008 Mar 25.

McCraith, S.; Holtzman, T.; Moss, B. & Fields, S. (2000). Genome-wide analysis of vaccinia virus protein-protein interactions. *Proc. Natl. Acad. Sci. USA*, 97, 9, (April 2000) 4879-4884

Ouzounis, C. A. & Kar_finalp P.D. (2000). Global properties of the metabolic map of escherichia coli. *Genome Res.* 10, 4, (April 2000) 568-576

Ptacek, J.; Devegan, G.; Michaud,; G. Zhu,; H. Zhu,; X. Fasolo,; J. & et al. (2005). Global analysis of protein phosphorylation in yeast. *Nature*, 438, 7068, (December 2005) 679-84

Rual, J. F.; Venkatesan, K.; Hao, T.; Hirozane-Kishikawa, T.; Crico, A.; Li, N.; Berriz, G. F. & et al. (2005).Towards a proteome-scale map of the human protein-protein interaction network. *Nature,* 437, 7062, (October 2005) 1173-1178

Sousa T.; Silva A., Neves A. (2004) Particle Swarm based data mining algorithms for classification tasks, *Parallel Computing J.*, Vol. 30, pp. 767-783

Thieffry, D.; Huerta, A.M.; Pérez-Rueda, E. & Collado-Vides, J. From specific gene regulation to genomic networks: a global analysis of transcriptional regulation in Escherichia coli. *Bioessays*, 20, 5, (1998 May) 433-40

Uetz, P.; Giot, L.; Cagney, G.; Mansfield,T.A.; Judson, R.S.; Knight, J.R.; & et al. (2000). A comprehensive analysis of protein-protein interactions in *Saccharomyces cerevisiae. Nature*, 403, 6770, (February 2000) 623-627

14

In Silico Identification of Plant-Derived Antimicrobial Peptides

Maria Clara Pestana-Calsa and Tercilio Calsa Jr.
Bioethanol Research Network of Pernambuco, Universidade Federal de Pernambuco, Brazil

1. Introduction

The widely available digital information about plant genomes and its products has triggered the use of bioinformatics and other *in silico* approaches over gene expression data, from genome to phenotype-based analysis methods. In such data universe, special attention has been given to a peptidic group of plant bioactive molecules, the antimicrobial peptides (AMP), usually small cysteine or glycine-rich peptides antagonistic to several pathogens and component of plant innate defense. Main classes of AMPs comprise defensins, thionins, lipid-transfer proteins, cyclotides, snakins and hevein-like, according to amino acid sequence homology.

Plant biodiversity for antimicrobial peptides search has led to increasing efforts on their identification and characterization, although such biodiversity still remains largely unexplored for drug development and other potential applications. As expected, crop species have been more frequently targeted for AMP research and application, mainly due to higher availability of molecular data (Pestana-Calsa et al., 2010). AMPs provide novel strategies not only in medicine but can potentially increase agricultural yields by phytopathogen or pest control. Those small peptides have wide-range inhibitory activity over phytopathogenic microorganisms (mostly fungi), as already comproved in enhanced crop resistance to pathogen attack through genetic breeding and transgenic manipulation (Terras et al., 1995).

Last decade advances in gene expression studies technology brought huge amounts of genomic, transcriptomic and proteomic datasets, a fruitful field for AMP prospection. Based on ethnobotanical or even laboratory information, literature concerning *in silico* analyses of plant antimicrobial peptides is much smaller than that focusing plant secondary metabolites. Additionally, bioinformatics tools are usually constrained to model species, whose "-omics" datasets are available, justifying many heterology-based studies for non-model species. Thus, data integration is highly desired and helpful to apply bioinformatics approaches aiming the *in silico* screening of genomic, transcriptional, proteomic and metabolomic datasets from cultivated or wild plant species. Preliminary analyses have been commonly linked to subsequent molecular techniques to identify and characterize AMP-coding genes, their products and their regulation, besides to validate putative functional aspects in a systems biology approach. For instance, in silico analyses of AMPs and respective coding genes determine the way by which different sequences can affect specific pathogens (Pestana-Calsa etal., 2010).

This chapter focuses on bioinformatics methods, usually associated to molecular biology tools, to prospect databases, identify and characterize putative or known AMPs, and discusses procedures for testing them *in vitro* and *in vivo*. Regarding applications, molecular data must be generated in large scale for a comprehensive set of plant species which have not been addressed up to date in AMP research field. "-Omics" based experimental procedures will then be more efficient and reliable, making easier the application of scientific knowledge from molecular biology and bioinformatics on medicine, agriculture and industry needs.

2. Bioinformatics on AMP identification

The huge amount of available information over plant secondary metabolites effects on human health has demanded the usage of bioinformatics tools on transcriptomic, proteomic and metabolomic profiles from different plant species, concerning single as well as mixtures of phytochemicals (Ulrich-Merzenich et al., 2007). Indeed, most results experimentally obtained for bioactive antimicrobial secondary metabolites might be also derived from correlated AMPs sinergy (Verpoorte at al., 2005). Nowadays increasing number of described plant antimicrobial peptides, in parallel to exponential growth in nucleotide and protein data for public access, allows several possibilities of potential identification of novel AMP. Preliminary analyses have been commonly followed by molecular methods to characterize AMP-coding genes and its regulation. Here, *in silico* studies of AMP and respective coding genes contributed to unravel their functional aspects (Hammami et al., 2009).

2.1 Search by genomics and transcriptomics

Searches in genome and transcript sequences datasets have shown to be a comprehensive and higher yield initial step for seeking candidate AMP-coding genes, mostly due to the typical large-scale coverage of the organism genetic potential in such analyses. *In silico* starting approaches constitute a way to quickly achieve reliable AMP-coding potential of the studied plant species, even if it requires further biological validation.

Genomic analyses through DNA sequencing and mapping have been useful to AMP prospection. Considering model organisms, most green plant species genomes harbor 15-50 defensin-coding genes, even if this number is an under-estimation according to some authors and may grow by further studies on several other plant species genome-wide expressed sequence tag (EST) libraries (Silverstein et al., 2005; 2007; Manners, 2007). In Arabidopsis, 317 defensin-like sequences could be assigned in the genome by hidden Markov models of in silico search (Silverstein et al., 2005).

Actually, a comprehensive search on standard publicly available gene expression databases for plant transcript sequences provides a relatively trustable picture of annotated and putative AMP- related sequences from plants (Table 1). Plant AMPs represent almost 16% of deposited AMP sequences and, considering all the organisms as well as just plants, lipid transfer proteins constitute the most abundant group, what may be an over-estimation (Pestana-Calsa et al., 2010) since they are not only involved in direct plant defense (Chen et al., 2008; Boutrot et al., 2008). Noteworthy, snakins and heveins appear as plant-representative peptides, while defensins and lipid transfer proteins have relatively similar distribution that when considering all organisms. In opposite, thionins accesses are about 4 times more abundant in plants than in all organisms dataset, and plant cycletides occur in about 3 times smaller frequency than for all organisms in GenBank, what may represent under-estimation.

AMP main group	All organisms	Green plants	(%)
Defensin	4,506	755	16.8
Thionin	1,318	780	59.2
Lipid transfer protein	54,485	13,139	24.1
Cycletides	10,497	591	5.6
Snakin-like	13	13	100.0
Hevein-like	1,434	1,043	72.7
Other	31,251	160	0.5
Total	103,504	16,481	15.9

Table 1. Number of AMP-related accesses in GenBank/NCBI/Entrez, identified as AMP main groups (from Pestana-Calsa et al., 2010).

Concerning reference available plant EST collection, the TIGR Plants Gene Indices (http://compbio.dfci.harvard.edu/tgi/plant.html) is useful to achieve distribution of AMP-related transcript sequences from crop and weed/pasture species (Figure 1). From the 14 botanical families represented by 34 species, ten are important commodities as soybean, maize, sugarcane, orange and coffee. It suggests grasses and solanaceous are snakins richer, while legumes (Graham et al., 2004) have less thionin coding genes.

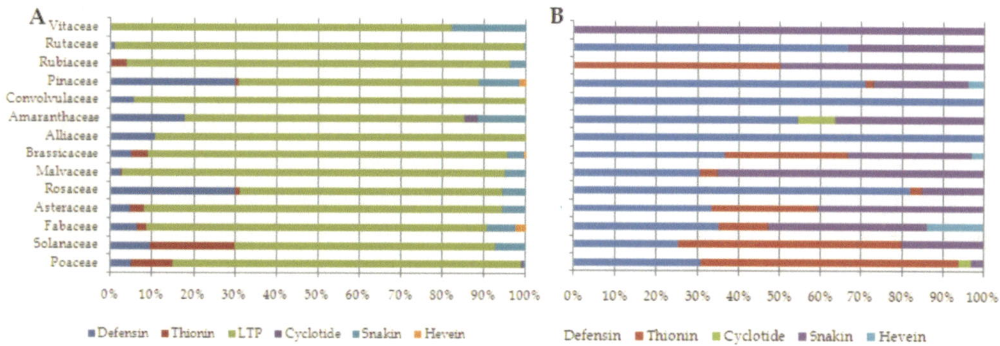

Fig. 1. Distribution of plant EST unigenes (clusters + singletons) from TIGR Gene Indices putatively annotated as AMPs. Frequencies were considered including (A) or not including (B) the putative lipid-transfer proteins (LTP) coding transcripts. Data in each botanical family derive from: Poaceae (*Hordeum vulgare, Zea mays, Oryza sativa, Secale cereale, Sorghum bicolor, Saccharum spp., Panicum virgatum, Festuca arundinacea, Triticum aestivum*); Solanaceae (*Capsicum annuum, Petunia hybrida, Solanum tuberosum, Nicotiana tabacum, Solanum lycopersicum*); Fabaceae (*Phaseolus vulgaris, Medicago truncatula, Phaseolus coccineus, Glycine max*); Asteraceae (*Lactuca sativa, Lactuca serriola, Helianthus annuus*); Rosaceae (*Malus_x_domestica, Prunus persica*); Malvaceae (*Theobroma cacao, Gossypium*); Brassicaceae (*Brassica napus*); Alliaceae (*Allium cepa*); Amaranthaceae (*Beta vulgaris*); Convolvulaceae (*Ipomoea nil*); Pinaceae (Pinus); Rubiaceae (*Coffea canephora*); Rutaceae (*Citrus sinensis*); and Vitaceae (*Vitis vinifera*). Original data from Pestana et al. (2010).

Such type of analysis usually constitutes an initial step in antimicrobial peptide research in plants. By allying *in silico* tools to predict potential AMP-coding genes with "wet"-bench procedures like PCR-based methods using specific designed oligonucleotides to amplify AMP-related regions from genomic DNA, novel grass and legume β-defensins have been recently identified from sugarcane (*Saccharum* spp.; Padovan et al., 2009) and cowpea (*Vigna unguiculata;* Padovan et al., 2010a). These novel peptides expression was co-related to fungal infection and drought responses, suggesting biotic and abiotic responsiveness for defensin activation.

Computational search matches have provided growing helpfulness for AMP genes identification in concluded and current DNA sequencing projects, as in large transcript profiling datasets, however the predicted defense-related genes must be functionally tested (Pestana-Calsa et a., 2010; Belarmino et al., 2010). In spite of rare use, in situ hybridization methods for AMP studies in plants keep very promising to analyze the spatial and temporal expression of such peptides in several organs and tissues, under a wide range of treatments (Tavares et al., 2008).

Genomics and transcriptomics platforms allow the development of strategies to identify distinct classes of antimicrobial peptides. In pepper (*Capsicum annuum*), an AMP was identified and named CaAMP1, which was isolated in a cDNA library from leaves inoculated by *Xanthomonas campestris* pv. *vesicatoria.* CaAMP1 expression was induced in leaves during pathogen infection and after abiotic stressing (Lee & Hwang, 2009). A cDNA sequence, named VvAMP1, was isolated from grape (*Vitis vinifera*) berries and coded for a 77 amino acid peptide homologous to defensins (de Beer & Vivier, 2008); also in grape, comprehensive *in silico* searches have successfully resulted in AMP candidates (Zamyatnin & Voronina, 2010).

Transcripts tag sequences from different large-scale gene expression analyses, like cDNA microarray and SAGE, have been assigned to plant AMP. For instance, cDNA microarray was used to validate regulation of known AMP genes after hormone signalling in defense responses (Wan et al., 2002). On the other hand, SAGE-derived approaches have resulted in suitable databases for AMP detection, as like in sugarcane (Calsa Jr. & Figueira, 2007) and soybean; in both, tens of transcripts have been putatively annotated as AMP-like and are under experimental validation (data not shown).

2.2 Search by proteomics

Plant genomic and post-genomic researches have generated large amounts of information that contributes to understand gene and protein expression profiles, besides their connection to biological processes. Proteomics results in important data over biological systems, because it produces information about proteins and peptides, the major functional and structural determinants of cells (Baginsky, 2009). Conventional two-dimensional gel electrophoresis (2-DE) to separate and visualize proteins, and mass spectrometry (MS) to identify proteins and peptides of interest, have been applied. Some 2-DE limitations concerning AMP analysis refer to retention and visualization on gel of target proteins with molecular weight lower than 10 KDa (Baggerman et al., 2004) and frequent production in small quantities by plant cells (Antunes, 2008).

Small protein and native peptide component of plant tissues is a still neglected proteomic area, and relatively few studies are available (Zhang et al., 2006). Peptidomics have improved the study of such small polypeptides, by coupling bi or multidimensional liquid chromatography to mass spectrometry, where high complexity or low concentrated samples can be efficiently separated via multidimensional liquid chromatography (MDLC), whose advantage over 2-DE is separation of complex mixtures by using multiple columns (Barbosa, 2008). Even so, conventional 2-DE followed by MS has also provided important antimicrobial subproteome/subpeptidome information associated to specific plant samples,

as soybean xylem sap (Djordjevic et al., 2007), and to plant-virus interaction, as from *Capsicum annuum* cv. Bungang (hot-pepper) infected by tobacco mosaic virus (TMV) (Lee e al., 2006). Differentially expressed pathogenesis-related proteins and peptides, as defensin, from sugar beet infected by beet necrotic yellow vein virus (BNYVV) were separated by MDLC, identified by MALDI-TOF/MS and annotated through homology-based search using amino acid sequence deduced from the MS/MS spectra (Larson et al., 2008).
An interesting perspective on antimicrobial peptides was developed to study the conservation of structural motifs in AMPs from phylogenetically distant organisms using proteomic bioinformatics datasets and tools, like PROSITE (ExPASy Proteomics Server, 2011), for multidimensional signature model for AMPs (Yount & Yeaman, 2004). Modeling is an attempt to unify structural domains in distinct classes of AMP.
Additionally, proteomics approaches have been successful in the identification of signalling elements that regulates defensins in plants (Widjaja et al., 2010).

2.3 AMP databases

Hundreds of known AMP sequences are available at UniProtKB/Swiss-Prot, while other repository collections are in other accessible links (Table 2). The major ones are ANTIMIC (Brahmachary et al., 2004), Antimicrobial Sequence database AMSDb (eukaryotic), Peptaibol (fungal, Whitmore & Wallace, 2004), APD (Wang & Wang, 2004) and APD2 (Wang et al., 2009). Among them, the PhytAMP (http://phytamp.pfba-lab.org), a database specific for plant AMPs, was developed providing access to informations regarding studies and applications for these peptides (Hammami et al., 2009), and organizing sequences and corresponding taxonomic, physicochemical, structural, taxonomical and publications over each deposited AMP.
Bioinformatics tools have been developed in agreement to "-omics" databases requirements. An example was the alignment of known antibacterial peptides aiming to detect preferential residues by artificial neural network, since certain amino acids are more frequent in some positions, particularly at the N or C terminus (Lata et al., 2007). Other successful advances resulted from enlargement of AMP catalogs, basically through clustering and alignment to previously annotated sequences; this approach has been quite useful on putative AMPs identification in comprehensive datasets (Fjell et al., 2207). Statistical modeling is applied over genomic, transcriptional and proteomic data in the aim to identify peptides, also domains from already functionally annotated proteins, which present significant motifs and/or structure match to AMP (Nagarajan et al., 2006).

Database	Web site (http://)	Organisms (Reference)
AMSDb	www.bbcm.units.it~tossi/pag1.htm	Eukaryotes (Tossi & Sandri, 2002)
ANTIMIC	research.i2r.a-star.edu.sg/Templar/DB/ANTIMIC	General (Brahmachary et al., 2004)
APD	aps.unmc.edu/AP/main.html	General (Wang & Wang, 2004)
APD2	aps.unmc.edu/AP/main.php	General (Wang et al., 2009)
APPDb	ercbinfo1.ucd.ie/APPDb/	general
Defensins	defensins.bii.a-star.edu.sg/	general
PenBase	penbase.immunaqua.com/	general (Gueguen et al., 2006)
Peptaibols	www.cryst.bbk.ac.uk/peptaibol	Fungi (Whitmore & Wallace, 2004)
PhytAMP	http://phytamp.pfba-lab.org	Plants (Hammami et al., 2009)
SAPD	oma.terkko.helsinki.fi:8080/~SAPD	General (Wade & Englund, 2002)

Table 2. URLs for main available databases for characterized antimicrobial peptides (modified from ExPASy Proteomics Server and Pestana-Calsa et al., 2010).

As expected, economically relevant crop plant species have commonly been the main target for antimicrobial peptides research and biotechnological application, due to associated agricultural and social demand and impact, but also because the higher availability of molecular data. In spite of such trend, several research efforts have pointed a huge potential for wild plant species prospection to achieve discovery of novel AMPs, as well as the identification of novel biological functions to known AMPs. Hence, plant biodiversity in different biomes and ecosystems is an outstanding promising focus on antimicrobial peptides investigation (Pieters & Vlietinck, 2005).

A very recent revised list of experimentally confirmed plant AMP, with physicochemical and antibacterial specificity is also available, comprising members of the main families: 23 defensins, eight hevein-like, six vicilin-like, five knottins, four cyclotides, two Impatiens-like, two sepherins, two snakins, one MBP-1 (from maize), one glutamate-rich, one glycine-rich, and other eight peptides not classified in these structural families (Pelegrini et al., 2011).

3. Structural analyses and function

An example of the lack in plant AMP cataloguing is the organized database for antibacterial phytochemical compounds hosted and managed by Kioto Encyclopedia of Genes and Genomes (KEGG Compound, Plant Secondary Metabolites), where the related biosynthetic enzymes and molecular targets in human organism are fully described. However, no plant-derived AMP or coding gene sequence is set, since the database allows the access only to human microbicidal and cytotoxic peptides.

Bioinformatics have been essential on plant AMP prospection and establishment of catalogues, which have increased in last decade (Hammami et al., 2009). Several sequence aligment tools, usually BLAST-based (www.ncbi.nlm.nih.gov/blast) are available, but other main technical approaches have included *in silico* structural peptide prediction and putative analyses of AMP-target molecule interactions. Last but not least, direct chemical isolation have also depended of *in silico* tools to achive more complete structural and functional characterization of isolated AMP.

3.1 Isolation

Complementarily to the several classes of chemical compounds synthesized by plants secondary metabolism, with proved antimicrobial effects, the characterization of plant AMPs function correlated to such effects is extremely useful to improve research in this area. Although very likely, supposing that AMPs are co-responsible or even responsible for antimicrobial activity of complex plant extracts relies on sub-fractioning and experimental evidence (Moreira et al., 2011). Direct isolation of plant AMPs has succeeded in several species (Kovaleva et al., 2009; Rogozhin et al., 2011), associated to further cDNA cloning and characterization as defensin and lipid transfer protein. Also, extraction and purification of candidate peptides, followed by sequencing and match to AMP databases, allowed the identification of novel promising antimicrobial molecules in wheat (Odintsova et al., 2009).

Efforts to prospect AMPs in plant crop species and many examples of direct isolation followed by functional assays and characterization are available. As alternative to fungicide usage to pathogen control in agriculture, several studies have focused on searching for plant proteins and peptides with antifungal activities (AFPs). Recently, two novel 10 and 15 kDa

AFPs were isolated from rosemary pepper (*Lippia sidoides* Cham.) flowers through Octyl-Sepharose hydrophobic column separation, and were able to inhibit the development of *Botrytis cinerea*, an economically harmful phytopathogen for many crops (Moreira et al., 2011). The N-termini sequences of these AFPs have homology with NBS-LRR R proteins, well known plant defense elements.

A 11,500 heterodimeric antifungal protein, named Pa-AFP1, highly similar to 2S albumin family, was purified by anionic exchange Q-Sepharose chromatography associated with HPLC reversed-phase C4 chromatography and structurally confirmed as dimer by MALDI-TOF spectra analyses (Ribeiro et al., 2011). It inhibits the growth of fungus *Colletotrichum gloeosporioides*, but no antibacterial nor anti-yeast activity was observed. an antiviral 2 KDa peptide was purified from sorghum seeds by gel filtration, ion exchange and high-performance liquid chromatography (HPLC), and showed strong inhibition of herpes simplex virus type 1 (HSV-1) and bovine herpes virus (BHV) replication (Camargo-Filho et al., 2007). On the other hand, two anti-yeast peptides were isolated from seeds of a phytobacterial-resistance pepper (*Capsicum annuum*) genotype and identified by amino acid sequencing (Ribeiro et al., 2007). Another research identified peptides with bactericidal activity from sesame (*Sesamum indicum*) kernel flour; one of them, with 5.8 KDa, showed activity only against *Klebsiella* sp., a Gram-negative bacterium causal of human urinary infection (Costa et al., 2007). More detailed structural and functional results were achieved for a cowpea seed γ-thionin/defensin, a wide-spectrum bactericide whose primary structure, mechanism of action and tissue localization during germination provided the understanding of these bioactive peptides in plant defense responses (Franco et al., 2006).

An example of how native or introduced plant biodiversity may be a fruitful option on AMP research is the recent growing number of such peptides identified in Brazilian species. Direct purification was achieved in originally African legume *Crotalaria pallida*, a widely dispersed weed in South America and abundant in drought and warm "caatinga" biome. A novel peptide structurally similar to defensin/2S-albumin was isolated from seeds and presented inhibitory effects over bacteria (Pelegrini et al., 2008). From seeds of guava (*Psidium guajava*) and passion fruit (*Passiflora edulis*), the antifungal and antibacterial peptides Pg-AMP1, passiflin and a 2S albumin-like were isolated. PgAMP1 comprises approximately 6 KDa of molecular mass and small amounts of a homodimer; amino acid sequencing indicated it belongs to glycine-rich plant protein family, being the first one having activity towards Gram-negative bacteria; instead passiflin and the 2S albumin-like peptide show high antifungal properties (Pelegrini et al., 2006; Lam & Ng, 2009).

Several plant species from Atlantic rainforest and "cerrado" biomes have been studied to confirm and explain their antimicrobial activity as transmitted popular medicine (reviewed in Pestana-Calsa et al., 2010). For instance, out from 32 plant species selected after field survey, extracts from 13 species presented antimicrobial activity against *Staphylococcus aureus*, but further analyses to identify potential peptides involved in such activity keep lacking (Brasileiro et al., 2006; Silva Jr. et al., 2009). Molecular and bioinformatics approaches could start from applied phytochemical researches similar to these and is very likely that known and novel AMPs could be found.

3.2 Structural and functional analysis

As any peptide, AMP function is strictly dependent of strucuture. Specifically for relatively abundant α-helical AMPs, structural features and physicochemical properties have been targeted to increase antimicrobial activity, usually by changing molecular size and charge, residues arrangement, hydrophobicity, amphipathicity and helix folding probability (Tossi et al., 2000; Tian et al., 2009). Antimicrobial peptides structure/activity ratio results in their molecular diversity, but they do share some common features, as the low molecular weight and the variable number of disulphide bond-cysteines residues stabilizing conserved scaffolds (Padovan et al., 2010c), which is used to group them into different structural classes, as depicted in Figure 2.

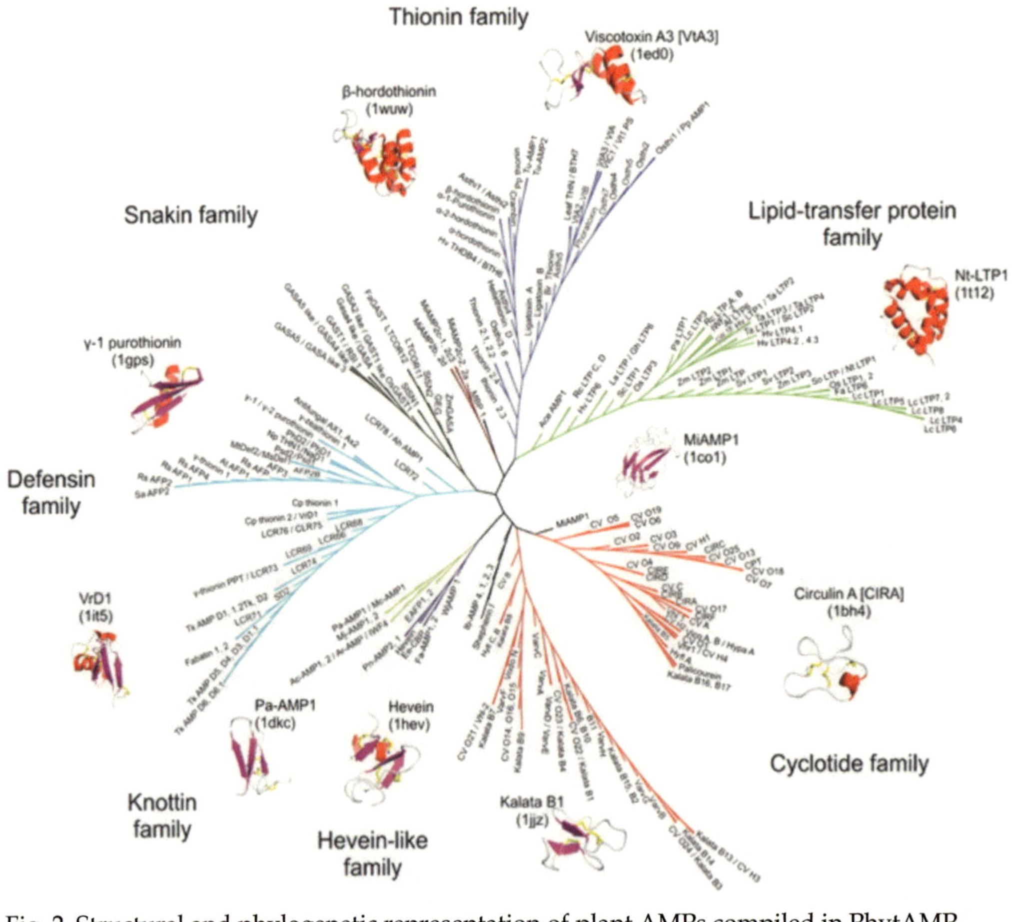

Fig. 2. Structural and phylogenetic representation of plant AMPs compiled in PhytAMP database. α-helices and β-sheets are respectively shown in red and purple (from original by Hammami et al., 2009)

Aiming to achieve more effective peptides, natural/original sequences may be re-arranged (Boman et al., 1989). Several protein databases and analyses tools (e.g. Swiss-Prot and links) are available, as softwares designed specifically to deal with predicting and modifying peptide structure and physicochemical improvement as ArgusLab, Anthewin and Peptool (detailed usage described by Wang, 2007; Hao et al., 2008). Simulations within *in silico* environments have been applied to test AMPs topology concerning biological effects and to unravel their probable mode of action, usually by interactions that destabilize lipid bilayers (e.g. maculatin, that changes the lysis mechanism depending on lipid structure of target membrane; Bond et al., 2008).

Sometimes, naturally occurring AMP structure is not the most effective. The influence of 2 disulfide bonds in a small β-sheet AMP (Ib-AMP1) from *Impatiens balsamina* seeds was verified as not essential for antimicrobial activity, because the synthetic linear analogs displayed by 4.8-fold higher inhibitory specificity than the wild-type peptide (Wang et al., 2009). Also the seed antimicrobial peptide Cy-AMP1 from *Cycas revoluta* was analyzed on its chitin-binding ability, a well-conserved feature in other AFP as knottin and hevein (Yokoyama et al., 2009): antifungal activity of the peptide was strongly reduced after variations in chitin-binding motifs.

Both MsDef1 and MtDef4 defensins, potent growth inhibitors of several filamentous fungi including *Fusarium graminearum*, induced plasma membrane permeabilization; however, MtDef4 is more efficient based on its unique γ-core motif, indicating that it defines specific antifungal properties of each defensin, and so may help de novo design of more effective AMP (Sagaram et al., 2011). The barley derived α-Hordothionin (αHTH), another membrane-permeabilizing peptide with broad-range antimicrobial activity, is supposed to act as small water-selective channel, through the αHTH double α-helix core when the peptide interacts with anions (Oard, 2011); conserved cysteine and tyrosine residues lined pore walls, resembling aquaporins that delivers water molecules to the lipid bilayer center, what may lead to localized membrane disruption.

The well known chitin-binding lectin hevein has served as template to synthetic mutant peptides that interact with chitin oligosaccharides (main components of fungi cell walls), but mutant AMP versions may present decrease in the association kinetics to target chito-oligosaccharides (Chávez et al., 2010). Such results, provided by nuclear magnetic resonance followed by image modelling analysis tools, pointed that mutant and parent Hev32 peptides three-dimensional structures were quite similar, including orientations of the three key Trp aromatic residues; hence, it was supposed that the mutant lower affinity relied on distinct topology orientation of key side chains and protein-sugar intermolecular essential hydrogen bonds.

Structural and functional design of AMPs from plants manipulated to be active in plants must attend to the agricultural demand for novel antimicrobial compounds, more specifically in plant disease control, with lower toxicity to consumers and environment (reviewed by Montesinos & Bardaji, 2008). In medicine and therapeutics for infectious diseases, AMPs effects on human cells can be widely verified in different gene expression levels, allowing to accurately confirming expected antimicrobial action without undesirable side effects (Ulrich-Merzenich et al., 2007). Also, synthetic peptides engineering have contributed with novel insights for antimicrobial drugs and treatments, by testing in vitro and in vivo several amino acid sequences putatively harbouring antagonistic effects to microorganisms (Choi & Moon, 2009).

Although every class or family of AMP has specific features in structure, activity and potential technological applications, few are so remarkably intriguing as the generally named cyclotides, very stable plant cyclic peptides which harbour many potential technological usages in pharmaceutical and agricultural strategies, according to their various bioactivities and fitness as protein-engineering templates (Henriques & Craik, 2010). Cyclotides compise the largest known family of head-to-tail cyclic peptides, have approximately 30 amino acid residues with a complex structure containing a circular peptide backbone and a cystine knot. They are mainly found in plants of Violaceae and

Rubiaceae families, and supposed to act mostly in plant protection. In addition to insecticidal properties, cyclotides may have cytotoxic, anti-HIV and antimicrobial effects, among other activities as to inhibit neurotensin binding ability (Gerlach et al., 2010). A cyclotide in alkaloid fraction from root bark of *Discaria americana* (Rhamnaceae) was isolated and structurally determined, but did not inhibit the growth of challenged bacteria significantly (Giacomelli et al., 2004). However, other cycletides extracted from *Scutia buxifolia*, also a Rhamnaceae family member, were much more efficient to inhibit bacterial cells than *Discaria*-derived cycletides, although no antifungal effect was observed (Morel et al., 2005).

Regarding prospection of structure-function clues, cyclic peptides isolated from Euphorbiaceae species have been intensively studied due to their rigid three-dimensional conformation, considered to be essential for bioactivity over lipid membranes (Barbosa et al., 2011). As example, cyclotide labaditin and derived synthetic open chain analogs were chemically and virtually analysed in comparison over their interaction with lipid bilayers, and results suggested the native labaditin had greater membrane insertion. A possible mechanism for this is based on initial hydrophobic interaction with the lipid membrane followed by conformational change, peptide adsorption and internalization; indeed, native labaditin reduced viability in Gram-positive bacteria (Barbosa et al., 2011).

Some plant AMPs have so many unique structural features which impair its insertion into any previous, well-characterized AMP family. This is the case of antimicrobial peptide Ib-AMP1, formed by a 20-residue disulfide-linked beta-sheet and usually found in the seeds of *Impatiens balsamina*. Using it as template molecule, synthetic analogs were obtained in order to check the relevance of 2 disulfide bonds on the antimicrobial activity and specificity (Wang et al., 2009).

Beyond conventionally detected cationic antimicrobial peptides, there are also anionic antimicrobial peptides/proteins (AAMPs), reported since 1980s and accepted as important components of innate immune systems of plants and animals (Harris et al., 2009). AAMPs present activity against bacteria, fungi, viruses and insects, but noteworthy their antimicrobial activity is believed to be secondary. Structures vary from alpha-helical peptides in amphibians to cyclic cystine knots in some plant peptides, and certain AAMPs are suggested to link metal ions forming cationic salt bridges with negatively charged lipids of microbial membranes. In bioinformatics context, softwares and analysis parameters must be adjusted and present enough flexibility to cope in a suitable manner with such huge structural and charge inversion, if compared to data obtained from "conventional" antimicrobial peptides dynamics and biological function.

3.3 Heterologous expression

Several AMPs, mostly defensins, have been studied through gene cloning and expression in heterologous systems (Kovaleva et al., 2011). In this context, perhaps one of the main contributions of bioinformatics is the codon-usage optimization for heterologous expression of AMPs, depending on host organism.

Heterologous expression of AMPs also refers to increasing interest for production of antimicrobial compounds such as the defensins, whose applications include health, agriculture and industry as targets. Specifically to protect food or bio-fuel crops, several strategies have been proposed and tested in order to isolate and produce defensins. Experimental viability is still mostly constrained in academic research, where defensins have been heterogously expressed in bacteria, yeasts, fungi and plants (Padovan et al.,

2010b); on the other hand, (bio)-chemical synthesis is not usual yet for commercial production purposes, and here the most striking challenge is to keep correct protein/peptide folding *in vitro*.

A novel defensin-like peptide, isolated from *Nicotiana megalosiphon*, NmDef02 was heterologously expressed in the yeast *Pichia pastoris*, and the purified recombinant protein was found to display antimicrobial activity in vitro against important plant pathogens. Constitutive expression of NmDef02 gene in transgenic tobacco and potato plants enhanced resistance against various plant microbial pathogens, including the oomycete *Phytophthora infestans*, causal agent of the economically important potato late blight disease, under greenhouse and field conditions (Portieles et al., 2010).

Other type of AMP, a defensin from cowpea seeds was heterologously assessed on its putative alpha-amylase inhibitory action probably involved in protection against pests (Dos Santos et al., 2010). Its cDNA was cloned into plasmidial expression vector and transformed into *Escherichia coli* cells; the recombinant peptide was then purified via affinity chromatography, identified by sequencing and submitted to alpha-amylase inhibition assay together with seeds-isolated defensin. Both peptides inhibited alpha-amylases from weevil (*Callosobruchus maculatus*) but were not able to inhibit mammalian alpha-amylases.

Heterologous expression, supported by comprehensive *in silico* prediction and peptide design tools, will probably keep being one of the most helpful tools for biological research over plant AMP, based on success in literature. However, large scale production of AMP establishment will depend on results of future studies where the main tasks shall be the engineering/re-design of more stable and self-folding peptides, and definition of optimized biotechnological procedure to cost-effectively produce the peptide, as molecular farming transgenic plants.

4. AMP-coding genes promoter analysis

The search on regulatory genomic regions, mainly promoter elements, has presented increasing usefulness to start unravelling the control mechanisms of activation of AMP-coding genes. These are known to be expressed in defense signalling against microbial pathogens, involving several transduction components that depend on the action of hormones as jasmonic acid (JA), ethylene (ET) and abscisic acid (ABA). An *in silico* approach with potential applicability on tracking regulatory pathways of AMP coding genes is their promoter sequence analysis.

It is known that JA and ET signaling pathways are synergistic for activation of AMPs, especially the defensin PDF1.2. The coding pdf1.2 is targeted and expressed after activation by ORA59 transcription factor, a APETALA2/Ethylene response Factor (AP2/ERF)-domain protein, which is dependent of JA and ET signaling pathways. The pdf1.2 promoter contains two GCC boxes that were confirmed to be the ligation site for ORA59 transcription factor, enabling PDF1.2 coding gene to respond simultaneously to both hormones (Zarei et al., 2011).

The characterization of tissue-specific and pathogen-inducible promoters is essential for localized expression of defense-related genes as AMP. Wheat and rice defensin genes expressed in early developing grain and during grain germination were compared regarding their promoters activity, through stable transformation with promoter-GUS reporter fusion constructs (Kovalchuk et al., 2010). Activity was detected mainly in ovary before and at anthesis in both transgenic cereal species, but differences concerning one or other species were

observed in the expression of transgenic constructs in reproductive tissues. Even so, wheat and rice promoters were strongly induced by wounding in leaf, stem and grain.
Specifically for plant lipid transfer proteins (LTPs), which are very unkown on antimicrobial function although being well-known in other cellular activities, the study of gene promoter sequence may reveal defense-related aspects. From a *Vitis vinifera* genomic library 2,100-bp fragment, the coding region and the promoter of a lipid transfer protein 1 (VvLTP1) was screened, revealing several defense-related *cis*-regulatory elements, like MYB-boxes (Laquitaine et al., 2006). The expression of VvLTP1 promoter-GUS fusion construct in *Arabidopsis thaliana* indicated the antifungal response of VvLTP1 from grape.

5. AMP from plant-related microorganisms

The growing number of distinct species whose genomes, transcripts, proteins and other molecular data are being deposited in public databases makes more reasonable to consider antimicrobial peptides produced by other organisms, specially in the cases where these species, although not plants, have strict ecological relationship with host or neighboring plants. Several examples have been described in literature normally including well-known or potential endophytes, which produce AMPs in a predictable symbiotic context. Obviously, the applications derived from this type of research are supposed to be useful within sustainable biological control of pathogens and pests.
The mycelium-forming actinomycetes of the genus Frankia (well known producers of bioactive compounds) are commonly found as symbionts in actinorhizal plants, performing facultative nitrogen-fixing. Bioinformatic analysis of the strains ACN14a, CcI3, and EAN1pec by genomes prediction and by intact cells MALDI-TOF allowed the identification of putative coding regions and molecules associated to cyclic peptides, siderophores, pigments, signaling molecules and specialized lipids, from which some cyclotides and lipid-transfer proteins are considered to be essential for host-endophyte recognizing and to inhibit other competitor microorganisms, as pathogens (Udwary et al., 2011).
Pyoverdines (PVDs), high affinity siderophores well studied in *Pseudomonas aeruginosa*, were searched *in silico* in *P. fluorescens* SBW25 (a plant growth-promoter endophyte) complete but not annotated genome, where 31 genes putatively involved in PVD biosynthesis, transport or regulation, could be identified (Moon et al., 2008). Since pyoverdine-mediated iron uptake is essential for this endophyte, structural analysis of its PVDs was achieved and defined it as a partly cyclic seven residue peptide backbone, which makes the bacteria able to utilize a wide variety of exogenous PVDs.
Another interesting example can be traced for alamethicin, a membrane-active AMP produced by root symbiont fungus *Trichoderma viride* that permeabilises plasma membrane, mitochondria and plastids of *in vitro* cultured plant cells by creating voltage-dependent pores (Aidemark et al., 2010). Cultured plant cells pre-treated with pathogen elicitors did not get resistant to alamethicin, while those treated with cellulase did; this suggested that different membrane lipid composition induced by cellulase may render the cells resistant to alamethicin, in a mechanism where possible cellulase-secreting pathogens (as several phytopathogenic fungi) would suffer alamethicin action in their membranes. Other fungus-derived AMPs have been described as candidates for production and biotechnological uses in plant protection, as the cystein-rich antifungal peptide AcAFP, secreted by *Aspergillus clavatus* (Skouri-Gargouri et al., 2010).

Development of applications derived from plant-related microorganisms AMP in biological control of pathogens and pests will depend on ecological modeling studies, even if it starts from a single AMP being produced by an endophyte but biologically active in the host plant.

6. Novel AMP functions in plants

Relatively recent studies have added new insights in plant-derived AMP functions, other than classical antimicrobial activity. In fact, the ubiquitous presence of AMPs in distantly related taxa allows the concept that novel functions for derived AMP genes and products may have arisen during each species evolutionary history. Literature is rich in well-established as well as still unkown biological functions for AMP in plants. In this context, even no-peptidic biomolecules could be considered in a wider view.

Perillic acid is a terpenoid plant extract with antiinfective and anticancer properties, and is a small cyclic molecule structurally similar to salicylic acid. It is known to cause large-scale membrane thinning, a clearly possible antimicrobial activity through a membrane-lytic mechanism very close to that of AMPs (Khandelia et al., 2010). Indeed, also subproteins or subpeptides may be relevant to antimicrobial activity *in silico* prospection. Plant-specific insert domain (PSI) is a region of about 100 amino acid residues, contained in several plant aspartic protease (AP) precursors. The PSI from potato aspartic protease 1 was purified after heterologous expression, and was able to kill pathogenic spores in a dose-dependent manner, without deleterious effect on host plant, and through lytic interaction with microbial cell wall/or membrane (Muñoz et al., 2010).

In roots of host legume plants, a complex and evolutionary successful symbiosis takes place with nitrogen-fixing *Rhizobium* bacteria. Surprisingly, the bacteria irreversible differentiation to bacterioid form is also dependent of plant factors, nodule-specific cysteine-rich (NCR) peptides, that are driven to bacterial membrane and cytosol (van de Velde et al., 2010). Since NCRs are similar to AMPs, it is very likely that the host plant adopted effectors from innate immune system for symbiosis, resulting in a control mechanism to the endosymbiotic bacteria cell fate.

Other proteins initially thought to be not related to AMPs have been well linked to regulation of defense mechanisms that include such peptides. Examples have been described where specific phosphatases are key responsive proteins to pathogen infection and induce plant defensins (Widjaja et al., 2010). In addition, some AMPs may be effectors as well as regulators of other molecules. Plant lipid transfer proteins (LTPs) are ubiquitous lipid-binding proteins involved in diverse stress responses; for example, 14 LTPs from *Tamarix hispida* Willd. were screened over possible functions in response to various abiotic stresses (Wang et al., 2009). Results showed that all 14 LTPs were expressed in roots, leaves and stems, in different levels according to the organ; also, some of them were induced by NaCl, PEG, $NaHCO_3$, $CdCl_2$ and ABA, suggesting novel roles beyond defense, and in abiotic stress tolerance. In flower buds of *Brassica campestris* L. ssp. *chinensis*, a putative LTP of 103 amino acids was characterized as being a membrane protein with a signal peptide at the N-terminus, and strongly related to pollen viability and male sterility (Tian et al., 2009).

In the endosperm cells undergoing programmed cell death, LTPs have been described as participating in recycling of endosperm lipids, or acting as protease inhibitors to protect growing cotyledons from released proteases (Eklund & Edqvist, 2003).

The mode of action of AMPs on external pathogen or pest cells seems to hide still unknow mechanisms: certain cyclotides are able to increase permeability across nematodes cuticle layers, suggesting that one action of the such AMPs involves the interaction with the lipid-rich epicuticle layer at the pathogen worm surface (Colgrave et al., 2010).

7. Towards biotechnological applications

Identification of plant defense genes against pathogens and environmental stresses provides novelties to plant breeding, also genetic transformation (Vidal et al., 2003), mainly due cross-activity of distinct organisms AMPs that has significant potential in phytopatology and plant resistance improvement. As already confirmed, insect-derived AMP gene is able to be expressed in plant genome and its product can be correctly sorted in plant cell or tissue. The metchnikowin, a 26-amino acid residue proline-rich AMP from *Drosophila melanogaster*, was used for resistance in barley against pathogenic fungi (Rahnamaeian et al., 2009). In an interfamily transfer, transgenic tobacco (Solanaceae) and peanut (Fabaceae) plants expressed a defensin from mustard (Brassicaceae), effective against to phytopathogenic fungi (Anuradha et al., 2008).

Many examples of structurally manipulated AMP followed by their transfer to crop plants resulted in increased resistance against phytopathogens: about 18 sequence-optimized AMPs have been transfected to plants with beneficial results for agricultural (Marcos et al., 2008; Jan et al., 2010; Ma et al., 2010; Eggenberger et al., 2011). Plant-derived AMPs are not just possible templates for bioactive molecule design, but candidates to nanotechnologies applied to crop protection as inner content of nanocapsules to be used in greenhouses or in the field to enhance plant resistance and or to control phytopathogens, pests or parasitic weeds (Perez-de-Luque & Rubiales, 2009; Pestana-Calsa et al., 2010; Imamura et al., 2010; Choi et al., 2009) or as integral domains fused to another host organ/tissue targeted protein to local delivery of AMPs in transgenics (Bryksa et al., 2010). Enginnered LTPs have the potential to be utilized as scaffolds to design hydrophobic ligand biosensors or to serve as drug carriers (Choi et al., 2007). Inclusion of plastid transformation technology to enhance yield of peptides accumulation in molecular farming approaches is also very expected in next few years (Oey et al., 2009), establishing another alternative to the production of next-generation antimicrobial peptides in plants, from plants or non-plant source organism.

Novel bioactive molecules produced by plant growth-stimulator endophytes have been comproved on suppressing the growth of bacterial and fungal plant pathogens, as the case of TOMM, a thiazole/oxazole-modified microcin produced by the soil bacterium *Bacillus amyloliquefaciens*. TOMM requires extensive posttranslational modification to become bioactive against other bacteria, involving host plant factors to achieve such activity (Kajula et al., 2010). How to solve this extensive need for proper peptide folding in other expression systems remains another good question for next years. Probably, answers will come with intensive *in silico* modeling to trial all theoretical three-dimensional intramolecular interactions.

By the way, AMP-target interaction has been a recurrent bioinformatic issue in human immunology and nutrition, concerning several allergenic AMPs. Immunodominance of certain T-cells epitopes have been screened by modeling, helping breeders to achieve the theoretical amino acid sequence of a given allergenic AMP that would have the lowest allergenic potential (Oseroff et al., 2010). AMPs have also been tested and used for medicine biotechnological applications, as cardiovascular functioning and diseases (Li, 2009).

A biotechnological focus on plant AMPs with potentially high impact in biofuels industry is the investigation of such peptides effects over bioethanol production, from alcoholic fermentation by yeasts as well as from promising cellulolytic filamentous fungi species. In both cases, it is likely that AMPs present in plant feedstock may still have inhibitory activity on these industrial microorganisms, and so, over industrial yield (Nierop et al., 2008). In the specific case of sugarcane and other potential crops for bioethanol production in Brazilian northeast, as sorghum, Opuntia and other 'caatinga'-adapted plant species, research efforts have been directed to quantify this inhibition (if significant) and to achieve alternative agronomical and/or breeding solutions to reduce it (Bioethanol Research Network of Pernambuco).

8. Conclusion

As presented, antimicrobial compounds have been relatively well studied, since a long period and usually from native traditional usage of plants. However, biological and chemical bioinformatics have focused phytochemicals prospection and effects, while plant antimicrobial peptides are left apart. Expansion of AMP prospection through *in silico* methodologies is in perfect adjustment to the huge amount of biological data still not screened. Several antimicrobial effects observed in some plant extracts may also be explained due to AMPs supposed to be in sample, but such valuable information still has to be generated in lab benches as well as in databases extensive computational analyses.
Plant biodiversity in natural and anthropical ecosystems provide almost infinite targets number to unravel novel AMP candidates. Achieving such results relies on the generation of molecular data from crop and wild plant species. "-Omics" based experiments will then be more profitable and reliable, making easier the application of scientific knowledge from molecular biology and bioinformatics to develop systems biology approaches in accordance to nowadays and future needs in medicine, agriculture and industry.

9. Aknowledgement

Authors are grateful to CNPq (Conselho Nacional de Desenvolvimento Científico e Tecnológico), FACEPE (Fundação de Apoio à Ciência e Tecnologia do Estado de Pernambuco), Bioethanol Research Network of Pernambuco and UFPE (Universidade Federal de Pernambuco) for financial support. Authors would like to thank collaborative researchers for critical reviewing of the manuscript.

10. References

Aidemark, M.; Tjellström, H.; Sandelius, A.S.; Stålbrand, H.; Andreasson, E.; Rasmusson, A.G.; Widell, S. (2010). *Trichoderma viride* Cellulase Induces Resistance To The Antibiotic Pore-Forming Peptide Alamethicin Associated With Changes In The Plasma Membrane Lipid Composition Of Tobacco BY-2 Cells. *BMC Plant Biol*, Vol.14, No.10, (2010), pp. 274, ISSN.

Antunes, P.W.P. (2008).Thesis, Universidade Federal de Viçosa (2008), pp.

Anuradha, T.S.; Divya, K.; Jami, S.K.; Kirti, P.B. (2008). Transgenic Tobacco And Peanut Plants Expressing A Mustard Defensin Show Resistance To Fungal Pathogens. *Plant Cell Rep*, Vol.27, No. ,(2008), pp.1777-1786, ISSN.

Baggerman, G.; Verleyen, P.; Clynen, E.; Huybrechts, J.; De Loof, A.; Schoofs, L. (2004). Peptidomics Rewiew. *J. Chromatogr. B. Analyt. Technol. Biomed. Life Sci*, Vol. 803, (2004), pp.3-16, ISSN.

Baginsky, S. (2009). Plant Proteomics: Concepts, Applications, And Novel Strategies For Data Interpretation. *Mass Spectrom*. Rev, (2009), Vol.28, pp.93-120, ISSN.

Barbosa, M.O. (2008). Thesis, Universidade Federal De Viçosa, 2008 Pp.

Barbosa, S.C.; Cilli E.M.; Dias, L.G.; Stabeli, R.G.; Ciancaglini, P. (2011). Labaditin, A Cyclic Peptide With Rich Biotechnological Potential: Preliminary Toxicological Studies And Structural Changes In Water And Lipid Membrane Environment. *Amino Acids*, Vol.40, No.1, (January 2011), pp.135-44, ISSN.

Belarmino, L.C.; Capriles, P.V.; Crovella, S.; Dardene, L.E.; Benko-Iseppon, A.M. (2010) EST-Database Search Of Plant Defensins - An Example Using Sugarcane, A Large And Complex Genome. *Curr Protein Pept Sci*, 2010 Vol. 1, No.11,(May 2010), pp.248-54, ISSN.

Bioethanol Research Network of Pernambuco, Brazil. http://www.cetene.gov.br/laboratorios/bioetanol.php (Accessed in April 3th, 2011).

Boman, H.G.; Wade, D.; Boman, I.A.; Wahlin, B.; Merrifield, R.B. (1989). Antibacterial And Antimalarial Properties Of Peptides That Are Cecropin-Melittin Hybrids. *FEBS Lett*, Vol. No.259, (1989), pp.103-106, ISSN.

Bond, P.J.; Parton, D.L.; Clark, J.F.; Sansom, M.S.P. (2008).Coarse-Grained Simulations Of The Membrane-Active Antimicrobial Peptide Maculatin 1.1. *Biophys. J*, Vol.95, (2008), pp.20083802–3815, ISSN.

Boutrot, F.; Chantret, N.; Gautier, M.F. (2008). Genome-Wide Analysis Of The Rice And Arabidopsis Non-Specific Lipid Transfer Protein (Nsltp) Gene Families And Identification Of Wheat Nsltp Genes By EST Data Mining. *BMC Genomics*. 2008 Feb 21; Vol.9, No., (February 2008), Pp.86, ISSN.

Brahmachary, M.; Krishnan, S.P.T.; Koh, J.L.Y.; Khan, A.; Seah, S.H.; Tan, T.W.; Brusic, V.; Bajic, V.B. (2004). ANTIMIC: A Database Of Antimicrobial Sequences. *Nucl. Acids Res*, Vol. 32, No., pp. 586-D589 (2004), ISSN.

Brasileiro, B.G.; Pizziolo, V.R.; Raslan, D.S.; Jamal, C.M.; Silveira, D. (2006). Antimicrobial And Cytotoxic Activities Screening Of Some Brazilian Medicinal Plants Used In Governador Valadares District. *Brazilian J. Pharmaceut. Sci*, Vol.42, No., (2006), Pp.195-202, ISSN.

Bryksa, B.C; Horimoto, Y.; Yada, R.Y. (2010). Rational Redesign Of Porcine Pepsinogen Containing An Antimicrobial Peptide. *Protein Eng. Des. Sel.*, Vol.23, No.9, (September 2010), pp.711-719, ISSN

Calsa Jr., T. & Figueira, A. (2007). Serial Analysis Of Gene Expression In Sugarcane (*Saccharum* Spp.) Leaves Revealed Alternative C4 Metabolism And Putative Antisense Transcripts. *Plant Mol. Biol*, Vol.63, No., (2007), pp.745-762, ISSN

Camargo-Filho, I.; Cortez, D.A.G.; Ueda-Nakamura, T.; Nakamura, C.V.; Dias-Filho, B.P. (2007). Antiviral Activity And Mode Of Action Of A Peptide Isolated From *Sorghum bicolor*. *Phytomedicine*, Vol.15, No., (2007), pp.202–208, ISSN.

Chávez, M.I.; Vila-Perelló, M.; Cañada, F.J.; Andreu, D.; Jiménez-Barbero, J. (2010). Effect Of A Serine-To-Aspartate Replacement On The Recognition Of Chitin

Oligosaccharides By Truncated Hevein. A 3D View By Using NMR. *Carbohydr Res*, Vol.345, No.10, (July 2010), pp.1461-1468, ISSN

Chen, L.; Zhang, S.; Illa, E.; Song, L.; Wu, S.; Howad, W.; Arús, P.; Van De Weg, E.; Chen, K.; Gao, Z. (2008). Genomic Characterization Of Putative Allergen Genes In Peach/Almond And Their Synteny With Apple. *BMC Genomics*, Vol.9, No, (November 2008), pp.543, ISSN

Choi, E.J.; Mao, J.; Mayo, S.L. (2007). Computational Design And Biochemical Characterization Of Maize Nonspecific Lipid Transfer Protein Variants For Biosensor Applications. *Protein Sci.*, Vol.16, No.4, (April 2007), pp.:582-528, ISSN

Choi, J.; Moon, E. (2009). Identification Of Novel Bioactive Hexapeptides Against Phytopathogenic Bacteria Through Rapid Screening Of A Synthetic Combinatorial Library. *J Microbiol Biotechnol.*, Vol.19, No.8, (August 2009), pp.792-802, ISSN

Choi, M.S.; Kim, Y.H.; Park, H.M.; Seo, B.Y.; Jung, J.K.; Kim, S.T.; Kim, M.C.; Shin, D.B.; Yun, H.T.; Choi, I.S.; Kim, C.K.; Lee, J.Y. (2009). Expression Of Brd1, A Plant Defensin From *Brassica Rapa*, Confers Resistance Against Brown Planthopper (*Nilaparvata Lugens*) In Transgenic Rices. *Mol Cells.*, Vol.28, No.2, (August 2009), pp.131-137, ISSN

Colgrave, M.L.; Huang, Y.H.; Craik, D.J.; Kotze, A.C. (2010). Cyclotide Interactions With The Nematode External Surface. *Antimicrob Agents Chemother*, Vol.54, No.5, (May 2010), pp.2160-2166, ISSN

Costa, F.T.; Maria-Neto, S.; Bloch Jr., C.; Franco, O.L. (2007). Susceptibility Of Human Pathogenic Bacteria To Antimicrobial Peptides From Sesame Kernels. *Curr. Microbiol.*, Vol.55, No., (2007), pp.162–166, ISSN

De Beer, A.; Vivier, M.A. (2008). Vv-AMP1, A Ripening Induced Peptide From *Vitis vinifera* Shows Strong Antifungal Activity. *BMC Plant Biol.*, Vol.8, No., (2008), pp.75, ISSN

Djordjevic, M.A.; Oakes, M.; Li, D.X.; Hwang, C.H.; Hocart, C.H.; Gresshoff, P.M. (2007). The *Glycine Max* Xylem Sap And Apoplast Proteome. *J Proteome Res.*, Vol.6, No.9, (September 2007), pp.3771-3779, ISSN

Dos Santos, I.S.; Carvalho, A. de O.; De Souza-Filho, G.A.; Do Nascimento, V.V.; Machado, O.L.; Gomes, V.M. (2010). Purification Of A Defensin Isolated From *Vigna unguiculata* Seeds, Its Functional Expression In *Escherichia Coli*, And Assessment Of Its Insect Alpha-Amylase Inhibitory Activity. *Protein Expr Purif.*, Vol.71, No.1, (May 2009), pp.8-15, ISSN

Eggenberger, K.; Mink, C.; Wadhwani, P.; Ulrich, A.S.; Nick, P. (2011). Using The Peptide BP100 As A Cell-Penetrating Tool For The Chemical Engineering Of Actin Filaments Within Living Plant Cells. *Chembiochem.*, Vol.12, No.1, (January 2011), pp.132-137, ISSN

Eklund, D.M.; Edqvist, J. (2003). Localization Of Nonspecific Lipid Transfer Proteins Correlate With Programmed Cell Death Responses During Endosperm Degradation In *Euphorbia Lagascae* Seedlings. *Plant Physiol.*, Vol.132, No.3, (July 2003), pp.1249-1259, ISSN

Expasy Proteomics Server, Prosite. Http://www.Expasy.Ch/Prosite (Accessed In March 4th, 2011).

Fjell, C.D.; Hancock, R.E.W.; Cherkasov, A. (2007). Amper: A Database And An Automated Discovery Tool For Antimicrobial Peptides. *Bioinformatics*, Vol.23, (2007), pp.1148–1155, ISSN.

Franco, O.L.; Murad, A.M.; Leite, J.R.; Mendes, P.A.M.; Prates, M.V.; Bloch Jr., C. (2006). Identification Of A Cowpea C-Thionin With Bactericidal Activity. *FEBS J.*, Vol.273, (2006), pp.3489–3497, ISSN

Gerlach, S.L.; Burman, R.; Bohlin, L.; Mondal, D.; Göransson, U. (2010). Isolation, Characterization, And Bioactivity Of Cyclotides From The Micronesian Plant *Psychotria Leptothyrsa. J Nat Prod.*, Vol.73, No.7, (July 2010), pp.1207-1213, ISSN

Giacomelli, S.R.; Maldaner, G.; Gonzaga, W.A.; Garcia, C.M.; Silva, U.F.; Dalcol, I.I.; Morel, A.F. (2004). Cyclic Peptide Alkaloids From The Bark Of *Discaria Americana. Phytochem.*, Vol.65, No., (2004), pp.933–937, ISSN

Graham, M.A.; Silverstein, K.A.; Cannon, S.B.; Vandenbosch, K.A. (2004). Computational Identification And Characterization Of Novel Genes From Legumes. *Plant Physiol.*, Vol.135, No.3, (July 2004), pp.1179-1197, ISSN

Gueguen, Y.; Garnier, J.; Robert, L.; Lefranc, M.P.; Mougenot, I.; De Lorgeril, J.; Janech, M.; Gross, P.S.; Warr, G.W.; Cuthbertson, B.; Barracco, M.A.; Bulet, P.; Aumelas, A.; Yang, Y.; Bo, D.; Xiang, J.; Tassanakajon, A.; Piquemal, D.; Bachere, E. Penbase, The Shrimp Antimicrobial Peptide Penaeidin Database: Sequence-Based Classification And Recommended Nomenclature. *Dev. Comp. Immunol.*, Vol.30, No., (2006), pp.283-288, ISSN

Hammami, R.; Ben-Hamida, J.; Vergoten, G.; Fliss, I. (2009). Phytamp: A Database Dedicated To Antimicrobial Plant Peptides. *Nucl. Ac. Res.*, Vol.37, (2009), pp.D963-D968, ISSN

Hao, G.; Shi, Y.H.; Han, J.H.; Li, Q.H.; Tanga, Y.L.; Lea, G.W. (2008). Design And Analysis Of Structure–Activity Relationship Of Novel Antimicrobial Peptides Derived From The Conserved Sequence Of Cecropin. *J. Pept. Sci.*, Vol.14, (2008), pp.290–298, ISSN

Harris, F.; Dennison, S.R.; Phoenix, D.A. (2009). Anionic Antimicrobial Peptides From Eukaryotic Organisms. *Curr Protein Pept Sci.*, Vol.10, No.6, (December 2009), pp.585-606, ISSN

Henriques, S.T.; Craik, D.J. (2009). Cyclotides As Templates In Drug Design. *Drug Discov Today*, Vol.15, No.1-2, (January 2010), pp.57-64, ISSN

Imamura, T.; Yasuda, M.; Kusano, H.; Nakashita, H.; Ohno, Y.; Kamakura, T.; Taguchi, S.; Shimada, H. (2009). Acquired Resistance To The Rice Blast In Transgenic Rice Accumulating The Antimicrobial Peptide Thanatin. *Transgenic Res.*, Vol.19, No.3, (June 2010), pp.415-424, ISSN

Jan, P.S.; Huang, H.Y.; Chen, H.M. (2010). Expression Of A Synthesized Gene Encoding Cationic Peptide Cecropin B In Transgenic Tomato Plants Protects Against Bacterial Diseases. *Appl Environ Microbiol.*, Vol.76, No.3, (February 2010), pp.769-775, ISSN

Kajula, M.; Tejesvi, M.V.; Kolehmainen, S.; Mäkinen, A.; Hokkanen, J.; Mattila, S.; Pirttilä, A.M. (2010). The Siderophore Ferricrocin Produced By Specific Foliar

Endophytic Fungi In Vitro. *Fungal Biol.*, Vol.114, No.2-3, (February 2010), pp.248-254, ISSN

Khandelia, H.; Witzke, S.; Mouritsen, O.G. (2010). Interaction Of Salicylate And A Terpenoid Plant Extract With Model Membranes: Reconciling Experiments And Simulations. *Biophys J.*, Vol.99, No.12, (December 2010), pp.3887-3894, ISSN..

Kioto Encyclopedia For Genes And Genomes, KEGG Compound, Plant Secondary Metabolites. Http://www.Genome.Jp/Kegg/Compound (Accessed In March 5th, 2009).

Kovalchuk, N.; Li, M.; Wittek, F.; Reid, N.; Singh, R.; Shirley, N.; Ismagul, A.; Eliby, S.; Johnson, A.; Milligan, A.S.; Hrmova, M.; Langridge, P.; Lopato, S. (2010). Defensin Promoters As Potential Tools For Engineering Disease Resistance In Cereal Grains. *Plant Biotechnol J.*, Vol.8, No.1, (January 2010), pp.47-64, ISSN

Kovaleva, V.; Krynytskyy, H.; Gout, I.; Gout, R. (2010). Recombinant Expression, Affinity Purification And Functional Characterization Of Scots Pine Defensin 1. *Appl Microbiol Biotechnol.*, Vol.89, No.4, (February 2011), pp.1093-1101, ISSN

Kovaleva, V.; Kiyamova, R.; Cramer, R.; Krynytskyy, H.; Gout, I.; Filonenko, V.; Gout, R. (2009). Purification And Molecular Cloning Of Antimicrobial Peptides From Scots Pine Seedlings. *Peptides*, (2009), Doi:10.1016/J.Peptides.2009.08.007.

Lam, S.K.; Ng, T.B. (2009). Passiflin, A Novel Dimeric Antifungal Protein From Seeds Of The Passion Fruit. *Phytomedicine*, Vol.16, (2009), pp.172-180, ISSN

Laquitaine, L.; Gomès, E.; François, J.; Marchive, C.; Pascal, S.; Hamdi, S.; Atanassova, R.; Delrot, S.; Coutos-Thévenot, P. (2006). Molecular Basis Of Ergosterol-Induced Protection Of Grape Against Botrytis Cinerea: Induction Of Type I LTP Promoter Activity, WRKY, And Stilbene Synthase Gene Expression. Mol Plant Microbe Interact., Vol.19, No.10, (October 2006), pp.1103-1112, ISSN

Larson, R.L.; Wintermantel, W.M.; Hill, A.; Fortis, L.; Nunez, A. (2008). Proteome Changes In Sugarbeet In Response To Beet Necrotic Yellow Vein Virus Infection. *Physiological Mol. Plant Pathol.*, Vol.72, (2008), pp.62-72, ISSN

Lata, S.; Sharma, B.K.; Raghavya, G.P.S. (2007). Analysis And Prediction Of Antibacterial Peptides. *BMC Bioinformatics*, Vol.8, (2007), pp.263, ISSN

Lee, B.J.; Kwon, S.J.; Kim, S.K.; Kim, K.J.; Park, C.J.; Kim, Y.J.; Park, O.K.; Paek, K.H. (2006). Functional Study Of Hot Pepper 26S Proteasome Subunit RPN7 Induced By Tobacco Mosaic Virus From Nuclear Proteome Analysis. *Biochem. Biophys. Res. Commun.*, Vol.351, (2006), pp.405–411, ISSN

Li Y. (2009). The Role Of Antimicrobial Peptides In Cardiovascular Physiology And Disease. *Biochem Biophys Res Commun.*, Vol.390, No.3, (December 2009), pp.363-367, ISSN

Ma, X.; Yang, X.; Zeng, F.; Yang, L.; Yu, D.; Ni, H. (2010). Physcion, A Natural Anthraquinone Derivative, Enhances The Gene Expression Of Leaf-Specific Thionin Of Barley Against *Blumeria Graminis*. *Pest Manag Sci.*, Vol.66, No.7, (July 2010), pp.718-724, ISSN

Manners, J.M. (2007). Hidden Weapons Of Microbial Destruction In Plant Genomes. *Genome Biol.*, Vol.8, (2007), pp.225, ISSN

Marcos, J.F.; Muñoz, A.; Pérez-Payá, E.; Misra, S.; López-García, B. (2008). Identification And Rational Design Of Novel Antimicrobial Peptides For Plant Protection. *Annu. Rev. Phytopathol.*, Vol.46, (2008), pp.273-301, ISSN

Moon, C.D.; Zhang, X.X.; Matthijs, S.; Schäfer, M.; Budzikiewicz, H.; Rainey, P.B. (2008). Genomic, Genetic And Structural Analysis Of Pyoverdine-Mediated Iron Acquisition In The Plant Growth-Promoting Bacterium *Pseudomonas fluorescens* SBW25. *BMC Microbiol.*, Vol.8, (January 2008), pp.7, ISSN

Moreira, J.S.; Almeida, R.G.; Tavares, L.S.; Santos, M.O.; Viccini, L.F.; Vasconcelos, I.M.; Oliveira, J.T.; Raposo, N.R.; Dias, S.C.; Franco O.L. (2011). Identification Of Botryticidal Proteins With Similarity To NBS-LRR Proteins In Rosemary Pepper (*Lippia Sidoides* Cham.) Flowers. *Protein J.*, Vol.30, No.1, (January 2011), pp.32-38, ISSN

Morel, A.F.; Maldaner, G.; Ilha, V.; Missau, F.; Silva, U.F.; Dalcol, I.I. (2005). Cyclopeptide Alkaloids From *Scutia Buxifolia* Reiss And Their Antimicrobial Activity. *Phytochem.*, Vol.66, (2005), pp.2571–2576, ISSN

Muñoz, F.F.; Mendieta, J.R.; Pagano, M.R.; Paggi, R.A.; Daleo, G.R.; Guevara, M.G. (2010). The Swaposin-Like Domain Of Potato Aspartic Protease (Stasp-PSI) Exerts Antimicrobial Activity On Plant And Human Pathogens. *Peptides*, Vol.31, No.5, (May 2010), pp.777-785, ISSN

Nagarajan, V.; Kaushik, N.; Murali, B.; Zhang, C.; Lakhera, S.; Elasri, M.O.; Deng, Y. (2006). A Fourier Transformation Based Method To Mine Peptide Space For Antimicrobial Activity. *BMC Bioinformatics*, Vol.7, (2006), S2.

Oard, S.V. (2011). Deciphering A Mechanism Of Membrane Permeabilization By A-Hordothionin Peptide. *Biochim Biophys Acta,* Vol.1808, No.6, (June 2011), pp.1737-1745, ISSN

Odintsova, T.I.; Vassilevski, A.A.; Slavokhotova, A.A.; Musolyamov, A.K.; Finkina, E.I.; Khadeeva, N.V.; Rogozhin, E.A.; Korostyleva, T.V.; Pukhalsky, V.A.; Grishin, E.V.; Egorov, T.A. (2009). A Novel Antifungal Hevein-Type Peptide From *Triticum Kiharae* Seeds With A Unique 10-Cysteine Motif. *FEBS J.*, Vol.276, (2009), pp.4266–4275, ISSN

Oey, M.; Lohse, M.; Scharff, L.B.; Kreikemeyer, B.; Bock, R. (2009). Plastid Production Of Protein Antibiotics Against Pneumonia Via A New Strategy For High-Level Expression Of Antimicrobial Proteins. *Proc. Nat. Acad. Sci. USA*, Vol.106, (2009), pp.6579–6584, ISSN.

Oseroff, C.; Sidney, J.; Kotturi, M.F.; Kolla, R.; Alam, R.; Broide, D.H.; Wasserman, S.I.; Weiskopf, D.; Mckinney, D.M.; Chung, J.L.; Petersen, A.; Grey, H.; Peters, B.; Sette, A. (2010). Molecular Determinants Of T Cell Epitope Recognition To The Common Timothy Grass Allergen. *J Immunol.*, Vol.185, No.2, (July 2010), pp.943-955, ISSN.

Padovan, L.; Crovella, S.; Tossi, A.; Segat, L. (2010). Techniques For Plant Defensin Production. *Curr Protein Pept Sci.*, Vol.11, No.3, (May 2010), pp.231-235, ISSN..

Padovan, L.; Scocchi, M.; Tossi, A. (2010). Structural Aspects Of Plant Antimicrobial Peptides. *Curr Protein Pept Sci.*, Vol.11, No.3, (May 2010)c, pp.210-219, ISSN.

Padovan, L.; Segat, L.; Tossi, A.; Antcheva, N.; Benko-Iseppon, A.M.; Kido, E.A.; Brandao, L.; Calsa Jr, T.; Crovella, S. (2009). A Plant-Defensin From Sugarcane (*Saccharum* spp.). *Protein Pept. Lett.*, Vol.16, (2009), pp.430-436, ISSN.

Padovan, L.; Segat, L.; Tossi, A.; Calsa Jr, T.; Kido, E.A.; Brandão, L.; Guimarães, R.L.; Pestana-Calsa, M.C.; Pandolfi, V.; Belarmino, L.C.; Benko-Iseppon, A.M.; Crovella, S. (2010a). Characterization Of A New Defensin From Cowpea (*Vigna unguiculata* (L.) Walp.). *Protein Pept Lett.*, Vol.17, No.3, (March 2010), pp.297-304, ISSN.

Pelegrini, P.B.; Del Sarto, R.P.; Silva, O.N.; Franco, O.L.; Grossi-De-Sa, M.F. (2011). Antibacterial Peptides From Plants: What They Are And How They Probably Work. *Biochemistry Research International*, Vol.2011, Article ID 250349, (2011), 9p., ISSN.

Pelegrini, P.B.; Murad, A.M.; Silva, L.P.; Santos, R.C.P.; Costa, F.T.; Tagliari, P.D.; Bloch Jr., C.; Noronha, E.F.; Miller, R.N.G.; Franco, O.L. (2008). Identification Of A Novel Storage Glycine-Rich Peptide From Guava (*Psidium guajava*) Seeds With Activity Against Gram-Negative Bacteria. *Peptides*, Vol.29, (2008), pp.1271–1279, ISSN.

Pelegrini, P.B.; Noronha, E.F.; Muniz, M.A.; Vasconcelos, I.M.; Chiarello, M.D.; Oliveira, J.T.; Franco, O.L. (2006). An Antifungal Peptide From Passion Fruit (*Passiflora edulis*) Seeds With Similarities To 2S Albumin Proteins. *Biochim. Biophys. Acta*, Vol.1764, (2006), pp.1141-1146, ISSN.

Perez-De-Luque, A.; Rubiales, D. (2009). Nanotechnology For Parasitic Plant Control. *Pest Manag. Sci.*, Vol.65, (2009), pp.540–545, ISSN.

Pestana-Calsa, M.C.; Ribeiro, I.L.; Calsa Jr., T. (2010). Bioinformatics-coupled molecular approaches for unravelling potential antimicrobial peptides coding genes in Brazilian native and crop plant species. *Curr Protein Pept Sci.*, Vol.11, No.3, (May 2010), pp.199-209, ISSN.

Pieters, L.; Vlietinck, A.J. (2005). Bioguided Isolation Of Pharmacologically Active Plant Components, Still A Valuable Strategy For The Finding Of New Lead Compounds? *J. Ethnopharmacol.*, Vol.100, (2005), pp.57-60, ISSN.

Portieles, R.; Ayra, C.; Gonzalez, E.; Gallo, A.; Rodriguez, R.; Chacón, O.; López, Y.; Rodriguez, M.; Castillo, J.; Pujol, M.; Enriquez, G.; Borroto, C.; Trujillo, L.; Thomma, B.P.; Borrás-Hidalgo, O. (2010). Nmdef02, A Novel Antimicrobial Gene Isolated From *Nicotiana Megalosiphon* Confers High-Level Pathogen Resistance Under Greenhouse And Field Conditions. *Plant Biotechnol J.*, Vol.8, No.6, (August 2010), pp.678-690, ISSN.

Rahnamaeian, M.; Langen, G.; Imani, J.; Khalifa, W.; Altincicek, B.; Von Wettstein, D.; Kogel, K.H.; Vilcinskas, A. (2009). Insect Peptide Metchnikowin Confers On Barley A Selective Capacity For Resistance To Fungal Ascomycetes Pathogens. *J. Exp. Botany*, Vol.60, (2009), pp.4105–4114, ISSN.

Ribeiro, S.M.; Almeida, R.G.; Pereira, C.A.; Moreira, J.S.; Pinto, M.F.; Oliveira, A.C.; Vasconcelos, I.M.; Oliveira J.T.; Santos, M.O.; Dias, S.C.; Franco O.L. (2010). Identification Of A Passiflora Alata Curtis Dimeric Peptide Showing Identity With 2S Albumins. *Peptides*, Vol.32, No.5, (May 2011), pp.868-874, ISSN.

Ribeiro, S.F.F.; Carvalho, A.O.; Cunha, M.; Rodrigues, R.; Cruza, L.P.; Melo, V.M.M.; Vasconcelos, I.M.; Melo, E.J.T.; Gomes, V.M. (2007). Isolation And Characterization Of Novel Peptides From Chilli Pepper Seeds: Antimicrobial Activities Against Pathogenic Yeasts. *Toxicon*, Vol.50, (2007), pp.600–611, ISSN.

Rogozhin, E.A.; Oshchepkova, Y.I.; Odintsova, T.I.; Khadeeva, N.V.; Veshkurova, O.N.; Egorov, T.A.; Grishin, E.V.; Salikhov, S.I. (2011). Novel antifungal defensins from *Nigella sativa* L. seeds. *Plant Physiol Biochem.*, Vol.49, No.2, (February 2011), pp.131-137, ISSN.

Sagaram, U.S.; Pandurangi, R.; Kaur, J.; Smith, T.J.; Shah, D.M. (2011). Structure-Activity Determinants In Antifungal Plant Defensins Msdef1 And Mtdef4 With Different Modes Of Action Against *Fusarium Graminearum*. *Plos One*, Vol.6, No.4, (April 2011), pp.E18550, ISSN..

Silva Jr., I.F.; Cechinel-Filho, V.C.; Zacchino, S.A.; Lima, J.C.S.; Martins, D.T.O. (2009). Antimicrobial Screening Of Some Medicinal Plants From Mato Grosso Cerrado. *Brazilian J. Pharmacognosy*, Vol.19, (2009), pp.242-248, ISSN.

Silverstein, K.A.T.; Graham, M.A.; Paape, T.D.; Bosch, K.A.V. (2005). Genome Organization Of More Than 300 Defensin-Like Genes In Arabidopsis. *Plant Physiol.*, Vol.138, (2005), pp.600–610, ISSN.

Silverstein, K.A.T.; Moskal Jr., W.A.; Wu, H.C.; Underwood, B.A.; Graham, M.A.; Town, C.D.; Vandenbosch, K.A. (2007). Small Cysteine-Rich Peptides Resembling Antimicrobial Peptides Have Been Underpredicted In Plants. *Plant J.*, Vol.51, (2007), pp.262–280, ISSN.

Skouri-Gargouri, H.; Jellouli-Chaker, N.; Gargouri, A. (2010). Factors Affecting Production And Stability Of The Acafp Antifungal Peptide Secreted By *Aspergillus clavatus*. *Appl Microbiol Biotechnol.*, Vol.86, No.2, (March 2010), pp.535-543, ISSN.

Tavares, L.S.; Santos, M.O.; Viccini, L.F.; Moreira, J.S.; Miller, R.N.G.; Franco, O. L. (2008). Biotechnological Potential Of Antimicrobial Peptides From Flowers. *Peptides*, Vol.29, (2008), pp.1842-1851, ISSN.

Terras, F.R.; Eggermont, K.; Kovaleva, V.; Raikhel, N.V.; Osborn, R.W.; Kester, A.; Rees, S.B.; Torrekens, S.; Van Leuven, F.; Vanderleyden, J.; Cammue, B.P.A.; Broekaert, W.F. (1995). Small Cystein-Rich Antifungal Proteins From Radish: Their Role In Host Defense. *Plant Cell*, Vol.7, (1995), pp.573-588, ISSN.

Tian, A.; Cao, J.; Huang, L.; Yu, X.; Ye, W. (2009). Characterization Of A Male Sterile Related Gene Bcmf15 From *Brassica campestris* Ssp. *chinensis*. *Mol Biol Rep.*, Vol.36, No.2, (February 2009), pp.307-314, ISSN.

Tian, Z.; Dong, T.; Teng, D.; Yang, Y.; Wang, J. (2009). Design And Characterization Of Novel Hybrid Peptides From LFB15(W4,10), HP(2-20) And Cecropin A Based On Structure Parameters By Computer-Aided Method. *Appl. Microbiol. Biotechnol.*, Vol.82, (2009), pp.1097–1103, ISSN.

Tossi, A.; Sandri, L. (2002). Molecular Diversity In Gene-Encoded, Cationic Antimicrobial Polypeptides. *Curr. Pharm. Des.*, Vol.8, (2002), pp.743-761, ISSN.

Tossi, A.; Sandri, L.; Giangaspero, A. (2000). Amphipathic, Alpha-Helical Antimicrobial Peptides. *Biopolymers*, Vol.55, (2000), pp.4–30, ISSN.

Udwary, D.W.; Gontang, E.A.; Jones, A.C.; Jones, C.S.; et al.; Moore, B.S. (2011). Comparative Genomic And Proteomic Analysis Of The Actinorhizal Symbiont Frankia Reveals Significant Natural Product Biosynthetic Potential. *Appl Environ Microbiol.*, (April 2011), Epub Ahead Of Print, ISSN.

Ulrich-Merzenich, G.; Zeitler, H.; Jobst, D.; Panek, D.; Vetter, H.; Wagner, H. (2007). Application Of The "Omic-" Technologies In Phytomedicine. *Phytomedicine*, Vol.14, (2007), pp.70–82, ISSN.

Van De Velde, W.; Zehirov, G.; Szatmari, A.; Debreczeny, M.; Ishihara, H.; et al.; Mergaert, P. (2010). Plant Peptides Govern Terminal Differentiation Of Bacteria In Symbiosis. *Science*, Vol327, No.5969, (February 2010), pp.1122-1126, ISSN.

Verpoorte, R.; Choi, Y.H.; Kim, H.K. (2005). Ethnopharmacology And Systems Biology: A Perfect Holistic Match. *J. Ethnopharmacol.*, Vol.100, (2005), pp.53-56, ISSN.

Vidal, J.R.; Kikkert, J.R.; Wallace, P.G.; Reisch, B.I. (2003). High-Efficiency Biolistic Co-Transformation And Regeneration Of Chardonnay (*Vitis vinifera* L.) Containing Npt-II And Antimicrobial Peptide Genes. *Plant Cell Rep.*, Vol.22, (2003), pp.252–260, ISSN.

Wade, D.; Englund, J. (2002). Synthetic Antibiotic Peptides Database. *Protein Pept. Lett.*, Vol.9, (2002), pp.53-57, ISSN.

Wan, J.; Dunning, F.M.; Bent, A.F. (2002). Probing Plant-Pathogen Interactions And Downstream Defense Signaling Using DNA Microarrays. *Funct. Integr. Genomics*, Vol.2, (2002), pp.259–273, ISSN.

Wang, C.; Yang, C.; Gao, C.; Wang, Y. (2009). Cloning And Expression Analysis Of 14 Lipid Transfer Protein Genes From *Tamarix hispida* Responding To Different Abiotic Stresses. *Tree Physiol.*, Vol.29, No.12, (December 2009), pp.1607-1619, ISSN.

Wang, P.; Bang, J.K.; Kim, H.J.; Kim, J.K.; Kim, Y.; Shin, S.Y. (2009). Antimicrobial Specificity And Mechanism Of Action Of Disulfide-Removed Linear Analogs Of The Plant-Derived Cys-Rich Antimicrobial Peptide Ib-AMP1. *Peptides*, Vol.30, No.12, (December 2009), pp.2144-2149, ISSN.

Wang, G. (2007). Tool Developments For Structure-Function Studies Of Host Defense Peptides. *Protein Pept. Lett.*, Vol.14, (2007), pp.57-69, ISSN.

Wang, G.; Li, X.; Wang, Z. (2009). APD2: The Updated Antimicrobial Peptide Database And Its Application In Peptide Design. *Nucl. Acids Res.*, Vol.37, (2009), pp.D933–D937, ISSN.

Wang, P.; Bang, J.K.; Kim, H.J.; Kim, J.K.; Kim, Y.; Shin, S.Y. (2009). Antimicrobial Specificity And Mechanism Of Action Of Disulfide-Removed Linear Analogs Of The Plant-Derived Cys-Rich Antimicrobial Peptide Ib-AMP1. *Peptides*, (2009), Doi:10.1016/J.Peptides.2009.09.020, ISSN.

Wang, Z.; Wang, G. (2004). APD: The Antimicrobial Peptidc Database. *Nucl. Acids Res.*, Vol.32, (2004), pp.D590-D592, ISSN.

Whitmore, L.; Wallace, B.A. (2004). The Peptaibol Database: A Database For Sequences And Structures Of Naturally Occurring Peptaibols. *Nucl. Acids Res.*, Vol.32, (2004), pp.D593-D594, ISSN.

Widjaja, I.; Lassowskat, I.; Bethke, G.; Eschen-Lippold, L.; Long, H.H.; Naumann, K.; Dangl, J.L.; Scheel, D.; Lee, J. (2010). A Protein Phosphatase 2C, Responsive To The Bacterial Effector Avrrpm1 But Not To The Avrb Effector, Regulates Defense Responses In Arabidopsis. *Plant J.*, Vol.61, No.2, (January 2010), pp.249-258, ISSN.

Yokoyama, S.; Iida, Y.; Kawasaki, Y.; Minami, Y.; Watanabe, K.; Yagib, F. (2009). The Chitin-Binding Capability Of Cy-AMP1 From Cycad Is Essential To Antifungal Activity. *J. Pept. Sci.*, Vol.15, (2009), pp.492–497, ISSN.

Yount, N.Y.; Yeaman, M.R. (2004). Multidimensional Signatures In Antimicrobial Peptides. *Proc. Natl. Acad. Sci. USA*, Vol.101, (2004), pp.7363–7368, ISSN.

Zamyatnin, A.A.; Voronina, O.L. (2010). Antimicrobial And Other Oligopeptides Of Grapes. *Biochemistry (Mosc).*, Vol.75, No.2, (February 2010), pp.214-223, ISSN.

Zarei, A.; Körbes, A.P.; Younessi, P.; Montiel, G.; Champion, A.; Memelink, J. (2011). Two GCC Boxes And AP2/ERF-Domain Transcription Factor ORA59 In Jasmonate/Ethylene-Mediated Activation Of The PDF1.2 Promoter In Arabidopsis. *Plant Mol Biol.*, Vol.75, No.4-5, (March 2011), pp.321-331, ISSN.

Zhang, K.; Mckinlay, C.; Hocart, C.H.; Djordjevic, M.A. (2006). The *Medicago truncatula* Small Protein Proteome And Peptidome. *J Proteome Res.*, Vol5, No.12, (December 2006), pp.3355-3367, ISSN.

15

Mining Effector Proteins in Phytopathogenic Fungi

Li Cheng-yun and Yang Jing
Key Laboratory of Agro-Biodiversity and Pest Management of Education Ministry of China, Yunnan Agricultural University, Kunming, Yunnan, China

1. Introduction

"Pathogen effector" has been increasingly used in the past decades in the plant-pathogen interactions (Hogenhout et al., 2009).Presently, the definition of pathogen effector commonly adopted the definition given by Sophien Kamoun, that is, effectors are 'molecules that manipulate host cell structure and function, thereby facilitating infection (virulence factors or toxins) and/or triggering defense responses (avirulence factors or elicitors)' (Kamoun, 2006). Plant and their related pathogen have coevolved for many millions of years, which resulted in evolving some resistance genes in plants to prevent or limit pathogen infection, and simultaneously pathogen also evolved some effector proteins to overcome plant defense as well as cause disease. Many plant pathogens secreted effector proteins into host cells to repress plant defense and contribute to pathogen colonization and breach (Birch et al., 2006;Chisholm et al., 2006; Grant et al., 2006; Huang et al., 2006 a, 2006 b; Jones and Dangl, 2006; Kamoun, 2006; O'Connell and Panstruga, 2006).

Oomycetes could cause many destructive plant diseases, like potato late blight that caused the Irish potato famine in the nineteenth century (Tyler, 2007). Oomycete effector proteins could be translocated into host cells with the help of RXLR and dEER motifs in the absence of the pathogen (Dou et al., 2008; Whisson et al., 2007).Oomycetes secreted effector proteins into the infection site. The effector proteins were categorized into two classes based on action site of effectors, and one was extracellular effectors acted in the apoplastic space where they interact with extracellular molecules of hosts. The other was cytoplasmic effectors that acted within the boundary of plant cell wall.

Plant fungal pathogen also secreted effector proteins into host cells where they incompatibly interacted with plants receptors encoded by major resistance genes, which rapidly triggered host defense response (Ellis et al., 2006; Tyler, 2002). Oomycete effectors play dual role in disease and plant defense, fungal effector proteins were no exception. For example, victorin of *Cochliobolus victoriae*, NIP1 of *R.secalis*, and AAL toxin of *Alternaria alternate* played the role in toxin(Lorang et al., 2007; Navarre and Wolpert, 1999; Rohe et al., 1995; Spassieva et al., 2002; van't Slot and Knogge, 2002; Wang et al., 1996, Wolpert et al., 2002), but some studies showed *NIP1* and *ToxA* also interacted with corresponding host-resistance or toxin-sensitivity genes, which resulted in *NIP1* and *ToxA* acted in the same way as the *Avr* genes (Schürch et al., 2004; Stukenbrock and McDonald, 2007). So, it is necessary to clarify fungal effector proteins incompatibly or compatibly interact with host receptors when they were transported into host cells. In the meanwhile, identification and functional assay of fungal effectors-encoding genes will contribute to discovering mechanism for interaction and coevolution of pathogens and plants. In this chapter, we will focus on our recent years' studies on mining, sequence characterization and functional analysis of secreted effector proteins in fungi and model plant of *Arabidopsis thaliana*.

2. Mining effector-encoding genes in *Magnaporthe grisea* genome database

Magnaporthe grisea is an ascomycete fungus and the causal agent of rice blast disease, which is the most destructive disease of rice-growing areas in the worldwide. The annual rice yield loss caused by blast disease is enough to feed about 60 million people (Ou, 1985).Whole-genome sequence indicated fungal and oomycete plant pathogen had large amounts of secreted proteins (Dean et al., 2005; Kämper et al., 2006; Tyler et al., 2006).*Magnaporthe grisea* genome sequence was available online, which facilitated to mining many novel effector-encoding genes. And some online software could be used to predict some features such as secretion, domain and homology of effector protein. This provided some evidence for next functional verification.

2.1 Predicting classically and non-classically secreted effector proteins in *M. grisea*

Secreted effector proteins are secreted from pathogen cell into extracelluar space. Secreted proteins were categorized into two classes based on their secreted pathway, one was classically secreted proteins, there was a signal peptide in N-terminal of proteins, and the other was non-classically secreted proteins, their secreted pathway was known as leaderless secretion (Nickel, 2003).

Classically secreted proteins in *M. grisea* were predicted through combined online software such as SignalP v3.0, TargetP v1.01, big-PI predictor and TMHMM v2.0 (http://www.cbs.dtu.dk/ services/), the determinant standard of classically secreted proteins conformed to the following four standards, the first standard is $L=-918.235-123.455\times(\text{Mean S score})+1983.44\times(\text{HMM score})$ and $L>0$ for predicting proteins with N-terminal signal peptide, the second one is proteins with signal peptides were transported via Sec pathway, the third one is no transmembrane, the fourth one is no GPI-anchor site (Samuel et al., 2003).

Total of 12,595 putative proteins including 1,486 small proteins from *M. grisea* database were predicted. Of which, 1,134 putative proteins were predicted for classically secreted proteins with N-terminal signal peptide. Their signal peptide length lied in between 15-45 amino acids. Here, we will center on small secreted proteins (amino acid length <100), there were 119 classically secreted proteins among 1,486 small proteins, we selected 45 putative secreted proteins-encoding genes among 119 genes as candidates in order to analysis their polymorphism in blast strains from Yunnan, China, the results showed that the most of genes distributed in 21 tested blast strains from Yunnan, which indicated high polymorphism and conservative in blast fungus strains. In addition to classically secreted proteins, non-classically secreted proteins were in further predicted using SecretomeP 2.0 Server (http://www.cbs.dtu.dk/services/).

2.2 Predicted features of secreted protein sequence

To conveniently identify function of predicted secreted effector proteins, sequence features of secreted effector proteins needed to be predicted. The gene sequence prediction began with the identification of regions of DNA that coded for expression of proteins. Whether there was intron or not in DNA sequence for eukaryotic genome. Molecular weight of immature protein was important for gene cloning and functional identification. For secreted proteins, subcellular location was needed to be predicted, which contribute to understand organelle in which secreted effector protein interacted with host receptors. In addition, prediction of protein domain was necessary to experimentally assay function of protein in future. For example, it was predicted that many effectors from plant pathogenic *Phytophthora* species had N-terminal motifs (RXLR-dEER) that were necessary to translocate these effectors into host cells (Jiang et al., 2008), along with the motif prediction, many experiments such as oomycete effectors were translocated into host cells and their function had been carried out, So, to some extent, domain or motif prediction experimentally facilitated function identification of effector proteins.

We predicted subcellular location and domain of the parts of secreted proteins. For example, MultiLoc/TargetLoc (http://www-bs.informatik.uni-tuebingen.de/Services/MultiLoc/)was used to predict subcellular location of 8 secreted proteins. The result showed that subcellular location of MGS4 was predicted to be cytoplasmic, and MGS7, MGS11, MGS42, MGS53, MGS60 and MGS174 were predicted to be extracellularly secreted. Theoretical pI of 7 secreted proteins showed that they were predicted to lie in between 7.69-8.13 except MGS4 was 5.71, and their theoretical molecular weight lied in between 8.50-10.70. Their domain prediction revealed MGS7 possessed a transmembrane region, and MGS53 had a domain of ZnF-C_2H_2, while other 5 proteins had no any typical domain except a signal peptide sequence contained in N-terminal of proteins. The 8 secreted proteins sequences were searched using BLASTN or BLASTX of NCBI nr database. No high similarity to sequences from other organism was found, which revealed the effector-encoding genes were novel. Collectively, we analyzed molecular weight, subcellular location, domain and homology of putative secreted proteins, which provide a base for their functional prediction and identification.

2.3 Analysis of a Host-Targeting Motif and its flanking sequence in the genome of *Magnaporthe grisea*

It is well-known that bacterial pathogens delivered effectors inside plant cells through the type III secretion system (TTSS). The motif is primarily found in the pathogenic protein of *Plasmodium falciparum*, termed RxLxE/D/Q, and its role is to target the host during the export of virulence proteins; this motif is conserved in *P. falciparum*(Marti et al., 2004), and is defined as host-targeting signal (HTS) or host-targeting motif (HTM). In *P. falciparum* genome, this motif is detected within the 60 amino acids downstream of the secretion signal sequence cleavage sites in approximately more than 400 proteins (Hiller et al., 2004). Subsequently, a series of findings suggested that effectors from oomycete such as *Hyaloperonospora parasotica, Phytophthora infestans, Phytophthora sojae,* and *Phytophthora ramorum* possessed the conserved motif, termed RxLR, located within the N-terminal 60 amino acids downstream of signal peptide cleavage sites, which is similar in sequence and position to the Plasmodium sp HTM, also the RxLR motif and HTM domains are functionally interchangeable (Haldar et al., 2006; Bhattacharjee et al., 2006). In summary, these findings indicated that these oomycetes shared conserved machinery for the transport of effectors. Whether did the motif of RxLx exist in effectors from *Magnaporthe grisea*? So, we predicted that the secretory proteins of *M. grisea* possessed the motif RxLx.

Here, we applied a tool of MEME (http://www.meme.sdsc.edu/) to analysis RxLx of 1,270 putative secretory proteins from the fifth edition of the rice blast fungus genome (http://www.broad.mit.edu/ annotation/fungi/magnaporthe). The results showed that 297 putative secreted proteins possessed the motif of RxLx, the motif located within the region of 100 amino acids downstream of the N-terminal signal sequence cleavage sites. The number of secretory proteins with RxLx motif was similar with those of Plasmodium and Oomycetes Host-targeted secretome, which indicated that the RxLx motif possibly function as transporting of secreted proteins of *M. grisea* into host cells. However, biological experiments are required for further verification.

Weblogo (http://weblogo.berkeley.edu/cache/fileDo8NeU.png) was used to analyze a sequence logo of the MEME motif and its surrounding region of 149 putative secretory protein sequences (Figure1). Arg(R) in position 1 and Leu (L) in position 3 were the most highly conserved residues in the motif RxLx. It also showed the lower but possible finite positional value, the other residues represented as 'x' in the linearized motif RxLx. By contrast, the E/D/Q residues in the 5 amino acids core of the Plasmodium HTM and the enrichment in E/D residues downstream and the highly conservation Arg(R) in position 4 in the motif RxLR that were required for function of motif RxLR were no positionally conserved in secretory proteins containing motif RxLx of *M.grisea*.

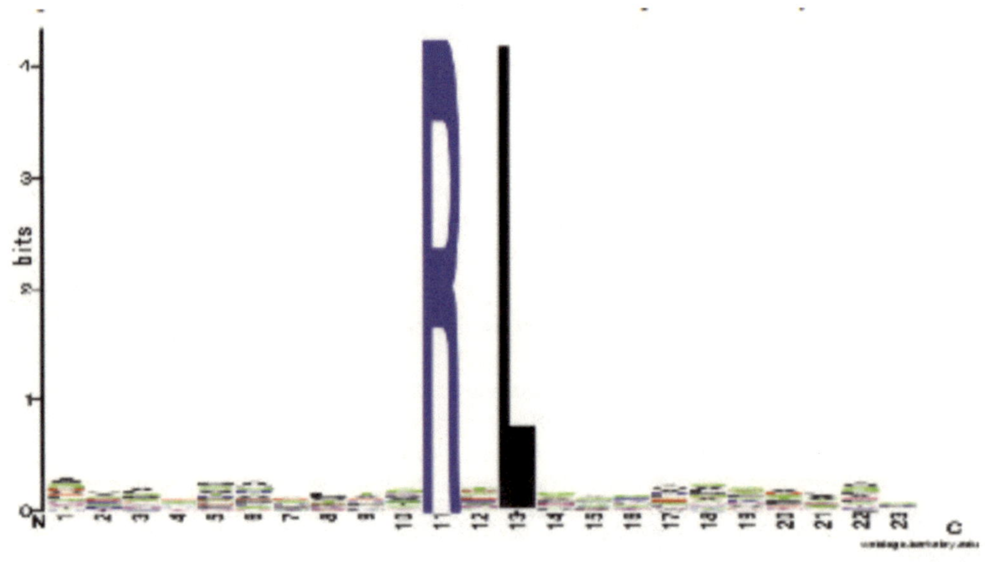

Fig. 1. Sequence logo derived from 149 predicted secreted proteins of *M. grisea*

2.4 Functional prediction and analysis of RxLx motif-containing secretory proteins of M. grisea genome

To analyze whether the RxLx motif-containing secretory proteins of *M. grisea* was involved in pathogenicity, we compared the predicted proteins with the PEDANT database. By manually comparing, there were putative ascribed functions of 62 RxLx motif-containing secretory proteins of *M. grisea* (Table 1).These proteins had putative functions such as proteins of *MGG_11036, MGG_09159, MGG_09248, MGG_08401, MGG_09143, MGG_08424, MGG_08537, MGG_07809,MGG_05479, MGG_06008* and *MGG_09460* were possibly related to cell wall degrading enzyme, proteins of *MGG_00922, MGG_09817, MGG_08436,*

Gene code	Description
MGG_00276	6-hydroxy-D-nicotine oxidase
MGG_00505	sun family protein
MGG_00671	arginase family protein
MGG_00922	aspartic proteinase
MGG_01085	conserved hypothetical protein
MGG_01195	conserved hypothetical protein
MGG_02853	nuclease S1 precursor
MGG_03029	neutral proteinase II
MGG_03276	Major allergen Asp f 2 precursor (Asp f II).
MGG_03476	protein-L-isoaspartate(D-aspartate) O-methyltransferase
MGG_03670	vacuolar subtilisin-like serine proteinase SPM1
MGG_03772	Bile-salt-activated lipase precursor
MGG_03995	carboxypeptidase S1
MGG_04825	probable endopolyphosphatase precursor
MGG_05164	probable membrane protein YML128c
MGG_05479	xylosidase : arabinofuranosidase
MGG_05529	Feruloyl esterase B precursor
MGG_05533	endochitinase class V precursor
MGG_05663	serine-type carboxypeptidase homolog precursor
MGG_05753	protein disulfide-isomerase precursor
MGG_05914	putative tyrosinase
MGG_06009	alpha-L-arabinofuranosidase
MGG_06303	epsilon-lactone hydrolase
MGG_06442	CATB protein
MGG_06538	blastomyces yeast phase-specific protein 1
MGG_07179	pepsin C precursor
MGG_07234	FK506-binding protein precursor (Peptidyl-prolyl *cis-* trans isomerase)
MGG_07331	probable GEL1 protein
MGG_07502	SCJ1 protein
MGG_07621	regulator of purine biosynthesis (adenine-mediated repression)
MGG_07809	cellulose 1,4-beta-cellobiosidase
MGG_08164	probable protein disulfide-isomerase precursor
MGG_08401	endoxylanase 11C
MGG_08795	hypothetical protein T10B9.2
MGG_09159	chitin deacetylase
MGG_09162	L-lactate dehydrogenase precursor
MGG_09351	aspartyl protease
MGG_11613	hypothetical protein B3E4.290
MGG_12799	preproalkaline protease

Table 1. In Silico annotation RxLx-containing secretary proteins of *M. Grisea*

MGG_03670 and *MGG_03029* were related to proteinase activity, MGG_06662, *MGG_00276, MGG_08528, MGG_10805, MGG_11286* and *MGG_10710* were related to oxidoreductase activity, proteins of *MGG_00238, MGG_10219* and *MGG_14395* were associated to reverse transcriptase, proteins of *MGG_08164, MGG_05753, MGG_02097* and *MGG_11485* were associated to post-translation modification and *MGG_09848* was related to energy activity, which suggested that seretory proteins with RxLx motif in *M.grisea* had diverse functions. Interestingly, among them, some proteins involved in multiple cellular activity, such as *MGG_08164* encoded disulfide isomerase-like protein that involved in cell rescue, defense, energy, development, cell fate and protein folding, modification, destination. In addition, endoxylanse, chitin binding protein, xylanase, pheromone precursor encoded by *MGG_08401, MGG_09159, MGG_09248, MGG_08424, MGG_07733*, respectively had been reported that they involved in the pathogenicity of the rice blast fungus. Cellobiose dehydrogenase, cutin hydrolase, endoglucanase, lipase, cellulose encoded by *MGG_11036, MGG_01943, MGG_08537, MGG_09839* and *MGG_07809*, respectively had previously been shown to involve in pathogenicity of other plant fungi (Tudzynski and Sharon, 2003; Mendgen et al, 1996).

3. Mining effector-encoding genes in other fungi genome database

Similarly, other fungi genome sequencing had been completed, many fungal genome sequences such as *Fusarium graminearum, Neurospora crassa, Saccharomyces cerevisiae* and *Ustilago maydis* were available online. We analyzed 10,082 proteins from *Neurospora crassa* genome database (http://www.broad.mit.edu/ftp/pub/annotation/neurospora/assembly3 /neurospora_3_protein.gz). There were 437 proteins with signal peptide among total of 10,082 proteins through combined software such as SignalP, TMHMM, TargetP and big-PI Predictior, their signal peptide length lied in between 15 and 59 amino acids. There were 205 predicted secreted proteins that had functional description among 437 proteins, their function mainly involved in diverse enzyme, cell energy, transition, cell recovered and defense mechanism. There were 284 secretory proteins through using software to predict 6,522 protein sequences of *U.maydis* of 284 proteins, 90 proteins contained functional description, and the minimum and maximum of open reading frame were 324 bp and 13,347 bp, respectively. The length range of signal peptides ranged from 16-42 amino acids and the average length was 23 amino acids. Among 284 secreted proteins, 56 proteins possessed the motif of RxLx that located within the region of 100 amino acids downstream of the N-terminal signal sequence cleavage sites.

Similarly, secreted proteins of *Saccharomyces cerevisia* were predicted. N-terminal amino acid sequences of 6,700 proteins of *S. cerevisiae* were available on http://supfam.mrc-lmb.cam.ac.uk/SUPERFAMILY/cgi-bin/gen_list.cgi?genome=sc. *Candida albicans* secretome data were from http://info.med.yale.edu/intmed/infdis/candida/Copyright (2003 John Wiley&Sons, Ltd.). Through the internet-based software such as SignalP v3.0, TargetP v1.01, Big-PI predictor and TMHMM v2.0 prediction for "typical" secretory proteins of *S. cerevisiae*, the 163 secretory ones among the 6 700 proteins were obtained. One hundred and sixty-three predicted secretory proteins were regarded as "Sec-type" signal peptides based on characteristics of four types signal peptides. C-domain of signal peptides of 163 secretory proteins was recognized and cleaved by type I SPases, and had common A-X-A motif, X stood for any amino acid residue. The length and types of amino acid residues of signal peptides were compared between 163 secretory proteins of *S. cerevisiae* and 283 ones of *C.*

albicans genome (Table 2 and Figure 2). The result showed that leucine, alanine, serine and valine were found, percentage of leucine was equally 18% in signal peptides of both the two eukaryotic secretomes, but percentage of alanine was 14% in signal peptides of *S. cerevisiae* secretory proteins and 11% in signal peptides of *C. albicans* secretory proteins. Signal peptides composed of a stretch of 19~21 residues in secretory proteins of both *S. cerevisiae* and *C. albicans*. 19 residues composed of signal peptide in *C albicans* secretory proteins had the highest frequency (16.8%), while 20 residues in *S. cerevisiae* secretory proteins had the highest frequency (19.0%).

Amino acid	Amount and frequency of single amino acid among signal peptide sequences			
	in *S.*	*cerevisiae* genome	in *C.*	*albicans* genome
A	805	14%	1233	11%
C	152	2%	167	1%
D	28	<1%	79	<1%
E	46	<1%	78	<1%
F	491	7%	844	8%
G	194	3%	385	3%
H	77	<1%	111	<1%
I	467	7%	1063	9%
K	201	3%	375	3%
L	1091	18%	2000	18%
M	350	6%	603	5%
N	126	2%	251	2%
P	94	1%	289	3%
Q	127	2%	235	2%
R	152	2%	231	2%
S	598	10%	1149	10%
T	420	7%	847	8%
V	978	8%	819	7%
W	81	1%	188	2%
Y	120	2%	253	2%

Table 2. Single amino acid frequency among predicted signal peptide sequence in *S. cerevisiae* genome and *C. albicans* genome

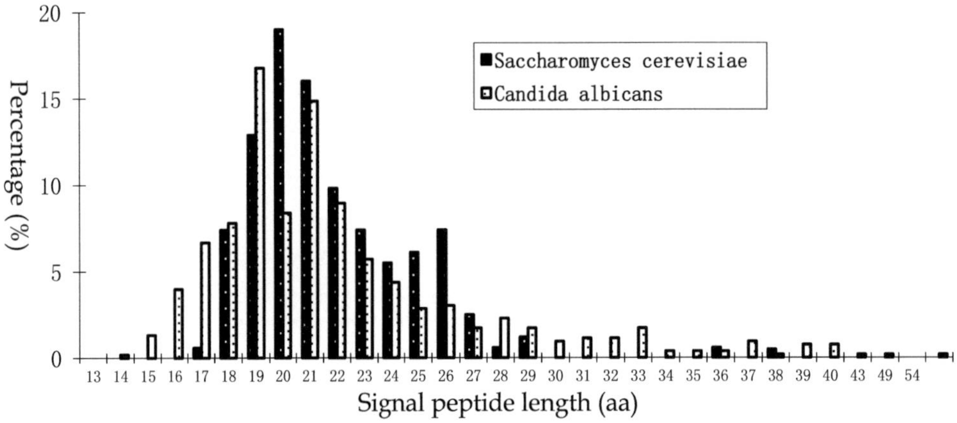

Fig. 2. Length distribution of predicted signal peptides in *S. cerevisiae* and *C. albicans*r

4. Mining effector-encoding genes in *Arabidopsis thaliana* genome database

Arabidopsis thaliana is the first plant genome that has been sequenced completely. At present, secretome of *Arabidopsis thaliana* and their gene function categorization have not been systematically reported. Here elementally report the prediction of secretome of *Arabidopsis thaliana* and function categorization of gene encoding predicted secretory protein. Combined the computer-based software such as SignalP v3.0, TargetP v1.01, big-PI predictor and TMHMM v2.0 was used to predict the secretome of *Arabidopsis thaliana*. The result showed that 282 secretory ones among 28,953 proteins were obtained, the proportion of predicted secretory proteins to total proteins in entire genome was 2.86 % (Table3). And sublocalization of 282 secretory proteins was further predicted through SubLoc v1.0, their sublocalization was cytoplasmic, extracellular, mitochondrial and nuclear, most of them were secreted into extracellular space, while a few of them were secreted into the other three sublocalization (Table 4).

Based on gene function category (http://webclu.bio.wzw.tum.de/genre/proj/uwe25/Search/Catalogs/searchCatFun.html?id=01), we categorized function of the putative secreted proteins from *Arabidopsis thaliana*. The result showed the most of gene encoding secreted proteins participated in cell metabolism, cell rescue and defense, cell transport, cell fate and storage protein. Among the genes participated in cell rescue and defense, the ratio of genes with functions of peroxidase, kinase, disease resistance protein, pathogenesis-related protein and leucine-rich repeat protein was 50.79%.Among the genes participated in cell metabolism, the ratio of genes with functions of hydrolase, lipase, carboxypeptidase, invertase, transferase, expansin and synthase was 70.33%. It was noteworthy that 724 secretary proteins (18.8 %) were found among 3,848 plant-specific proteins in *Arabidopsis thaliana* genome. Among 2,800 Arabidopsis genes that were reproducibly regulated in response to bacterial pathogen inoculation, 132 genes were potential secretary proteins. These results implied that many plant specific biochemical processes, including pathogen responsive genes were carried out at extracellular space. Prediction of *Arabidopsis thaliana* secretome by the aid of the related computer-based software will accelerate to the experimentally functional study of secretome.

No. of chromosome	Prediction of SignalP v3. 0	Prediction of TMHMM v2.0	Prediction of TargetP v1.01				Prediction of big-PI predictor
			S	M	C	-	
I	553	274	202	23	35	14	196
II	513	248	185	27	23	13	174
III	633	274	55	29	42	148	55
IV	497	234	191	11	23	9	188
V	711	307	228	17	41	21	215
Total	2907	1337	861	107	164	205	828

Note：“S” mean protein with secretory pathway；“M” indicated mitochondrial targeting protein; “C” indicated a chloroplast transit protein; “-” indicated any other location

Table 3. Prediction result through the computer-based software, the SignalP v3.0, TargetP v1.01, big-PI predictor and TMHMM v2.0

No. of chromosome	Amount of ORF encoding secretory proteins	Amount of total ORFs encoding proteins	Percentage (%)
I	196	7494	2.62
II	174	4589	3.79
III	55	5742	0.96
IV	188	4407	4.27
V	215	6721	3.20
Summary	828	28953	2.86

Table 4. Chromosomal distribution of secretory protein in *A. thaliana*

5. Expression pattern of effector protein-encoding genes from *M. grisea*

Many studies have used quantitative polymerase chain reaction (PCR) to evaluate fungal growth during the infection process (Hu et al., 1993; Mahuku et al., 1995; Groppe and Boller, 1997; Judelson and Tooley, 2000).Therefore, we detected the expression pattern of candidate novel genes *MGNIP10, MGNIP18, MGNIP24, MGNIP34, MGNIP38, MGNIP53, MGNIP74, MGNIP97* and *MgNIP04* in different isolates from Yunnan, China, the same isolate grown under nitrogen-starvation medium and complete medium and different time points when Lijiangxintuanheigu challenged with blast fungus using real-time fluorescence quantitative PCR.

All expression level of candidate genes were normalized by *actin* housekeeping gene and quantified by both the comparative threshold method and standard curve method. The results showed that expression level of all candidate genes were significantly different in isolates of 94-64-1b Y99-63, 95-23-4a, Y98-16 and 94-64-1b. When two isolates of Y98-16 and Y99-63 grown under complete medium and nitrogen-starvation medium, relative expression quantity of genes was different. And expression of more genes was detected when two isolates grew under nitrogen starvation for 24 h, comparing with when the two isolates grew under complete medium (Figure 3).

We detected expression level of all candidate genes at 24 hpi, 48 hpi, 72 hpi, 96 hpi and 168 hpi, the result revealed that all genes expression level apparently up-regulated, and the expression level achieved the maximum at 48hpi, the expression level had been decreasing after 72hpi (Table 5 and Figure 4).

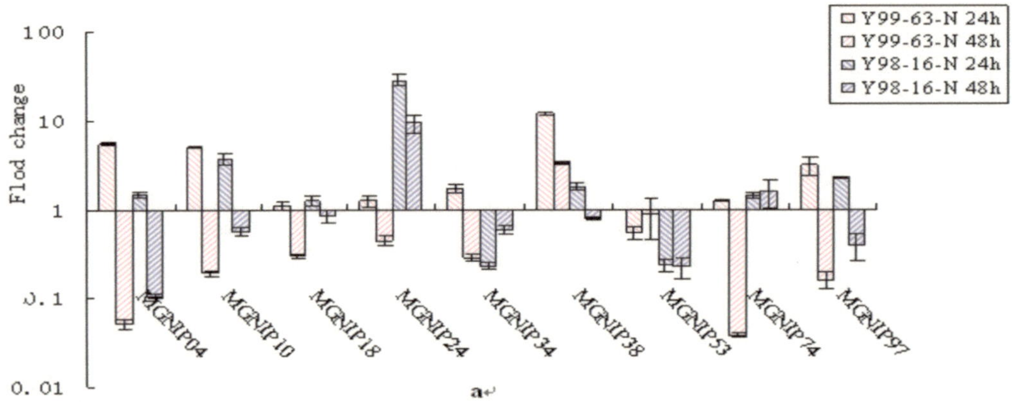

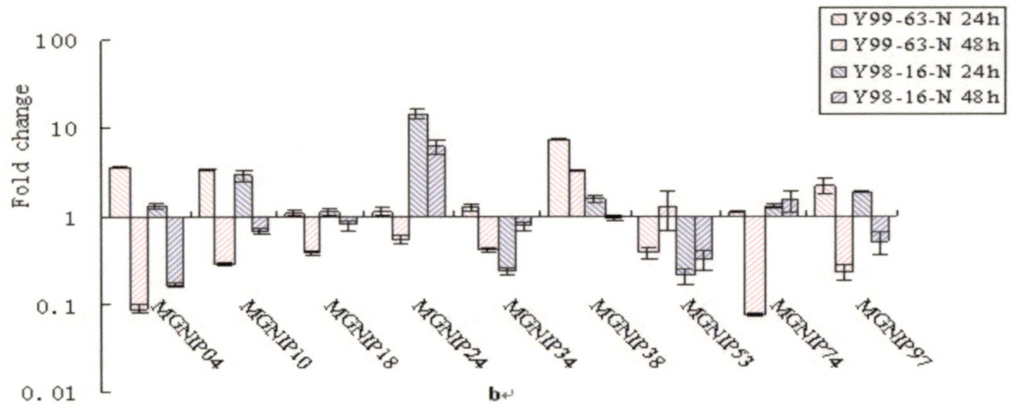

Fig. 3. Expression pattern of some predicted effector protein-encoding genes from *M. Grisea*

a: Relative expression quantity of target gene of Y99-63 and Y98-16 cultured in different mediums by $2^{-\Delta\Delta Ct}$ method

b: Relative expression quantity of target gene of Y99-63 and Y98-16 cultured in different mediums by standard curve method

Targeted gene \ Isolates	Y99-63	24 h-I		48 h-I		72 h-I		96 h-I		168 h-I	
	Control	$2^{-\Delta\Delta Ct}$ Fold change	Standard curve Fold change	$2^{-\Delta\Delta Ct}$ Fold change	Standard curve Fold change	$2^{-\Delta\Delta Ct}$ Fold change	Standard curve Fold change	$2^{-\Delta\Delta Ct}$ Fold change	Standard curve Fold change	$2^{-\Delta\Delta Ct}$ Fold change	Standard curve Fold change
MGNIP04	1	37602±6231	4800±854	71139±19038	7660±1802	54388±14248	7101±1626	44878±10524	6812±1360	243±76.8	291±237
MGNIP10	1	4.88±0.22	2.24±0.08	8.92±1.04	3.63±0.37	0.92±0.04	0.6±0.02	0.29±0.01	0.25±0.00	0.1±0.01	0.13±0.02
MGNIP18	1	1799±201	334±28.1	2386±743	414±95.7	359±36.3	95±7.37	76±4.36	29±1.26	5±0.48	3±0.24
MGNIP24	1	236±37.5	77±15.49	287±14.8	77±3.26	68±9.84	25±2.95	19±0.89	10±0.36	1±0.06	0.6±0.04
MGNIP34	1	2012±68.3	376±11.7	5401±297	879±44.2	773±8.74	192±1.99	88±1.51	32±0.51	3±0.11	2±0.08
MGNIP38	1	1323±121	263±21.4	1959±512	354±79.7	282±20.7	78±4.98	100±17.0	36±5.35	8±0.57	5±0.35
MGNIP53	1	4117±306	776±56	12888±1242	2198±209	1931±87.3	501±22.5	277±14.2	101±5.10	3±0.27	3±0.22
MGNIP74	1	9±2.00	5±0.82	17±5.22	8±1.87	5±0.39	3±0.19	3±0.17	2±0.09	0.1±0.01	0.1±0.01
MGNIP97	1	634±98.7	135±18.5	1825±385	325±59.7	337±34.3	89±7.86	130±9.00	45±2.65	3±0.19	2±0.13

Table 5. Relative expression quantity of target gene in infected rice leaves at different stages post inoculation

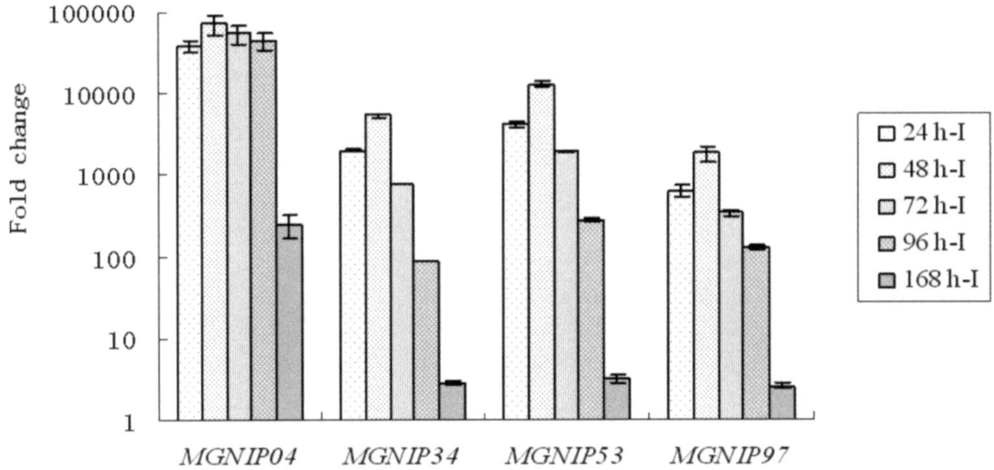

(a)

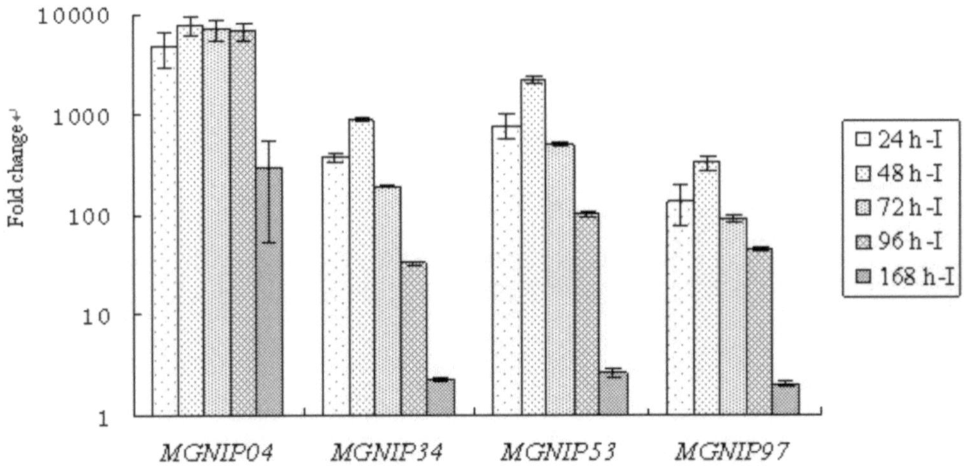

(b)

Fig. 4. Expression pattern of some predicted effector protein-encoding genes in infected rice leaves
a: Relative expression amount of target gene in infected rice leaves at different stages post inoculation by $2^{-\Delta\Delta Ct}$ method
b: Relative expression amount of target gene in infected rice leaves at different stages post inoculation by stand curve method

6. Pathogenicity analysis of secreted protein from the rice blast grown under nitrogen-starvation medium

Fungi maintained their cell living and even growth through material reutilization when they were in nutrition-stress environment. Some research showed that expression quantity of

pathogenicity-related genes increased when rice blast strains grew under nitrogen-starvation medium, which enhanced the pathogenicity of blast strains (Talbot et al., 1997).
The two isolates of Y99-63 and Y98-16 were from Yunnnan, China. And virulence test of two isolates of Y98-16 and Y99-63 on rice isogenic lines of IRBL1-24 had been previously performed in our lab, and virulence of Y99-63 was more intensive than Y98-16. To analyze the virulence of extracellularly secreted proteins on rice varieties such as susceptible variety of Lijiangxintuanheigu, resistant variety of Tetep and rice isogenic lines of IRBL1-24, we separated the extracellularly secreted proteins when Y98-16 and Y99-63 grew under nitrogen starvation for 48h, and man-made wounded rice leaves were inoculated with extracellularly secreted proteins. The result showed that necrosis speck occurred around the wounded leaves and wounded stems of rice when secreted proteins were inoculated on leaves or stems for 48h, and speck diameter of leaves or stems treated with secreted proteins was 2 to 4 folds larger than leaves or stems treated with sterilized water.
We compared difference of extracellularly secreted proteins from Y99-63 and Y98-16 growing under nitrogen-starvation medium for 48h using two-dimensional electrophoresis technology. The result showed that more proteins spots were detected from Y99-63 growing under nitrogen-starvation medium than Y98-16. And pI and molecular weight of secreted proteins had an apparent difference between Y99-63 and Y98-16.

7. Summary

In this chapter, we have showed how mine the secreted proteins from fungi and plant, and how predicted the some features of secreted proteins such as domain, pI, molecular weight and sequence similarity. And simultaneously, we also introduced some experiments centered on expression pattern of secreted protein-encoding genes, and pathogenicity analysis of secreted proteins from the rice blast strains grown under nitrogen-starvation medium.

8. Acknowledgments

This work was supported by the National Basic Research Program (No. 2011CB100400) from The Ministry of Science and Technology of China and the National Natural Science Funds, China (30860161), respectively.

9. References

Birch, P.R.J., Rehmany, A.P., Pritchard, L., Kamoun, S. & Beynon, J.L. (2006). Trafficking Arms: Oomycete Effectors Enter Host Plant Cells. Trends in Microbiology 14, 8–11.

Chisholm, S.T., Coaker, G., Day, B. & Staskawicz, B.J. (2006). Host–Microbe Interactions: Shaping the Evolution of the Plant Immune Response. Cell 124, 803–814.

Dean, R.A, Talbot, N.J., Ebbole, D.J., Farman, M.L., Mitchell, T.K., Orbach, M.J., Thon, M., Kulkarni, R., Xu, J.R. et al. (2005). The Genome Sequence of the Rice Blast Fungus *Magnaporthe grisea*. Nature 434, 980–986.

Dou, D., Kale, S.D., Wang, X., Jiang, R.H.Y., Bruce, N.A., Arredondo, F.D., Zhang, X. & Tyler, B.M. (2008). RXLR-mediated entry of Phytophthora sojae Effector Avr1b into

Soybean Cells does not Require Pathogen-encoded Machinery. Plant Cell 20, 1930–1947.

Ellis, J., Catanzariti, A.M. & Dodds P. (2006). The Problem of How Fungal and Oomycete Avirulence Proteins Enter Plant Cells. Trends Plant Sci. 11, 61–63.

Grant, S.R., Fisher, E.J., Chang, J.H., Mole, B.M. & Hardham, A.R. (2002). Subterfuge and Manipulation: Type Effector Proteins of Phytopathogenic Bacteria. Annu. *Rev. Microbiol.* 60:425-429

Groppe, K. & Boller, T.(1997). PCR Assay Based on a Microsatellite Containing for Detection and Quantification of Epichoe endophytes in Grass Tissue. Appl Environ Microbiol 63:1543–1550.

Haldar, K., Kamoun, S., Hiller, N.L., Bhattacharje, S. & van Ooij, C. (2006). Common Infection Strategies of Pathogenic Eukaryotes. Nature Reviews Microbilolgy. 4:1-12.

Hiller, N.L., Bhattacharjee, S., van Ooij, C., Liolios, K., Harrison, T., Lopez-Estrano, C. & Haldar, K. (2004).A Host-targeting Signal in Virulence Proteins Reveals a Secretome in Malarial Infection. Science. 306:1934-1937,

Hogenhout, S.A., Van der Hoorn, R.A.L., Terauchi, R. & Kamoun, S. (2009). Emerging Concepts in Effector Biology of Plant-associated Organisms. *Molecular Plant – Microbe Interactions* 22, 115–122.

Hu, X., Nazar, R.N. & Robb, J.(1993). Quantification of Verticillium Biomass in Wilt Disease Development. Physiol Mol Plant Pathol 42:23–36.

Huang, G., Allen, R., Davis, E.L., Baum, T.J. & Hussey, R.S. (2006a). Engineering Broad Root-knot Resistance in Transgenic Plants by RNAi Silencing of a Conserved and Essential Root-knot Nematode Parasitism gene. *Proc. Natl. Acad. Sci.* USA103:14302-14306.

Huang, G., Dong, R., Allen, R., Davis, E.L., Baum, T.J.& Hussey, R.S.(2006b). A Root-knot Nematode Secretory Peptide Functions as a Ligand for a Plant Transcription Factor. Mol.Plant Microbe. Interact.19:463-470.

Jiang, R.H.Y., Tripathy, S., Govers, F. & Tyler, B.M. (2008). RXLR Effector Reservoir in Tow Phytophthora Species is dominated by a Single Rapidly Evolving Superfamily with More Than 700 Members. PNAS.105:4874-4879

Jones, J.D. & Dangl, J.L. (2006). The Plant Immune System. Nature. 444: 323–329.

Judelson, H.S., Tooley, P.W. (2000). Enhanced Polymerase Chain Reaction Methods for Detecting and Quantifying Phytophthora infestans in Plants. Phytopathology 90:1112–1119.

Kamoun, S. (2006). A Catalogue of the Effector Secretome of Plant Pathogenic Oomycetes. Annual Review of Phytopathology. 44: 41–60.

Kämper, J., Kahmann, R., Bolker, M., Ma, L.J., Brefort, T., Saville, B.J., Banuett, F., Kronstad, J.W., Gold, S.E. et al. (2006). Insights from the Genome of the Biotrophic Fungal Plant Pathogen Ustilago maydis. Nature. 444:97–101.

Lorang, J.M., Sweat, T.A. & Wolpert, T.J. (2007). Plant Disease Susceptibility Conferred by a "Resistance" Gene. Proc. Natl. Acad. Sci. U.S.A.104:14861-14866

Mahuku, G.S., Goodwin, P.H. & Hall, R. (1995). A Competitive Polymerase Chain Reaction to Quantify DNA of Leptosphaeria maculans During Blackleg Development in Oilseed Rape. Mol Plant-Microbe Interact 8:761–767.

Marti, M., Good, R.T., Rug, M., Knuepfer, E., & Cowman, A.F. (2004).Targeting Malaria Virulence and Remodeling Proteins to the Host Erythrocyte. Science. 306:1930-1933.

Mendgen, K., Hahn, M. & Deising, H. (1996).Morphogenesis and Mechanisms of Penetration by Plant Pathogenic Fungi. Annu.Rev.Phytopahol. 34:367-368

Navarre, D.A. & Wolpert TJ. (1999). Victorin Induction of an Apoptotic/Senescence-Like Respinse in Oats. Plant Cell.11:237-250

Nickel, W. (2005). Unconventional Secretory Routes: Direct Protein Export across the Plasma Membrane of Mammalian Cells. Traffic 6, 607–614.

O'Connell, R.J. & Panstruga, R. (2006). Tete a tete inside a Plant Cell: Establishing Compatibility between Plants and Biotrophic Fungi and Oomycetes. New Phytol.171:699-718.

Ou, SH. (1985). Rice Disease. Kew, England: Commonwealth Mycological Institute.p97-184

Rohe, M., Gierlich, A., Hermann, H., Hahn, M., Schmidt, B., Rosahl, S.& Knogge, W. (1995). The Race-specific Elicitor, NIP1, from the Barley Pathogen, Rhynchosporium secalis, Determines Avirulence on Host Plants of the Rrs1 Resistance Genotype. EMBO14:4168-4177

Lee, S.A., Wormsley, S., Kamoun, S., Lee, A.F., Joiner, K. & Wong, B. (2003). An Analysis of the *Candida albicans* Genome Database for Soluble Secreted Proteins Using Computer-based Prediction Algorithms. *Yeast*, 20:595-610.

Schürch, S., Linde, C.C., Knogge, W., Jackson, L.F.& McDonald, B.A. (2004). Molecular Population Genetic Analysis Differentiates Two Virulence Mechanism of the Fungal Avirulence Gene NIP1, Mol. Plant-Microbe Interact.17:1114-1125

Spassieva, S.D., Markham, J.E. & Hille, J. (2002). The Plant Disease Resistance Gene Asc-1 Prevents Disruption of Sphingolipid Metabolism during AAL-toxin-induced Programmed Cell Death. Plant J.32:561-572

Stukenbrock, E.H.& McDonald, B.A. (2007). Geographical Variation and Positive Diversifying Selection in the Host Specific Toxin SnToxA. Mol. Plant Pathol. 8:321-323

Talbot NJ ' McCafeny HRK, Ma M et a1. (1997). Nitrogen Starvation of the Rice Blast Fungus *Magnaporthe grisea* may Act as an Environmental Cue for Disease Symptom Expression. Physiological and Molecular Plant Pathology.50:179-195

Tudzynski, P. & Sharon, A.(2003).Fungal Pathogenicity Genes. Appl Mycol Biotechnol.3:187-212.

Tyler BM.(2002). Molecular Basis of Recognition between Phytophthora Species and Their Hosts. Annu. Rev. Phytopathol. 40:137–167

Tyler BM. (2007). Phytophthora sojae: root rot pathogen of soybean and model oomycete. Mol. Plant Pathol. 8:1–8.

Tyler, B.M., Tripathy, S., Zhang, X.M., Dehal, P., Jiang, R.H.Y., Aerts, A., Arredondo, F.D., Baxter, L., Bensasson, D. et al. (2006). Phytophthora Genome Sequences Uncover Evolutionary Origins and Mechanisms of Pathogenesis. Science 313, 1261–1266.

van't Slot, K.A.E. & Knogge, W. (2002). A Dual Role for Microbial Pathogen-derived Effector Proteins in Plant Disease and Resistance. Crit.Rev. Plant Dis.39:471-482

Wang, H., Li, J., Bostock, R.M. & Gilchrist, D.G. (1996). Apoptosis: A New Model for Perception of Plant Pathogen Effectors. Plant Cell20:2009-2017

Whisson, S.C., Boevink, P.C., Moleleki, L., Avrova, A.O., Morales, J.G., Gilroy, E.M., Armstrong, M.R., Grouffaud, S., van West, P., Chapman. S. et al. (2007). A Translocation Signal for Delivery of Oomycete Effector Proteins into Host Plant Cells. Nature 450, 115–119.

Wolpert, T.J., Dunkle, L.D. & Ciuffetti, L.M.(2002). Host-selective Toxin and Virulence Determinants: What's in a Name? Annu. Rev. Phytopathol.40:251-285

16

Immuno-Modulatory Effects of Phytomedicines Evaluated Using Omics Approaches

Shu-Yi Yin and Ning-Sun Yang
Agricultural Biotechnology Research Center, Academia Sinica, Taipei
Taiwan, R.O.C

1. Introduction

The advent of the omics area has created new research systems, including genomics, proteomics, metabolomics, as well as the associated bio-informatics science, and databases. At the same time, progress in traditional western medical research has reached a bottleneck, as single compound drugs are costly to create, synthesize, or engineer. If we are to see real and sustained progress in research, we must find approaches to utilize traditional remedies to develop advanced medicines.

Instrumental systems for transcriptome, including the microarray of different messenger RNA, microRNA and other non-gene sequence-related RNA products, have already been developed; functional genomics studies using these systems have been remarkably successful. However, the candidate genes involved in specific functions often need further verifications for revealing their roles in the signal pathways. The difficulty may also arise from the high variety and seemingly unrelated responsive genes and complex signaling or regulatory systems involved.

Proteomic analysis has its disadvantages, although two-dimensional (2-D) gels can display viable candidate proteins for study. Up to two thousand proteins of biological systems can often be analyzed in sensitive 2-D gel systems. There are other proteins, such as cytokines or chemokines of most leukocyte cells are expressed at relatively low levels, and often are not detectable by 2-D gels. More sensitive methods, such as LC/MS and other fractionation systems, need to be used for such cases.

Metabolomics faces an even greater challenge: a 2-D or one run display of the components of a metabolome has not been defined and cannot be systematically evaluated. Therefore, sequential analyses, e.g. the LC/MS followed by NMR, were developed to address the "overall" or more comprehensive picture of metabolomes.

Several phyto-medicinal studies, including some in traditional Chinese herbal medicine (TCM), have been considered as metabolome investigations. New strategies employing omics approaches may be especially useful for phytomedicinal research, as conventional phytomedicines often employ multiple components and they often are believed to interact with multiple molecular targets related to cellular and physiological (e.g., immune-modulatory) effects. In order to successfully evaluate the effects of phytomedicines, various omics approaches are being systematically combined. New computational and cross-disciplinary analyses will be required for most experimental biology studies. Some examples of systematic and technical considerations, in terms of research into the immune-modulatory and anti-inflammatory effects using the omics approaches, are addressed in this brief review.

1.1 Importance of systems biology and bioinformatics

Scientists investigate medicinal plants in search of regulatory genes and metabolites that can affect, modulate or upgrade the biological and metabolic processes, which in turn can confer specific physiological or pharmacological functions. Recently, various high-output technologies, including genomics, transcriptomics, proteomics and metabolomics, are employed in such research effort [1-3]. Bioinformatics and systems biology approaches are considered by many as needed to organize, manage, process, and understand the vast amounts of data obtained in various omics studies [4-8]. In addition, systems biology is aimed at understanding complex biology by integrating omics data from various sources for network analysis, for evaluating the holistic system as a whole, as experimental results from omics studies are most often not obtained or isolated as a single set of data points or evens [9]. By analyzing the omics data, bioinformatics tools can help upgrade new approaches for classifying and authenticating potential medicinal plants, identifying new bioactive phytochemicals or compounds, and even improving medicinal plant species or cultivars that can tolerate stressful environmental challenges.

The human immune system, as we currently conceptualize it, is under the tight control of a complex network of regulatory genes, RNAs, modulatory proteins and stimulatory metabolites. Past studies have often focused on understanding the roles of specific genes in immune responses. To associate expression changes with immunological conditions such as suppression, cancer, or autoimmunity, we can investigate the interrelationship of the up- and down-regulation of genes or proteins patterns. Using microarray analysis and comparative genomics, Hutton et al. [10] have identified genes and their regulatory elements responsible for maintenance, differentiation, and the general functioning of specific immune systems. In addition, most of the expression pattern of genes is related to the biological role and effects of the products of genes, and a similar statement may be made for protein expression [11]. Taken together, evaluation of gene and protein expression profiles may lead us to identify links between specific genes or proteins and the associated specific immuno-modularory effects. Moreover, omics technologies may also be employed to address our views of the often-used concepts in immunology, such as: molecular dynamics in response to specific stimulations or alterations of the molecular state of targeted specific cells, in the hypothesis-driven research approach [12]. For instance, in the drug discovery process, pharmaceutical companies have used various microarray systems as screening tools to eliminate compounds that have molecular indications of toxicities before preclinical and clinical testing [13]. In basic research, omics technologies have continually improved our understanding on how drugs can regulate the immune system as well as of a variety of issues in mechanistic or hypothesis-driven research [14-17]. The data obtained from these studies not only may have significant impact on the future directions of those specific lines of research but also may improve our understanding of the specific immuno-modulatory regulation of given drugs.

Bioinformatics is the application of computational tools for biological sciences; its major aim is the management and interpretation of biological data [18]. It has been an essential tool for fully integrating and multi-disciplinary understanding the processes in various biological areas [19]. Among them, understanding omics data requires both common statistical and machine-learning methods, because the data are usually in high-dimensional form and complexity. On the other hand, as compared with other biomedical and agricultural areas, the study of omics and its use for research into medicinal plants are still in its infant stage. Given the demand for studies on immuno-modulatory effects of herbal medicines, this chapter introduces and summarizes the applications of some omics approaches and specific bioinformatics tools for investigating phytomedicines.

1.2 Omics technologies

The technology platforms generally used in systems biology research, including transcriptomics, proteomics and metabolomics, have enabled us to study living systems from a holistic or integrative perspective through revealing profiles of multitudinous biochemical components (Figure 1); it also opens up a unique opportunity to reinvestigate phytomedicines [20]. The revolution of genomics research and technology development has yielded complete or draft DNA sequence maps for a spectrum of species including human, mouse and a serious of model organisms. Having the genomic data available, many new 'drug-able' targets based on transcriptomics study have been identified, opening up new insights into explanations of biological systems at a global scale. Additionally, through proteomics, we are witnessing the development of wonderful and multi-application tools for studying various signaling or mechanism systems at the level of proteins and protein-protein interactions [20, 21]. In the meantime, studies on glycol-biology and bioactive polysaccharides are making great leaps in glycomics research; similarly, studies on regulation and metabolic control of a spectrum of lipids are creating new approaches for "lipidomics". The recent wave of data from genomics and proteomics has precipitated the measurement of increasingly a group or spectrum of elements to provide a systems approach, especially at the level of metabolites and for the field of metabolomics.

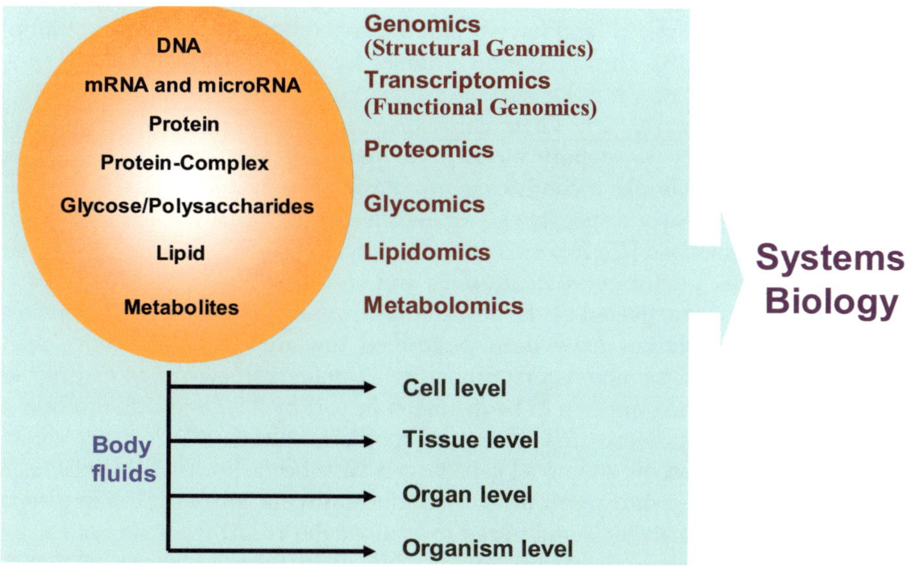

Fig. 1. The different levels of measurement in a systems biology approach.

Although omics are defined in several different ways today, in our opinion these systems provide "an integrated approach to study biological systems not only for the intracellular, but also at the cellular, organic and the whole body or organismic levels or networks, through measuring and integrating the genomic, proteomic and metabolic data" in a global consideration manner (Figure 1) [20, 22, 23].

In the search for new phyto-medicines, the necessary purification of single active components has been in general successful, whereas synergetic effects of mixtures of components (e.g., crude plant extracts) remain difficult to evaluate. Utilizing omics technology, scientists hope to develop methods and models to detect and observe the effects of complex mixtures such as various plant tissue extracts traditionally used in herbal medicines. This applies especially to approaches employing metabolomics which adress comprehensive phytochemical profiling, bioactivity phenotyping, sophisticated bio-organic chemistry instrumentation, and new cross-talk experimental designs. This scenario needs not only advancement in natural product research, but also revolutionary strategy in the development of molecular pharmacology-based herbal medicines [20].

1.3 Phytomics

The term "Phytomics" has been previously created to the "omics-based approach" for studying chemical compositions in plant (Kung PC et al., 2003), specifically: using bioinformatics and/or statistics to address qualitative and quantitative aspects of chemical compositions or profiles of the plant metabolites of our interest; or to develop databases for addressing such aspects [24].

2. Transcriptomics study on medicinal plant research

2.1 Application of DNA microarrays in toxicogenomics, pharmacogenomics and functional genomics studies of bioactivities from medicinal plants

Recent advances in genomics-based identification of responsive gene clusters, gene families or gene polymorphisms associated, with immune system dysfunction have helped to address some basic issues in immunology, and have begun to expand our understanding of immune-related disease processes [13]. The application of omics technologies in toxicological research (toxicogenomics) provided new insights into mechanisms of action, as well as data likely to be useful for risk assessment [13, 25]. Gene chips or microarrays are already employed in immunotoxicology research to identify biochemical pathways that are altered by specific chemical exposures. For example, trichothecene mycotoxin deoxynivalenol has been shown in mice to modulate splenic early responsive genes, which are functionally related to immunity, inflammation and chemotaxis [15, 26], indicating the importance of innate immune systems, including macrophages, granulocytes, neutrophils and various soluble mediators released in the inflammatory response activated by the hexachlorobenzene treatment. For basic research, a number of mechanistic studies have been performed towards gaining a comprehensive understanding of the immunomodulatory properties of potential new drugs or drug leads. Thymic atrophy, for instance, appears to be mediated in part by 2,3,7,8-tetrachlorodibenzo-*p*-dioxin (TCDD)-induced apoptosis [27]. Using an apoptosis-specific cDNA array combined with promoter analyses, specific and novel gene targets have been shown to enhance negative selection in the thymus and thus result in TCDD-induced thymic atrophy [16]. In a separate study, cDNA microarray analyses were utilized to evaluate the TCDD regulation of Fas ligand (FasL) promoter activity through modulation via NF-κB in thymic stromal cells and the subsequent initiation of the apoptotic pathway in thymic T cells [28].

During the past decade there has been a paradigm shift from utilizing single-target drugs to multi-target drugs [29, 30]. The concept of multi-targeted therapy was once believed to better represent the conventional herbal medicine treatments that often employ multi-component plant tissue extracts as natural products mixtures. However, very few phyto-medicinal products have clear or systematic documentations comparable to that of chemically synthesized drugs as single chemical compound. This situation has hampered our ability to predict precise or specific molecular targets, signaling or action mechanisms of activity, and possible side effects of "herbal drug" products [30]. With these requirements for botanical and clinical uses, a validated genomics and metabolomics approach in combination can be applied to quantify specific chemical markers and, subsequently, to obtain chemically standardized extracts [31]. In addition, researchers have witnessed a wide range of molecular mechanisms governing various cellular and tissue behaviors. The genomics approach with integrations of large and diverse sources of gene, protein and metabolite expression information will assist in making comprehensive and integrated predictions about the pharmacological effects of plant natural products [32].

While numerous laboratories use genomics in their investigation of underlying mechanisms of immunotoxicity, few have employed genomic analyses as a screening tool. Many differentially expressed genes are known to play a role in apoptosis, host defense, cell growth and differentiation, and trafficking of specific cells in body fluid systems. In the spleen, these may include the up-regulation of IL-18, lymphotoxin B receptor, and colony-stimulating factor receptor, and down-regulation of RANTES and histocompatibility antigens [15, 33-35]. In the thymus, gene changes included the down-regulation of nuclear factor of activated T cells, interferon gamma receptor, and T cell transcription factor 7, and the up-regulation of caspase 1 and ApoE. These findings are consistent with alterations previously observed in specific immune functions [34, 36] and could further expand our knowledge at gene regulation level.

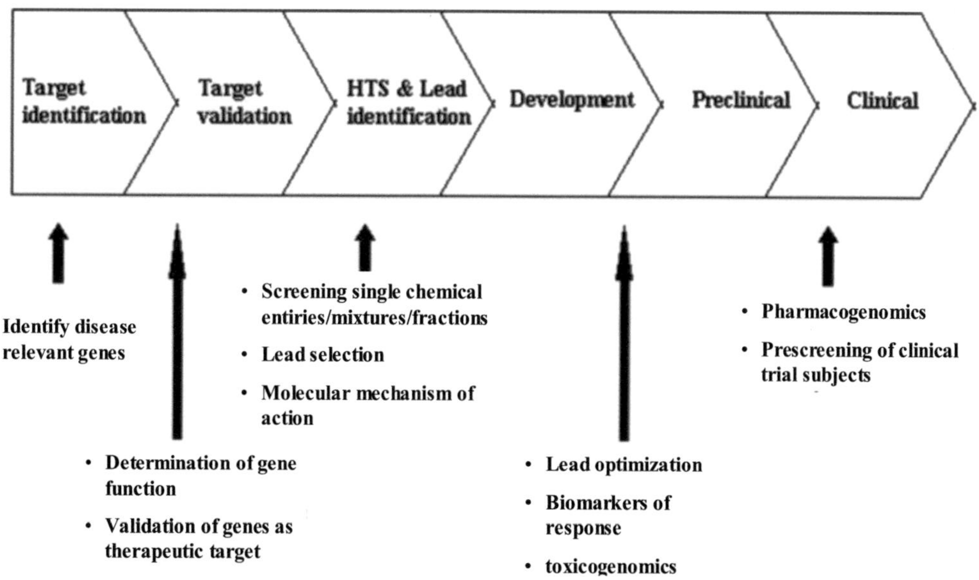

Fig. 2. DNA microarray applications in natural product drug discovery and development [37].

Applications of DNA microarray technologies in herbal drug research may be classified into three major areas. Firstly, it can be used in pharmacodynamics to aid the discovery of new diagnostic indicators and biomarkers for therapeutic response, elucidation of molecular mechanisms of herbal action, its formulations or its phytochemical components, and in identification and validation of new molecular targets for herbal drug development (Figure 2) [30], [37]. Secondly, it is applicable in toxicogenomics for predicting side effects of a medicinal herb or phytomedicine lead drug during preclinical activity and safety studies, conferring drug safety or resistance [38]. Thirdly, it is useful for botanical or plant identification and authentication of crude plant materials as part of an effort and regulatory system for standardization and quality control [39]. Given these considerations, DNA microarrays may thus offer powerful predictive functions at different stages of a typical drug/phytomedicine discovery pipeline.

2.2 Immuno-modulatory effects of different phyto-compounds/candidate phytomedicines

With the increased demand for validated herbal products for medicinal use comes the need to better understand the molecular mechanisms of their biological activities. Although many reputed herbal drugs are investigated at the molecular level, it remains difficult to realize the exact targets of individual phytochemical components and how these molecules together or independently can contribute to specific immuno-modulatory effects. Here, we discuss findings from some of the recent studies on microarray-based gene expression aimed at elucidating immune-regulatory mechanisms of pure phytochemicals as well as specific herbal extracts.

2.2.1 Purified compounds or specific phytochemical groups

The Chinese medicinal herb root *Tripterygium hypoglaucum* has been subjected to cDNA microarrays containing 3000 human genes (derived from a leukocyte cDNA library) in order to study its role in apoptosis-inducing activity of plant alkaloids. Apoptosis induced by these *T. hypoglaucum* alkaloids was shown to be mediated through c-myc and NF-kappa B signaling pathways [40]. In an animal model of aged rat, gene chip (Rat Genome U34A) analysis was applied to evaluate the gene regulatory pattern of *Epimedium* flavonoids in immune homeostasis. *Epimedium* flavonoids were found to reverse the "abnormal" or aging changes, allowing reconstruction of a beneficial equilibrium in gene expression and thus further remodeling of the immunohomeostasis in the aged rat [41]. Taken together, results from these studies indicate that the expression pattern characterized by up-regulation of specific apoptosis-promoting genes and down-regulation of certain apoptosis-inhibiting genes can be considered as important genomic background of an immunohomeostasis imbalance [30].

A traditional Chinese medicinal (TCM) herb prescription, Si-Jun-Zi decoction (SJZD), has been administered in a clinical setting to patients with disorders of the digestive system. Previous studies have indicated that the polysaccharides of SJZD are active components of the phyto-extract mixture in improving gastrointestinal function and immunity [42]. SJZD polysaccharides also had a protective effect and enhanced re-epithelialization on wounded IEC-6 cells. To further elucidate this effect at the molecular level, an oligonucleotide microarray was employed to study differential gene expression of SJZD-treated IEC-6 cells. There was, indeed, increased expression of genes encoding for ion channels and transporters, known as critical to cell migration and restoration of wounded cells, suggesting a mechanism for re-epithelialization as well as improved immunity [43]. These studies demonstrate the useful approach of functional genomics for research into modernization of TCM.

Shikonin and its derivatives, from the TCM-claimed medicinal herb *Lithospermum erythrorhizon*, have been shown to possess numerous beneficial pharmacological properties, including anti-inflammatory and antitumor properties [44, 45]. In our previous report, shikonin was shown to confer a potent bioactivity on suppression of TNF-α promoter activity [46]. Additionally, shikonin was found to mediate cytokine expression through inactivation of the RNA-activated protein kinase (PKR) pathway [46]. It was also suggested from the reports that regulation of TNF-α pre-mRNA splicing may constitute a promising target for future anti-inflammatory application [47]. Moreover, the functional genomic (DNA microarray) analysis on the cellular immunological effects of shikonin effectively distinguished the complex and specific bioactivities of this phyto-compound in human monocytes [48]. Further, ubiquitin pathway regulator, e.g., Rad23A, was also identified as possible key regulators for this shikonin effect [48]. A transcriptomics approach has therefore been instrumental in screening immune-modulatory effects of noteworthy phytocompounds. These studies have set useful examples for future systematization of key traditional herbal medicine-derived phytomedicines.

2.2.2 Medicinal herbal extracts

Screening of the human genome for TNF-α-inducible genes has been used to identify the anti-inflammatory effects of 5-Loxin, a standardized Boswellia serrata extract, in microvascular endothelial cells [49, 50]. It was shown that 113 out of the 522 TNF-α-induced genes were responsive to 5-Loxin treatment. These genes are directly or apparently related to inflammation, cell adhesion, and proteolysis. These robust 5-Loxin-sensitive candidate genes were subjected to further evaluation for molecular signaling, and this processing led to the suggestion of the primary 5-Loxin-sensitive TNF-α-inducible pathways. Mechanistically, 5-Loxin can completely inhibit VCAM-1 expression, and TNF-α can cause inflammation by strongly up-regulating the expression of this adhesion molecule VCAM-1. [49].

Recently, genomics analysis has also evolved for evaluation of the efficacy of bioactive chemicals in natural health products as therapeutics, for instance, with regard to the alleviation of specific inflammatory activities in human airway epithelial cells [33]. In addition, one application of gene expression profiling in this research field may be the growing appreciation for the multiple and pivotal roles played by various dendritic cells (DCs) in initiating and regulating a spectrum of immune responses. These cells are responsible for recognizing and processing various antigens and their ultimate presentation to specific immune cell (e.g., T cells) systems [51-53]. It has been well established that DCs present in the epidermis (Langerhans cells (LCs)) are required for the presentation of chemical allergens at the skin's surface, as well as for skin sensitization [54, 55]. Investigations by Enk and Katz [56] have revealed that topical exposure of mice to chemical allergens, but not to a non-sensitizing skin irritant, caused numerous changes in expression of cytokines and chemokines by LC and local epidermal cells. Among these changes recorded following allergen treatment was a rapid increase in LC expression of mRNA for interleukin-1β (IL-1β), a cutaneous cytokine necessary for the regulation of LC function and for skin sensitization [56-59]. These results concluded that changes in the expressions of IL-1β by LC in response to chemical allergens might hence provide a practical and efficient *in vitro* approach for identifying skin-sensitizing chemicals [60].

Genome-wide analysis has been adopted as a less selective approach for measuring the holistic or global changes in gene expression [17, 61-64]. It can be anticipated that the activation and functional maturation of DCs, as drastic cellular activities, are likely to be associated with changes in the levels of a spectrum of gene expressions that are responsible for: (a) intracellular metabolic processes; (b) control of cell motility (including those regulating intercellular communication and interactions with the tissue matrix); (c) cytokine and chemokine production, and (d) cell growth regulation and survival [17, 61]. Results of our previous studies on the immune-modulatory effects of a phytocompound mixture, extracted from the butanol fraction (BF) of a stem and leaf (S+L) extract of *Echinacea Purpurea* ([BF/S+L/Ep]), suggest that [BF/S+L/Ep] can effectively modulate DC mobility *in vivo* and related cellular physiology in the mouse immune system. In addition, [BF/S+L/Ep] modulated cell adhesion-, cell mobility-, cytokine- and NF-κB signaling- related activities in primary cultures of mouse DCs [17]. Similar study of Wang et al [60], have further shown that genes expressed in [BF/S+L/Ep]-treated human DCs revealed a key-signaling network involving a number of immune-modulatory molecules and lead to the activation of a downstream molecule, adenylate cyclase 8. These examples show that genomics approaches can be usefully employed for predicting candidate target molecules in future translational studies of phytochemicals, phytocompound mixtures, and medicinal herbal extracts.

2.3 Use of cDNA microarray/ expression sequence tags (ESTs) for evaluating bioactivities of medicinal plants

A transcriptome is the set of all detectable RNA molecules, including mRNA, tRNA, rRNA, and non-coding RNAs (e.g., siRNA, microRNA) produced in a group of test cells or tissues. By using the advanced transcriptomics, an organism's entire transcriptome can now be effectively analyzed for many experimental systems. Technically, transcriptomics is a technology to reveal genome-wide gene expression profiles, patterns, integrated or segregated features or networks describing a global view or analysis of gene expression activities of the genome at the mRNA or regulatory RNA levels. These technologies comprise cDNA-AFLP, SAGE, cDNA microarray (or gene chip), oligonucleotide-microarray, and microRNA microarray [2]. Microarrays also have been used to detect gene expression changes of medicinal plants in a variety of developmental stages, geographic locations, natural growth environments, and/or cultivation conditions [2]. In phytomics studies, studies have aimed to identify the responsive genes that are regulated by active medicinal compounds, anti-pathogen infection, or adaptation to harsh environment [65].

To design appropriate probe sequences for a DNA microarrays efficiently, we need to consider the genome sequence information for a specific organism in its entirety or with a definable set or subset. However, since only very limited genomes of medicinal plants have currently been sequenced, one alternative is to gather the necessary transcriptome information, by generating or making use of existing expression sequence tags (ESTs) [66, 67]. Increasing numbers of EST libraries from medicinal plants such as *Panax quinquefolius* [68], *Huperzia serrata* [69], *P. Notoginseng* [70], *Rehmannia glutinosa* [71], and *Catharanthus roseus* [72] have been recently obtained. An automatic system for large scale EST sequence retrieval, assembly, and functional and pathway analyses has been established [73]. This system has been successfully applied to analyzing both plant [74] and animal EST sequences [73, 75]. These EST and annotation systems have provided a good foundation for design of suitable arrays for representative genomes or focused transcriptomics, hence providing valuable information for genomic research into phytomedicine.

3. Proteomics studies on the research into medicinal plants

3.1 Use and advancement of analytical and instrumentation systems: Two-dimensional gel electrophoresis (2-DE), electrospray ionization, matrix-assisted laser desorption/ionization and surface-enhanced laser desorp

The Nobel Prize in Chemistry for 2002 was shared between scientists from two research expertise: mass spectrometry (MS) and nuclear magnetic resonance (NMR). These revolutionary breakthroughs have allowed chemical biology to become one of the most significant scientific disciplines in recent years. Scientists can now rapidly and reliably identify most proteins in a relatively small sample and readily produce three-dimensional display and/or images of expressed protein molecules with highly resolution. With these advancements, various experimental approaches and technologies were developed to obtain a better understanding of proteins and their regulatory effects on molecular and cellular functions of various biological systems [76, 77]. Among them, technologies including two-dimensional gel electrophoresis (2-DE) analysis [78, 79], matrix-assisted laser desorption/ionization (MALDI)- time-of-flight (TOF) [80] and Surface-Enhanced Laser Desorption/Ionization (SELDI)-TOF MS [81] have been broadly used in proteomics studies on the research of medicinal plants.

3.2 Application of proteomics for research into traditional herbal medicine

Proteomics technologies were applied to simultaneously study the function, organization, diversity, and the dynamic variety of total or a subset of proteins at the cellular or tissue levels [21]. The current integrative approach used in proteomics is in line with the practice and holistic philosophy of traditional Chinese medicine (TCM). Recent advances in multidimensional liquid chromatography, coupled with free-flow electrophoresis and capillary electrophoresis-based separation techniques, make it possible in separation of hundreds or even thousands of protein components in some medical plants [82, 83]. We may able now to explore an increased understanding of such complex mixtures and the reputed medicinal effects at the cellular and molecular levels through proteomics studies; it holds a key to the big demand for modernization and internationalization of a number of traditional phyto-medicines [83]. In this article, some of the proteomics approaches in TCM research and development are addressed, highlighting the application in mechanistic investigation of specific phytomedicines.

Panax ginseng and *Panax quinquefolius* are two of the valued herbs widely used in TCM. Conventional separation methods were unable to distinguish the different plant parts (main root, lateral roots, rhizome head and epidermal tissues) between these two species. On the other hand, when 2-DE maps were employed, plant tissue samples containing distinct or common protein species (spots) can be easily discriminated or distinguished. Clearly, these potential protein biomarkers may also facilitate the identification processes for various medicinal plants that may be difficult to identify morphologically or anatomically [84].

Numerous herbal medicines have been reported to have immunomodulatory and anti-tumor effects in cancer cells [85-87]. Recent biological and pharmaceutical researches have shown that diosgenyl saponins may exert a large variety of biological functions, with a potential for use in cancer chemoprevention [88]. By using 2-DE, tryptic in-gel digestion and MALDI-TOF MS analysis, Wang *et al.* [89] suggested that dioscin, a saponin extracted from *Polygonatum zanlanscianense* Pamp., exhibited cytotoxicity towards human myeloblast leukemia HL-60 cells. This proteomics analysis also revealed that the expression of mitochondria-associated proteins was substantially altered in HL-60 cells upon dioscin treatment, suggesting that mitochondria were the major cellular and organelle target of dioscin cytotoxicity. Moreover, the results indicated that other pathways were likely also involved in detected dioscin cytotoxicity, including phosphorylation-based cellular signaling, RNA-related protein synthesis, and oxidative stress processes. The study demonstrated the benefits of using a proteomics approach in anticancer phytomedicine research [90].

4. Metabolomics study on the research of medicinal plants

Metabolomics, including both targeted and global metabolite profiling strategies, is rapidly becoming a popular and powerful approach of choice across a broad range of medical and biological sciences including systems biology, drug discovery, and molecular and cell biology [24]. Specifically for human metabolites, it is believed that at least 3,000 metabolites that are essential for normal growth and development (primary metabolites) and >2000 secondary metabolites that are not essential for growth and development but may help fight off infection and other forms of stress on the body [91]. In addition, metabolomics are now being generally considered a vital component of the systems biology approach, in which it can reflect and connect the genotypes with diverse yet specific phenotypes of specific types of cells, tissues, or organs [91]. Within the past decade, the number of publications of metabolomics-related research articles has increased from roughly 40 in 2002 to 100, 170, 200 and >250 articles in the years 2004, 2005, 2006 and 2007, respectively. Now it is estimated that >300 articles, with a general aim or study on metabolomics were published annually in 2010. Owing to its remarkable versatility, metabolomics is rapidly becoming a universal tool and key component in medical research [24]. Combined with genomics and proteomics technologies, systems biology research using metabolomics investigates characteristic molecular signatures for disease diagnosis, prognosis, and therapeutics [92]. This section reviews the recent developments in technology platforms and experimental approaches for metabolomics studies in the research of immunomodulatory properties of potential medicinal plants.

4.1 Use of GC-MS, LC- MS, FT-IR and NMR technologies

Currently, the term 'metabolomics' often can be used interchangeably with "metabolite profiling" because the type of one-step, two dimensional exhibition analysis used in genomics and proteonomics experiments is not possible at the present time, as the complexity of chemicals in most biological systems, especially in plants, is highly diversified and can be enormous [93]. The two basic approaches in metabolomics can be classified the targeted- and the global metabolite analyses. Targeted metabolite analysis, (or metabolite profiling), as the name implies, targets mainly a subset of metabolites in test sample, instead of a complete, global metabolome analysis, often by using a particular set of analytic technique(s) such as gas chromatography-mass spectrometry (GC-MS) and liquid chromatography–mass spectrometry (LC-MS), and yields an estimate of quantity [94]. Metabolomics approaches using GC-MS, LC-MS, or 2D NMR are effective tools for quality control of medicinal plants or herbal medicine products [95, 96]. As shown in Figure 3 [24], key aspects of the technology were assembled in many research institutions as "core labs/facilities" in the metabolomics approach for herbal medicine or other integrated research interest. Various other technical systems, methodologies or techniques, including thin layer chromatography (TLC), Fourier transform infrared spectroscopy (FT-IR), Raman spectroscopy and NMR [97-99] are also important research facilities in the metabolite analysis arsenal.

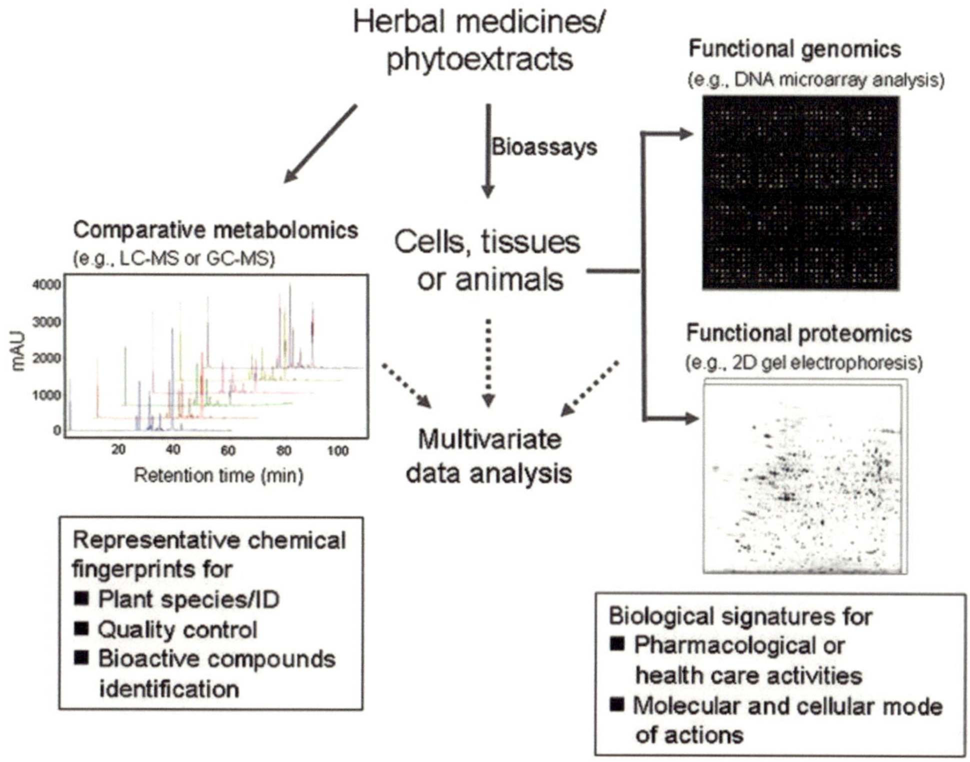

Fig. 3. Key features of metabolomics technologies employed for research into phytomedicines [24].

Mass spectrometry is currently the most broadly applied technology in metabolomic studies. Among the variety of MS techniques, GC-MS has long been popularly used in metabolite profiling of plant extracts [100, 101]. Rapid, high-resolution 2D GC x GC-TOF MS has been employed in the phenotyping of natural rice variants [102] as well as for efficient quality control or analysis of herbal medicines [95]. Recently, capillary electrophoresis-MS has also been developed as a metabolomics tool, capable of simultaneously analyzing over 1,000 charged chemical species, a technique that is expected to create a number of obvious applications in processing and characterization of various biological samples [95]. A shotgun approach using MALDI-TOF/TOF MS has recently been established for rapid analysis of negatively charged metabolites in mammalian tissues to: (a) facilitate the detection of low-abundant metabolites such as cAMP, cGMP, and IP3; and (b) discriminate isomeric molecular species [103]. In addition, novel instrumentation/equipment set ups developed recently, such as Fourier Transform Ion Cyclotron Resonance Mass Spectrometry (FT-MS), represents a quantum jump in the new capabilities of mass spectrometers for metabolite analysis. Due to the exceptionally high resolution of these instruments, metabolites with mass differences of less than 2 ppm can now be separated on a chromatographic time scale (Van der Greef *et al.*, 2004). The accurate results obtained can help reveal elemental compositions, which often enable unequivocal metabolite identification.

Remarkable recent developments in analytical biochemistry regarding the detection and characterization of compounds with small molecular mass, such as MS and high-field NMR coupled with user-friendly multivariate statistics, have led to highly efficient systems for comprehensive analysis of the metabolite data matrices generated by metabolomics experiments [104]. One-dimensional (1D) NMR spectrometry has shown its capability for high-output analysis and classification of chemically similar groups of test samples. At the same time, the large numbers of overlapping peaks generated by such method may also hinder in some case accurate identification of specific metabolites. Recently, a replacement for the 1D ^{1}H NMR spectroscopic technology also has been developed: a two-dimensional (2D) ^{1}H–^{13}C NMR strategy (fast metabolite quantification, FMQ, by NMR), was developed for analyzing metabolites as multivariate statistical objects [105].

The new 'hyphenated' techniques that combine in assay sequence various forms of liquid chromatography with NMR, such as HPLC-SPENMR, have effectively improved the sensitivity of NMR analyses and can be employed to characterize both high- and low-abundant metabolites in complex crude plant extracts [106, 107].

4.2 Metabolomics research in medicinal chemistry studies

Diverse secondary plant metabolites are believed to have evolved through continuous interactions with challenging and predominantly hostile environments, including both abiotic and biotic stresses. When these features are coupled with characteristic species and agronomic differences, various phyto-chemicals as secondary metabolites generally can confer various specific bioactivities related to their biochemical structures [108]. These bioactivities apparently can help the host plants to defend specific plant pathogens and to reduce a spectrum of abiotic stresses, e.g., drought, heat and saline conditions. Interestingly, these secondary plant metabolites often were also found to confer potent and valuable bioactivities for defending human sickness, including viral, cancerous and inflammatory diseases. Some well-known cancer chemotherapeutic drugs have been initially derived from plant secondary metabolites, such as paclitaxel (taxol), camptothecin (irinotecan, topotecan), and podophyllotoxins (etoposide, teniposide) [24, 109]. Recent re-recognization of the vast potential of plant secondary metabolites or natural products to serve as lead compounds for drug discovery and development, or as various general health care products, has renewed a lot of interest in pharmaceutical and nutraceutical research. *De novo* combinational chemistry has so far produced only a very limited number of novel drugs, the natural products and their derivatives are still considered by many scientists to be the primary source of leads for drug development [110]. In this area, the use of whole plants or their extracts as medicines gave way to the isolation of active phyto-compounds, beginning in the early 19th century with the isolation of morphine from opium. In such a reductionist approach, however, single active phytocompounds may often be not identifiable because of their low abundance in test plant extracts, or alternatively, a spectrum of pharmacological efficacy traditionally observed arises only as a synergistic action of the multiple but specific ingredients present in a single plant or even from a multiple medicinal plant formulation, as in TCM [111, 112].

To efficiently link the flood of experimental data and specific metabolites or general metabolite profiles information to biology and metabolism study systems, traditional bioinformatics is being combined with cheminformatics to generate a basic computational infrastructure for analysis of metabolomics [113, 114]. A number of metabolomics databases, some based on both chemical and biological/biochemical data, have been made publicly available [114]. The Human Metabolome Database (HMDB) is currently the largest and most complete database in breadth and depth, offering spectral, physico-chemical, clinical, biochemical, genomic, and metabolism information for a library of >2500 known human metabolites [115, 116]. Other databases include the BioMagResBank (BMRB) with an emphasis on NMR data (>270 pure compounds), the Madison Metabolomics Consortium Database (MMCD) which presents MS and/or NMR data on more than 10,000 metabolites [117], and the Golm Metabolome Database (GMD) which has been specifically designed for plant research and utilizes GC–MS data [118]. Additionally, Wishart [113] has reviewed the development of algorithms and innovations in informatics concerning data reduction, normalization, and alignment that offer sufficient biological insight into metabolic profiles.

4.3 Metabolomics approach applied to research into immuno-modulatory effects of phytomedicine

It is now generally accepted that chronic inflammation is a key factor in the development of many types of cancers. Natural products, especially from plants, were once popular choices in cancer therapeutics based on their immunosuppressive or anti-inflammatory effects [110, 119-121]. Recently, metabolomics has been effectively used to characterize and monitor carcinogenesis activities in mouse models [122]. In addressing oncology metabolomics, NMR was used to target biomarkers for prostate cancer by analyzing metabolites with anti-inflammatory effects in the development and progression of this cancer for better future management [123, 124]. This metabolomics approach has also been successfully implemented to monitor the metabolism in human brain, liver tumors, lymphomas, and colon cancers [125].

5. Comparative and bioinformatics tools for omics studies

5.1 Ingenuity (http://www.ingenuity.com/)

Functional genomics experimental approaches were employed in our previous studies on the modulatory effect of *Echinacea* plant extracts (e.g., the butanol-fractionated Leaf and Stem tissue extract designated as BF/S+L/Ep) on both mouse and human DCs [17, 63, 64]. Using the same defined phytochemical extracts in the study, we analyzed the genome-wide transcriptional response in the context of known functional activities and interrelationships among specific protein molecules and/or different cell phenotypes. Ingenuity systems, a structured network knowledge-based approach, provided us good tools and insight into the regulation of bone marrow-derived dendritic cell activities relevant to the body's immune system. Figure 5 shows candidate molecular networks revealed by clustering analysis of the representative genes involved in the BMDC response to [BF/S+L/Ep] treatment [17]. The prototypical cell was constructed from 37 representative genes that responded to treatment with [BF/S+L/Ep] *in vitro* from 4 hours to 12 hours. Genes whose expression was upregulated (more than doubled) are indicated in red, and those whose expression was downregulated (to less than half) are shown in green. Selected regions of the network highlight three groups of genes. Group 1: Immune response-related genes. Group 2: Adhesion molecules and cytoskeleton; cell movement related genes. Group 3: Cell cycle, cell proliferation and apoptosis related genes. Gene networks were analyzed using the Ingenuity Pathways program.

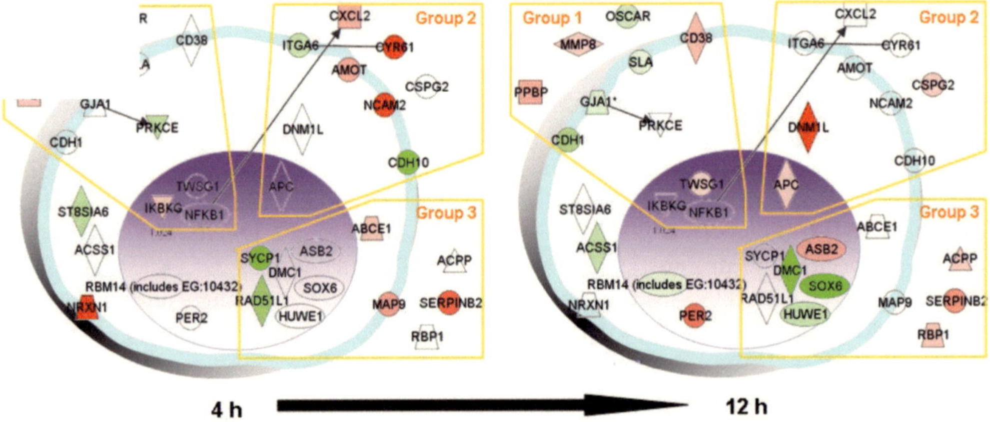

Fig. 5. Pathway analysis of representative genes that responded to [BF/S+L/Ep] treatment [17].

5.2 Metacore™ (http://www.genego.com/metacore.php)

MetaCore™ is another integrated knowledge database and software suite for pathway analysis of experimental data and gene lists. In the research of phytomedicines, it has also been used to evaluate the possible hierarchical control of microRNA expression from mouse tissues in order to identify trends of miRNA and mRNA expressions in response to targeted phytomedicinal treatment. Utilizing Metacore software, a prototypical network was constructed from 6 representative microRNAs that responded to treatment *in vivo* with specific phyto-chemicals (Figure 6). All selected microRNAs were found to be down-regulated to less than half of the untreated levels, and are shown with blue circles.

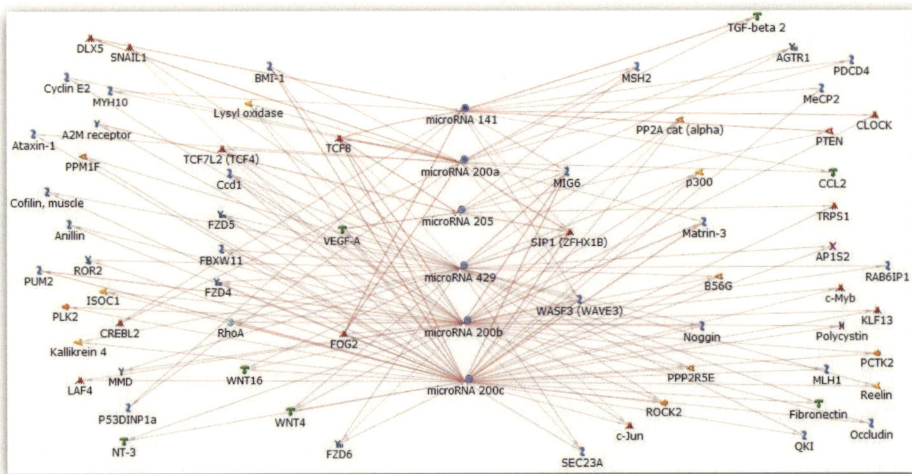

Fig. 6. Pathway/network analysis of representative microRNAs which are responsive *in vivo* to a specific single phytocompound treatment in inflammed mouse tissues.

Specifically, connections (hits) within 6 microRNAs were employed as the parameter for this specific search. Arrows indicate the cross talks among five key molecules/pathways, (TCF8, VEGF-A, FOG2, MIG6, SIP1 and WASF3), and there are postulated to be regulated by treatment with a specific phyto-chemical from TCM formulation.

5.3 TRANTHPATH (http://www.gene-regulation.com/index.html)

TRANSPATH, a database system about gene regulatory networks, combines encyclopedic information on signal transduction with tools for visualization and analysis. By integrating with TRANSFAC, a database about transcription factors and their DNA binding sites, TRANSPATH can predict putative signaling pathways from ligand to target genes and their products, which may themselves be involved in a regulatory action.

For studying specific imumodulatory effect of herbal medicine, the possible signaling pathways, networks or potential interactions among the responsive genes/target molecules in DCs treated with *Echinacea* extracts [BF/S+L/Ep] was assessed by using such Transpath software. This bioinformatics analysis has predicted a key-signaling network involving a number of immune-modulatory molecules leading to the activation of a very important downstream regulatory molecule, the adenylate cyclase 8, effectively in regulating cAMP levels in mammalian cells. This analysis indicated two postulated key molecules/pathways, Adenylate cyclase (AC8) and calmudulin (CaM), responsive to the *Echinacea* extracts (Figure 7).

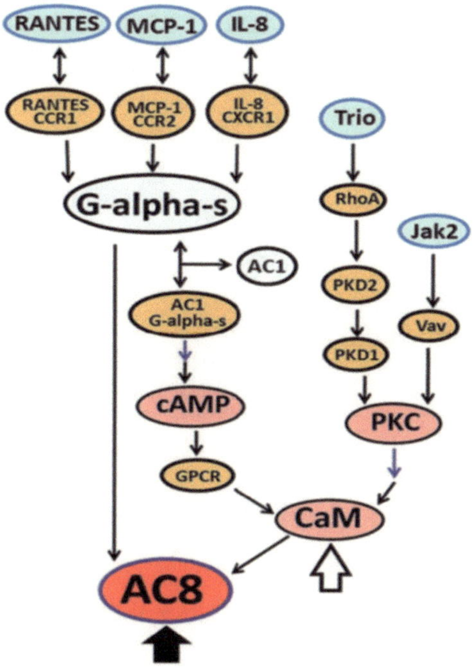

Fig. 7. Bioinformatics analysis of [BF/S+L/Ep] bioactivity and the underlying candidate molecular signaling networks in human DCs. The 20 genes that were up- or down-regulated at least 5-fold over controls were analyzed. Specifically, connections (hits) within 7 genes were employed as the parameter for the current search [63].

5.4 KEGG (http://www.kegg.jp/kegg/)

KEGG (*Kyoto Encyclopedia of Genes and Genomes*) is a multi-functional bioinformatics resource for linking genomes to metabolic activities. It consists of 16 main databases and has been widely used as a reference knowledge base for biological interpretation of large-scale datasets generated by sequencing and other high-throughput experimental technologies. Among these databases, the KEGG DRUG database contains crude drugs (consisting of multiple chemical compounds) and formulas (consisting of multiple crude drugs) in the Traditional Chinese Medicine (TCM). In addition, KEGG PATHWAY and KEGG ENVIRON are also being organized to interpret and correlate relationships between genomic and chemical information of various natural products/metabolites from plants. For example, the biosynthetic pathway of stilbenoids, a group of phenolic compounds, was provided for revealing specific molecular interaction and different reaction networks (Figure 8). Although the knowledge on biosynthetic pathways of plant natural products is in general largely incomplete, the genomic information is expected to provide clues to missing enzymes and reactions for biosynthesis of specific plant secondary metabolites, the source for future modernized phytomedicines, either as pure compounds, fractionated phytochemical mixtures, or as crude plant extracts. Moreover, the genomic information may also uncover the architecture of biosynthetic pathways for generating chemical diversity of natural products.

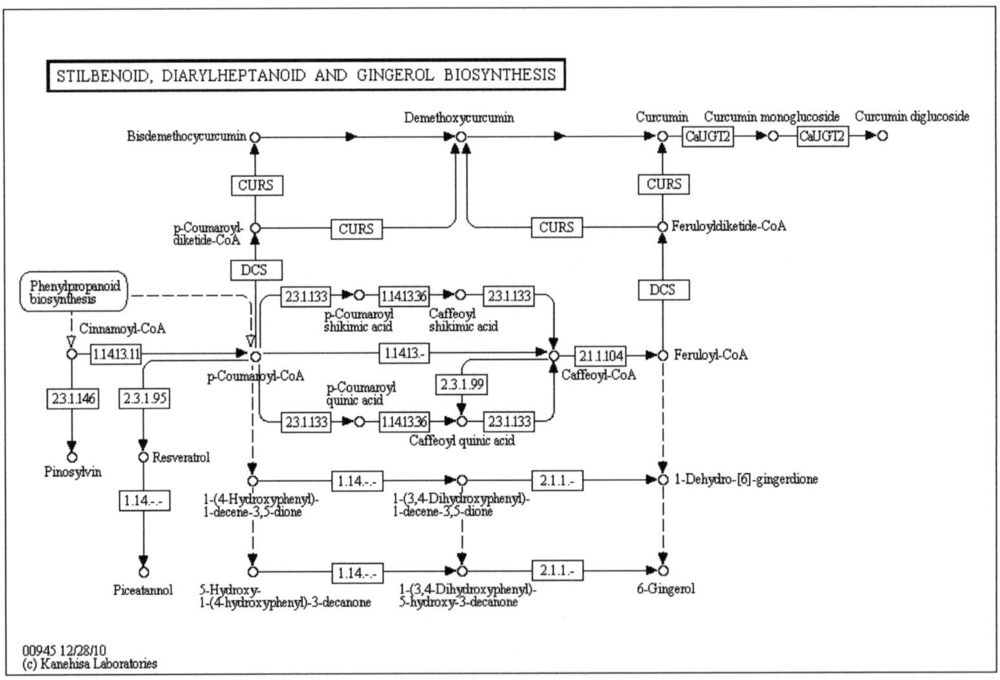

Fig. 8. Stilbenoid, diarylheptanoid and gingerol biosynthetic networks cited in KEGG. (Adopted from http://www.kegg.jp/kegg/pathway/map/map00945.html)

6. Challenges and perspectives

Traditionally, the pharmaceutical research and industries have focused on evaluating or monitoring individual gene, proteins as the target or basis for identifying new drugs. The quest for single molecules to modify single key factors in a disease process is now recognized as may not be able to provide a solution for a spectrum of diseases in which multiple cell types, target molecules and/or multiple pathways are known or believed to contribute to the diseases. Herbal extracts/mixtures as conventional phytomedicines may represent the combinational chemistry of the nature of traditional medicines, and encompass a vast repertoire of chemical entities that may have anecdotally and empirically found through long human culture history to confer a complex and yet integrated effect on numerous cellular components and functions, effecting a medicinal activity. Various traditional herbal drugs may thus have good potential for re-invention and newly found use in the multi-target approach in treating various diseases. However, such potential of herbal drugs is undermined by difficulties in standardization, and the pharmacodynamics and pharmacokinetics studies of these multicomponent plant extract mixtures. Microarray analysis of gene expression profiles may be useful for elucidating such complex molecular mechanisms and networks underlying the multi-target pharmacological functions of herbal extracts and phytomedicine mixtures. Research into the patterns of gene expression at a range of stages during the treatment process may reveal key targets and mechanisms and help to identify biomarkers of either adverse- or favorable response. A positive correlation between the transcriptional response induced by a putative or candidate herbal drug and the database profile of an existing pharmaceutical or therapeutic agent as a single chemical may provide us insight into the target specificity, mechanism of action, as well as in facilitating analysis of signaling pathways downstream of the specific target. This information could in turn be used to interpret possible bioactivity, function or effectiveness of test phytomedicines. In addition, various DNA, RNA or protein microarrays may also be used for bioactivity-guided fractionation of herbal extracts, thereby narrowing in the active principles delivering the desired or observed effect. Microarrays may also improve the power for selection of biological targets and lead compounds up or down the drug discovery pipeline. Once useful transcriptome or/and proteome data from herbal drug candidates can be correlated with *in vivo* bioactivity or preclinical or "existing clinical" (as in some TCM) response outcomes (biomarkers) in defined biological systems, the best candidates can then be selected for further drug development [30].

Although some DNA microarrays have already offered impressive potential for pharmacodynamics and toxigenomics applications, they are still being considered as in an exploratory stage and the data obtained from them will need validation by other biological experiments. Bioinformatics and statistical tools have a major role to play in analysis of the microarray results, whereby data from multiple experiments can and may need to be integrated to address complex biological activities, functions or effects. Another factor currently limiting microarray application is the cost of this technology [30]. The challenge we face today is to develop or construct standardized, sensitive, reproducible microarray platforms, databases and visualization methods for expression profiles that are affordable to most research scientists. With the use or development of improved, uniform and sophisticated experimental designs, data management systems [126, 127], statistical tools and upgraded algorithms for data analysis [128, 129], DNA microarrays hopefully can be more optimally used in herbal drug research. In spite of the vast potential offered by microarray and the related functional genomics and proteomics technology, the importance of integrating various *in vitro* biological assays, cell culture-based and *in vivo* animal experimental systems cannot be ignored. Comprehensive strategy integrating information from diverse scientific experiments and technologies are expected to benefit and lead to molecule and cell evidence-based phytomedicines.

The integration of information from genomics, proteomics, and metabolomics is hoped to provide solid evidence-based rationales for systematic development of various modern phytomedicines, on top of the foundations of various traditional medicine cultures. The search for specific, active single phytocompound may also be expedited when various metabolomics approaches are combined with a comprehensive array of bioactivity assay systems using standardized and normalizable mammalian cell, tissue and animal models. Wherea a "complete metabolome-exhibition" system is currently not available, HPLC-, GC- and LC/MS-based metabolite-profiling systems, alone or in combination, may already offer a good description or authentication tool for comparative and qualitative analyses and definition of the unique, distinctive, or combinational profile features of the conventional herbal medicine formulations, as elegantly demonstrated recently by W. Lam et al (2010) [112]. These and the improved or newly developed metabolomics technologies in linkage may also be usefully applied to discovery and development of new phytomedicines, as single phyto-chemicals or their mixtures, or as fractions or the whole preparation of the crude extracts of various medicinal plant tissues. Our challenges together, as scientists and health care-takers are to coordinate and integrate our intellectual thrusts, talents and efforts to address and target specific medical and medicinal research areas, e.g., for anti-inflammation and related chronic or cancerous diseases, for future research and development of advanced phytomedicines, may be to be pursued more effectively as an international program.

7. References

[1] E. Fridman, E. Pichersky, Metabolomics, genomics, proteomics, and the identification of enzymes and their substrates and products, Curr Opin Plant Biol 8 (2005) 242-248.
[2] D.J. Lockhart, H. Dong, M.C. Byrne, M.T. Follettie, M.V. Gallo, M.S. Chee, M. Mittmann, C. Wang, M. Kobayashi, H. Horton, E.L. Brown, Expression monitoring by hybridization to high-density oligonucleotide arrays, Nat Biotechnol 14 (1996) 1675-1680.
[3] A. Pandey, M. Mann, Proteomics to study genes and genomes, Nature 405 (2000) 837-846.
[4] L. Lederman, Bioinformatics and systems biology, Biotechniques 46 (2009) 501-503.
[5] Y.L. Yip, The promise of systems biology in clinical applications. Findings from the Yearbook 2008 Section on Bioinformatics, Yearb Med Inform (2008) 102-104.
[6] F.A. Middleton, C. Rosenow, A. Vailaya, A. Kuchinsky, M.T. Pato, C.N. Pato, Integrating genetic, functional genomic, and bioinformatics data in a systems biology approach to complex diseases: application to schizophrenia, Methods Mol Biol 401 (2007) 337-364.
[7] N. Rapin, C. Kesmir, S. Frankild, M. Nielsen, C. Lundegaard, S. Brunak, O. Lund, Modelling the human immune system by combining bioinformatics and systems biology approaches, J Biol Phys 32 (2006) 335-353.
[8] T. Yao, Bioinformatics for the genomic sciences and towards systems biology. Japanese activities in the post-genome era, Prog Biophys Mol Biol 80 (2002) 23-42.
[9] L.B. Ray, L.D. Chong, N.R. Gough, Computational biology, Sci STKE 2002 (2002) eg10.
[10] J.J. Hutton, A.G. Jegga, S. Kong, A. Gupta, C. Ebert, S. Williams, J.D. Katz, B.J. Aronow, Microarray and comparative genomics-based identification of genes and gene regulatory regions of the mouse immune system, BMC Genomics 5 (2004) 82.

[11] L.M. Staudt, P.O. Brown, Genomic views of the immune system*, Annu Rev Immunol 18 (2000) 829-859.
[12] N.J. Davies, M.G. Tadesse, M. Vannucci, H. Kikuchi, V. Trevino, D. Sarti, I. Dragoni, A. Contestabile, E. Zanders, F. Falciani, Making sense of molecular signatures in the immune system, Comb Chem High Throughput Screen 7 (2004) 231-238.
[13] L.A. Burns-Naas, R.J. Dearman, D.R. Germolec, N.E. Kaminski, I. Kimber, G.S. Ladics, R.W. Luebke, J.C. Pfau, S.B. Pruett, "Omics" Technologies and the Immune System (a) , (b), Toxicol Mech Methods 16 (2006) 101-119.
[14] S.D. Holladay, L.V. Sharova, K. Punareewattana, T.C. Hrubec, R.M. Gogal, Jr., M.R. Prater, A.A. Sharov, Maternal immune stimulation in mice decreases fetal malformations caused by teratogens, Int Immunopharmacol 2 (2002) 325-332.
[15] S. Kinser, Q. Jia, M. Li, A. Laughter, P. Cornwell, J.C. Corton, J. Pestka, Gene expression profiling in spleens of deoxynivalenol-exposed mice: immediate early genes as primary targets, J Toxicol Environ Health A 67 (2004) 1423-1441.
[16] M.T. Fisher, M. Nagarkatti, P.S. Nagarkatti, Combined screening of thymocytes using apoptosis-specific cDNA array and promoter analysis yields novel gene targets mediating TCDD-induced toxicity, Toxicol Sci 78 (2004) 116-124.
[17] S.Y. Yin, W.H. Wang, B.X. Wang, K. Aravindaram, P.I. Hwang, H.M. Wu, N.S. Yang, Stimulatory effect of Echinacea purpurea extract on the trafficking activity of mouse dendritic cells: revealed by genomic and proteomic analyses, BMC Genomics 11 612.
[18] M. Swindells, M. Rae, M. Pearce, S. Moodie, R. Miller, P. Leach, Application of high-throughput computing in bioinformatics, Philos Transact A Math Phys Eng Sci 360 (2002) 1179-1189.
[19] M.G. Kann, Advances in translational bioinformatics: computational approaches for the hunting of disease genes, Brief Bioinform 11 96-110.
[20] M. Wang, R.J. Lamers, H.A. Korthout, J.H. van Nesselrooij, R.F. Witkamp, R. van der Heijden, P.J. Voshol, L.M. Havekes, R. Verpoorte, J. van der Greef, Metabolomics in the context of systems biology: bridging traditional Chinese medicine and molecular pharmacology, Phytother Res 19 (2005) 173-182.
[21] W.C. Cho, Application of proteomics in Chinese medicine research, Am J Chin Med 35 (2007) 911-922.
[22] R.J. Lamers, J. DeGroot, E.J. Spies-Faber, R.H. Jellema, V.B. Kraus, N. Verzijl, J.M. TeKoppele, G.K. Spijksma, J.T. Vogels, J. van der Greef, J.H. van Nesselrooij, Identification of disease- and nutrient-related metabolic fingerprints in osteoarthritic Guinea pigs, J Nutr 133 (2003) 1776-1780.
[23] B.A. t Hart, J.T. Vogels, G. Spijksma, H.P. Brok, C. Polman, J. van der Greef, 1H-NMR spectroscopy combined with pattern recognition analysis reveals characteristic chemical patterns in urines of MS patients and non-human primates with MS-like disease, J Neurol Sci 212 (2003) 21-30.
[24] L.F. Shyur, N.S. Yang, Metabolomics for phytomedicine research and drug development, Curr Opin Chem Biol 12 (2008) 66-71.
[25] V.V. Barnatskii, V.D. Grigor'eva, S.B. Pershin, N.A. Derevnina, E.B. Gontar, [Influence of combined rehabilitation treatment including novel non-pharmacological technologies on immune system of patients with seronegetive spondylarthritis], Vopr Kurortol Fizioter Lech Fiz Kult (2005) 20-24.

[26] J. Ezendam, F. Staedtler, J. Pennings, R.J. Vandebriel, R. Pieters, J.H. Harleman, J.G. Vos, Toxicogenomics of subchronic hexachlorobenzene exposure in Brown Norway rats, Environ Health Perspect 112 (2004) 782-791.

[27] M.J. Rhile, M. Nagarkatti, P.S. Nagarkatti, Role of Fas apoptosis and MHC genes in 2,3,7,8-tetrachlorodibenzo-p-dioxin (TCDD)-induced immunotoxicity of T cells, Toxicology 110 (1996) 153-167.

[28] I.A. Camacho, N. Singh, V.L. Hegde, M. Nagarkatti, P.S. Nagarkatti, Treatment of mice with 2,3,7,8-tetrachlorodibenzo-p-dioxin leads to aryl hydrocarbon receptor-dependent nuclear translocation of NF-kappaB and expression of Fas ligand in thymic stromal cells and consequent apoptosis in T cells, J Immunol 175 (2005) 90-103.

[29] C.G. Wermuth, Multitargeted drugs: the end of the "one-target-one-disease" philosophy?, Drug Discov Today 9 (2004) 826-827.

[30] P. Chavan, K. Joshi, B. Patwardhan, DNA microarrays in herbal drug research, Evid Based Complement Alternat Med 3 (2006) 447-457.

[31] S. Amagaya, A. Iizuka, B. Makino, M. Kubo, Y. Komatsu, F.C. Cheng, T.I. Ruo, T. Itoh, K. Terasawa, General pharmacological properties of Sho-seiryu-to (TJ-19) extracts, Phytomedicine 8 (2001) 338-347.

[32] K.K. Ahmed, A.C. Rana, V.K. Dixit, B.G. Shivananda, Internet--implications for the future of phytopharmacological research, Indian J Exp Biol 41 (2003) 1233-1238.

[33] S. Katz, R. Harris, J.T. Lau, A. Chau, The use of gene expression analysis and proteomic databases in the development of a screening system to determine the value of natural medicinal products, Evid Based Complement Alternat Med 3 (2006) 65-70.

[34] J.P. Luyendyk, W.B. Mattes, L.D. Burgoon, T.R. Zacharewski, J.F. Maddox, G.N. Cosma, P.E. Ganey, R.A. Roth, Gene expression analysis points to hemostasis in livers of rats cotreated with lipopolysaccharide and ranitidine, Toxicol Sci 80 (2004) 203-213.

[35] S.B. Pruett, C. Schwab, Q. Zheng, R. Fan, Suppression of innate immunity by acute ethanol administration: a global perspective and a new mechanism beginning with inhibition of signaling through TLR3, J Immunol 173 (2004) 2715-2724.

[36] P. Mondola, F. Santangelo, C. Falconi, A. Belfiore, The serum apo B and apo E in rats following cholesterol diet and thymus treatment, Horm Metab Res 19 (1987) 407-410.

[37] P.A. Clarke, R. te Poele, R. Wooster, P. Workman, Gene expression microarray analysis in cancer biology, pharmacology, and drug development: progress and potential, Biochem Pharmacol 62 (2001) 1311-1336.

[38] D.J. Crowther, Applications of microarrays in the pharmaceutical industry, Curr Opin Pharmacol 2 (2002) 551-554.

[39] M.I. Klapa, J. Quackenbush, The quest for the mechanisms of life, Biotechnol Bioeng 84 (2003) 739-742.

[40] W.J. Zhuang, C.C. Fong, J. Cao, L. Ao, C.H. Leung, H.Y. Cheung, P.G. Xiao, W.F. Fong, M.S. Yang, Involvement of NF-kappaB and c-myc signaling pathways in the apoptosis of HL-60 cells induced by alkaloids of Tripterygium hypoglaucum (levl.) Hutch, Phytomedicine 11 (2004) 295-302.

[41] Y. Chen, Z.Y. Shen, W.H. Chen, [Molecular mechanism of epimedium flavonoids in immune homeostasis remodeling in aged rats revealed by lymphocyte gene expression profile], Zhongguo Zhong Xi Yi Jie He Za Zhi 24 (2004) 59-62.

[42] C.Q. Hu, X.G. Chen, C.M. Li, G.J. Chen, Q. Zhao, Effect of "Si Jun Zi (four gentlemen) Tang" decoction on monoamine transmitter in brain of reserpinized mice, J Tradit Chin Med 3 (1983) 33-35.
[43] L. Liu, L. Han, D.Y. Wong, P.Y. Yue, W.Y. Ha, Y.H. Hu, P.X. Wang, R.N. Wong, Effects of Si-Jun-Zi decoction polysaccharides on cell migration and gene expression in wounded rat intestinal epithelial cells, Br J Nutr 93 (2005) 21-29.
[44] X. Chen, L. Yang, J.J. Oppenheim, M.Z. Howard, Cellular pharmacology studies of shikonin derivatives, Phytother Res 16 (2002) 199-209.
[45] K. Nakaya, T. Miyasaka, A shikonin derivative, beta-hydroxyisovalerylshikonin, is an ATP-non-competitive inhibitor of protein tyrosine kinases, Anticancer Drugs 14 (2003) 683-693.
[46] V. Staniforth, S.Y. Wang, L.F. Shyur, N.S. Yang, Shikonins, phytocompounds from Lithospermum erythrorhizon, inhibit the transcriptional activation of human tumor necrosis factor alpha promoter in vivo, J Biol Chem 279 (2004) 5877-5885.
[47] S.C. Chiu, N.S. Yang, Inhibition of tumor necrosis factor-alpha through selective blockade of Pre-mRNA splicing by shikonin, Mol Pharmacol 71 (2007) 1640-1645.
[48] S.C. Chiu, S.W. Tsao, P.I. Hwang, S. Vanisree, Y.A. Chen, N.S. Yang, Differential functional genomic effects of anti-inflammatory phytocompounds on immune signaling, BMC Genomics 11 513.
[49] S. Roy, S. Khanna, H. Shah, C. Rink, C. Phillips, H. Preuss, G.V. Subbaraju, G. Trimurtulu, A.V. Krishnaraju, M. Bagchi, D. Bagchi, C.K. Sen, Human genome screen to identify the genetic basis of the anti-inflammatory effects of Boswellia in microvascular endothelial cells, DNA Cell Biol 24 (2005) 244-255.
[50] G.B. Singh, C.K. Atal, Pharmacology of an extract of salai guggal ex-Boswellia serrata, a new non-steroidal anti-inflammatory agent, Agents Actions 18 (1986) 407-412.
[51] N. Nakamura, [Antigen presentation and immune induction by dendritic cells], Tanpakushitsu Kakusan Koso 53 (2008) 2263-2268.
[52] Y. Jin, L. Fuller, G. Ciancio, G.W. Burke, 3rd, A.G. Tzakis, C. Ricordi, J. Miller, V. Esquenzai, Antigen presentation and immune regulatory capacity of immature and mature-enriched antigen presenting (dendritic) cells derived from human bone marrow, Hum Immunol 65 (2004) 93-103.
[53] U. Yrlid, M. Svensson, C. Johansson, M.J. Wick, Salmonella infection of bone marrow-derived macrophages and dendritic cells: influence on antigen presentation and initiating an immune response, FEMS Immunol Med Microbiol 27 (2000) 313-320.
[54] P.R. Bergstresser, G.B. Toews, J.W. Streilein, Natural and perturbed distributions of Langerhans cells: responses to ultraviolet light, heterotopic skin grafting, and dinitrofluorobenzene sensitization, J Invest Dermatol 75 (1980) 73-77.
[55] M. Cumberbatch, J.C. Hope, R.J. Dearman, S.J. Hopkins, I. Kimber, Migration of interleukin-6 producing Langerhans cells to draining lymph nodes following skin sensitization, Adv Exp Med Biol 378 (1995) 531-533.
[56] A.H. Enk, S.I. Katz, Early molecular events in the induction phase of contact sensitivity, Proc Natl Acad Sci U S A 89 (1992) 1398-1402.
[57] G.M. Halliday, H.K. Muller, Sensitization through carcinogen-induced Langerhans cell-deficient skin activates specific long-lived suppressor cells for both cellular and humoral immunity, Cell Immunol 109 (1987) 206-221.

[58] I. Kimber, M. Cumberbatch, Dendritic cells and cutaneous immune responses to chemical allergens, Toxicol Appl Pharmacol 117 (1992) 137-146.
[59] I. Kimber, M. Cumberbatch, R.J. Dearman, M. Bhushan, C.E. Griffiths, Cytokines and chemokines in the initiation and regulation of epidermal Langerhans cell mobilization, Br J Dermatol 142 (2000) 401-412.
[60] S. Casati, P. Aeby, D.A. Basketter, A. Cavani, A. Gennari, G.F. Gerberick, P. Griem, T. Hartung, I. Kimber, J.P. Lepoittevin, B.J. Meade, M. Pallardy, N. Rougier, F. Rousset, G. Rubinstenn, F. Sallusto, G.R. Verheyen, V. Zuang, Dendritic cells as a tool for the predictive identification of skin sensitisation hazard, Altern Lab Anim 33 (2005) 47-62.
[61] W.D. Pennie, I. Kimber, Toxicogenomics; transcript profiling and potential application to chemical allergy, Toxicol In Vitro 16 (2002) 319-326.
[62] C.A. Ryan, G.F. Gerberick, L.A. Gildea, B.C. Hulette, C.J. Betts, M. Cumberbatch, R.J. Dearman, I. Kimber, Interactions of contact allergens with dendritic cells: opportunities and challenges for the development of novel approaches to hazard assessment, Toxicol Sci 88 (2005) 4-11.
[63] C.Y. Wang, V. Staniforth, M.T. Chiao, C.C. Hou, H.M. Wu, K.C. Yeh, C.H. Chen, P.I. Hwang, T.N. Wen, L.F. Shyur, N.S. Yang, Genomics and proteomics of immune modulatory effects of a butanol fraction of echinacea purpurea in human dendritic cells, BMC Genomics 9 (2008) 479.
[64] C.Y. Wang, M.T. Chiao, P.J. Yen, W.C. Huang, C.C. Hou, S.C. Chien, K.C. Yeh, W.C. Yang, L.F. Shyur, N.S. Yang, Modulatory effects of Echinacea purpurea extracts on human dendritic cells: a cell- and gene-based study, Genomics 88 (2006) 801-808.
[65] M. Carles, M.K. Cheung, S. Moganti, T.T. Dong, K.W. Tsim, N.Y. Ip, N.J. Sucher, A DNA microarray for the authentication of toxic traditional Chinese medicinal plants, Planta Med 71 (2005) 580-584.
[66] K.L. Wan, J.M. Blackwell, J.W. Ajioka, Toxoplasma gondii expressed sequence tags: insight into tachyzoite gene expression, Mol Biochem Parasitol 75 (1996) 179-186.
[67] J. Liu, C. Hara, M. Umeda, Y. Zhao, T.W. Okita, H. Uchimiya, Analysis of randomly isolated cDNAs from developing endosperm of rice (Oryza sativa L.): evaluation of expressed sequence tags, and expression levels of mRNAs, Plant Mol Biol 29 (1995) 685-689.
[68] S.L. Chen, Y.Q. Sun, J.Y. Song, Y. Li, C.J. Li, S.N. Hu, X.W. Li, H. Yao, X.W. Zhang, [Analysis of expressed sequence tags (EST) from Panax quinquefolium root], Yao Xue Xue Bao 43 (2008) 657-663.
[69] H. Luo, C. Sun, Y. Li, Q. Wu, J. Song, D. Wang, X. Jia, R. Li, S. Chen, Analysis of expressed sequence tags from the Huperzia serrata leaf for gene discovery in the areas of secondary metabolite biosynthesis and development regulation, Physiol Plant 139 1-12.
[70] F. He, Y. Zhu, Y. Zhang, Identification and characterization of differentially expressed genes involved in pharmacological activities of roots of Panax notoginseng during plant growth, Plant Cell Rep 27 (2008) 923-930.
[71] P. Sun, Y. Guo, J. Qi, L. Zhou, X. Li, Isolation and expression analysis of tuberous root development related genes in Rehmannia glutinosa, Mol Biol Rep 37 1069-1079.

[72] A.K. Shukla, A.K. Shasany, M.M. Gupta, S.P. Khanuja, Transcriptome analysis in Catharanthus roseus leaves and roots for comparative terpenoid indole alkaloid profiles, J Exp Bot 57 (2006) 3921-3932.
[73] Y. Deng, Y. Dong, V. Thodima, R.J. Clem, A.L. Passarelli, Analysis and functional annotation of expressed sequence tags from the fall armyworm Spodoptera frugiperda, BMC Genomics 7 (2006) 264.
[74] V.K. Thara, A.R. Seilaniantz, Y. Deng, Y. Dong, Y. Yang, X. Tang, J.M. Zhou, Tobacco genes induced by the bacterial effector protein AvrPto, Mol Plant Microbe Interact 17 (2004) 1139-1145.
[75] E.V. Boyko, C.M. Smith, V.K. Thara, J.M. Bruno, Y. Deng, S.R. Starkey, D.L. Klaahsen, Molecular basis of plant gene expression during aphid invasion: wheat Pto- and Pti-like sequences are involved in interactions between wheat and Russian wheat aphid (Homoptera: Aphididae), J Econ Entomol 99 (2006) 1430-1445.
[76] W.C. Cho, Contribution of oncoproteomics to cancer biomarker discovery, Mol Cancer 6 (2007) 25.
[77] W.C. Cho, C.H. Cheng, Oncoproteomics: current trends and future perspectives, Expert Rev Proteomics 4 (2007) 401-410.
[78] P.H. O'Farrell, High resolution two-dimensional electrophoresis of proteins, J Biol Chem 250 (1975) 4007-4021.
[79] J. Klose, Protein mapping by combined isoelectric focusing and electrophoresis of mouse tissues. A novel approach to testing for induced point mutations in mammals, Humangenetik 26 (1975) 231-243.
[80] S.A. Smith, T.A. Blake, D.R. Ifa, R.G. Cooks, Z. Ouyang, Dual-source mass spectrometer with MALDI-LIT-ESI configuration, J Proteome Res 6 (2007) 837-845.
[81] L.H. Cazares, B.L. Adam, M.D. Ward, S. Nasim, P.F. Schellhammer, O.J. Semmes, G.L. Wright, Jr., Normal, benign, preneoplastic, and malignant prostate cells have distinct protein expression profiles resolved by surface enhanced laser desorption/ionization mass spectrometry, Clin Cancer Res 8 (2002) 2541-2552.
[82] M. Gao, C. Deng, S. Lin, F. Hu, J. Tang, N. Yao, X. Zhang, Recent developments and contributions from Chinese scientists in multidimensional separations for proteomics and traditional Chinese medicines, J Sep Sci 30 (2007) 785-791.
[83] J.K. Wen, M. Han, [Application of genomics and proteomics in study of traditional Chinese medicine], Zhong Xi Yi Jie He Xue Bao 2 (2004) 323-325.
[84] J.H. Lum, K.L. Fung, P.Y. Cheung, M.S. Wong, C.H. Lee, F.S. Kwok, M.C. Leung, P.K. Hui, S.C. Lo, Proteome of Oriental ginseng Panax ginseng C. A. Meyer and the potential to use it as an identification tool, Proteomics 2 (2002) 1123-1130.
[85] W.C. Cho, K.N. Leung, In vitro and in vivo anti-tumor effects of Astragalus membranaceus, Cancer Lett 252 (2007) 43-54.
[86] W.C. Cho, K.N. Leung, In vitro and in vivo immunomodulating and immunorestorative effects of Astragalus membranaceus, J Ethnopharmacol 113 (2007) 132-141.
[87] H.N. Koo, H.J. Jeong, I.Y. Choi, H.J. An, P.D. Moon, S.J. Kim, S.Y. Jee, J.Y. Um, S.H. Hong, S.S. Shin, D.C. Yang, Y.S. Seo, H.M. Kim, Mountain grown ginseng induces apoptosis in HL-60 cells and its mechanism have little relation with TNF-alpha production, Am J Chin Med 35 (2007) 169-182.

[88] M.J. Liu, Z. Wang, Y. Ju, J.B. Zhou, Y. Wang, R.N. Wong, The mitotic-arresting and apoptosis-inducing effects of diosgenyl saponins on human leukemia cell lines, Biol Pharm Bull 27 (2004) 1059-1065.
[89] Y. Wang, Y.H. Cheung, Z. Yang, J.F. Chiu, C.M. Che, Q.Y. He, Proteomic approach to study the cytotoxicity of dioscin (saponin), Proteomics 6 (2006) 2422-2432.
[90] W.C. Cho, [Research progress in SELDI-TOF MS and its clinical applications], Sheng Wu Gong Cheng Xue Bao 22 (2006) 871-876.
[91] B, Linda, Metabolomics: working toward personalized medicine, FDA Consum 39(2005)28-33.
[92] O. Fiehn, J. Kopka, P. Dormann, T. Altmann, R.N. Trethewey, L. Willmitzer, Metabolite profiling for plant functional genomics, Nat Biotechnol 18 (2000) 1157-1161.
[92] U.M. Malyankar, Tumor-associated antigens and biomarkers in cancer and immune therapy, Int Rev Immunol 26 (2007) 223-247.
[94] L.W. Sumner, P. Mendes, R.A. Dixon, Plant metabolomics: large-scale phytochemistry in the functional genomics era, Phytochemistry 62 (2003) 817-836.
[95] R. t'Kindt, K. Morreel, D. Deforce, W. Boerjan, J. Van Bocxlaer, Joint GC-MS and LC-MS platforms for comprehensive plant metabolomics: repeatability and sample pre-treatment, J Chromatogr B Analyt Technol Biomed Life Sci 877 (2009) 3572-3580.
[96] Z.D. Zeng, Y.Z. Liang, F.T. Chau, S. Chen, M.K. Daniel, C.O. Chan, Mass spectral profiling: an effective tool for quality control of herbal medicines, Anal Chim Acta 604 (2007) 89-98.
[97] S.Y. Yang, H.K. Kim, A.W. Lefeber, C. Erkelens, N. Angelova, Y.H. Choi, R. Verpoorte, Application of two-dimensional nuclear magnetic resonance spectroscopy to quality control of ginseng commercial products, Planta Med 72 (2006) 364-369.
[98] H.H. Draisma, T.H. Reijmers, F. van der Kloet, I. Bobeldijk-Pastorova, E. Spies-Faber, J.T. Vogels, J.J. Meulman, D.I. Boomsma, J. van der Greef, T. Hankemeier, Equating, or correction for between-block effects with application to body fluid LC-MS and NMR metabolomics data sets, Anal Chem 82 1039-1046.
[99] B. Biais, J.W. Allwood, C. Deborde, Y. Xu, M. Maucourt, B. Beauvoit, W.B. Dunn, D. Jacob, R. Goodacre, D. Rolin, A. Moing, 1H NMR, GC-EI-TOFMS, and data set correlation for fruit metabolomics: application to spatial metabolite analysis in melon, Anal Chem 81 (2009) 2884-2894.
[100] N.J. Serkova, J.L. Spratlin, S.G. Eckhardt, NMR-based metabolomics: translational application and treatment of cancer, Curr Opin Mol Ther 9 (2007) 572-585.
[101] E.C. Horning, M.G. Horning, Metabolic profiles: gas-phase methods for analysis of metabolites, Clin Chem 17 (1971) 802-809.
[102] L. Pauling, A.B. Robinson, R. Teranishi, P. Cary, Quantitative analysis of urine vapor and breath by gas-liquid partition chromatography, Proc Natl Acad Sci U S A 68 (1971) 2374-2376.
[103] M. Kusano, A. Fukushima, M. Kobayashi, N. Hayashi, P. Jonsson, T. Moritz, K. Ebana, K. Saito, Application of a metabolomic method combining one-dimensional and two-dimensional gas chromatography-time-of-flight/mass spectrometry to metabolic phenotyping of natural variants in rice, J Chromatogr B Analyt Technol Biomed Life Sci 855 (2007) 71-79.

[104] G. Sun, K. Yang, Z. Zhao, S. Guan, X. Han, R.W. Gross, Shotgun metabolomics approach for the analysis of negatively charged water-soluble cellular metabolites from mouse heart tissue, Anal Chem 79 (2007) 6629-6640.
[105] J.C. Lindon, E. Holmes, J.K. Nicholson, Metabonomics in pharmaceutical R&D, FEBS J 274 (2007) 1140-1151.
[106] I.A. Lewis, S.C. Schommer, B. Hodis, K.A. Robb, M. Tonelli, W.M. Westler, M.R. Sussman, J.L. Markley, Method for determining molar concentrations of metabolites in complex solutions from two-dimensional 1H-13C NMR spectra, Anal Chem 79 (2007) 9385-9390.
[107] M. Lambert, J.L. Wolfender, D. Staerk, S.B. Christensen, K. Hostettmann, J.W. Jaroszewski, Identification of natural products using HPLC-SPE combined with CapNMR, Anal Chem 79 (2007) 727-735.
[108] C. Clarkson, D. Staerk, S.H. Hansen, P.J. Smith, J.W. Jaroszewski, Discovering new natural products directly from crude extracts by HPLC-SPE-NMR: chinane diterpenes in Harpagophytum procumbens, J Nat Prod 69 (2006) 527-530.
[109] N. Schauer, A.R. Fernie, Plant metabolomics: towards biological function and mechanism, Trends Plant Sci 11 (2006) 508-516.
[110] B. Singh, T.K. Bhat, Potential therapeutic applications of some antinutritional plant secondary metabolites, J Agric Food Chem 51 (2003) 5579-5597.
[111] D.J. Newman, G.M. Cragg, Natural products as sources of new drugs over the last 25 years, J Nat Prod 70 (2007) 461-477.
[112] E.M. Williamson, Synergy and other interactions in phytomedicines, Phytomedicine 8 (2001) 401-409.
[113] W. Lam, S. Bussom, F. Guan, Z. Jiang, W. Zhang, E.A. Gullen, S.H. Liu, Y.C. Cheng, The four-herb Chinese medicine PHY906 reduces chemotherapy-induced gastrointestinal toxicity, Sci Transl Med 2 45ra59.
[114] D.S. Wishart, Current progress in computational metabolomics, Brief Bioinform 8 (2007) 279-293.
[115] V. Shulaev, Metabolomics technology and bioinformatics, Brief Bioinform 7 (2006) 128-139.
[116] D.S. Wishart, Human Metabolome Database: completing the 'human parts list', Pharmacogenomics 8 (2007) 683-686.
[117] D.S. Wishart, D. Tzur, C. Knox, R. Eisner, A.C. Guo, N. Young, D. Cheng, K. Jewell, D. Arndt, S. Sawhney, C. Fung, L. Nikolai, M. Lewis, M.A. Coutouly, I. Forsythe, P. Tang, S. Shrivastava, K. Jeroncic, P. Stothard, G. Amegbey, D. Block, D.D. Hau, J. Wagner, J. Miniaci, M. Clements, M. Gebremedhin, N. Guo, Y. Zhang, G.E. Duggan, G.D. Macinnis, A.M. Weljie, R. Dowlatabadi, F. Bamforth, D. Clive, R. Greiner, L. Li, T. Marrie, B.D. Sykes, H.J. Vogel, L. Querengesser, HMDB: the Human Metabolome Database, Nucleic Acids Res 35 (2007) D521-526.
[118] J.L. Markley, M.E. Anderson, Q. Cui, H.R. Eghbalnia, I.A. Lewis, A.D. Hegeman, J. Li, C.F. Schulte, M.R. Sussman, W.M. Westler, E.L. Ulrich, Z. Zolnai, New bioinformatics resources for metabolomics, Pac Symp Biocomput (2007) 157-168.
[119] J. Kopka, N. Schauer, S. Krueger, C. Birkemeyer, B. Usadel, E. Bergmuller, P. Dormann, W. Weckwerth, Y. Gibon, M. Stitt, L. Willmitzer, A.R. Fernie, D. Steinhauser, GMD@CSB.DB: the Golm Metabolome Database, Bioinformatics 21 (2005) 1635-1638.

[120] J.H. Kempen, E. Daniel, J.P. Dunn, C.S. Foster, S. Gangaputra, A. Hanish, K.J. Helzlsouer, D.A. Jabs, R.O. Kacmaz, G.A. Levy-Clarke, T.L. Liesegang, C.W. Newcomb, R.B. Nussenblatt, S.S. Pujari, J.T. Rosenbaum, E.B. Suhler, J.E. Thorne, Overall and cancer related mortality among patients with ocular inflammation treated with immunosuppressive drugs: retrospective cohort study, BMJ 339 (2009) b2480.

[121] A. Martinez, A. Castro, I. Dorronsoro, M. Alonso, Glycogen synthase kinase 3 (GSK-3) inhibitors as new promising drugs for diabetes, neurodegeneration, cancer, and inflammation, Med Res Rev 22 (2002) 373-384.

[122] J.M. Stewart, L. Gera, D.C. Chan, P.A. Bunn, Jr., E.J. York, V. Simkeviciene, B. Helfrich, Bradykinin-related compounds as new drugs for cancer and inflammation, Can J Physiol Pharmacol 80 (2002) 275-280.

[123] J.L. Griffin, Understanding mouse models of disease through metabolomics, Curr Opin Chem Biol 10 (2006) 309-315.

[124] K.W. Jordan, L.L. Cheng, NMR-based metabolomics approach to target biomarkers for human prostate cancer, Expert Rev Proteomics 4 (2007) 389-400.

[125] L.L. Cheng, M.A. Burns, J.L. Taylor, W. He, E.F. Halpern, W.S. McDougal, C.L. Wu, Metabolic characterization of human prostate cancer with tissue magnetic resonance spectroscopy, Cancer Res 65 (2005) 3030-3034.

[126] J.L. Griffin, R.A. Kauppinen, Tumour metabolomics in animal models of human cancer, J Proteome Res 6 (2007) 498-505.

[127] A. Brazma, P. Hingamp, J. Quackenbush, G. Sherlock, P. Spellman, C. Stoeckert, J. Aach, W. Ansorge, C.A. Ball, H.C. Causton, T. Gaasterland, P. Glenisson, F.C. Holstege, I.F. Kim, V. Markowitz, J.C. Matese, H. Parkinson, A. Robinson, U. Sarkans, S. Schulze-Kremer, J. Stewart, R. Taylor, J. Vilo, M. Vingron, Minimum information about a microarray experiment (MIAME)-toward standards for microarray data, Nat Genet 29 (2001) 365-371.

[128] G.A. Churchill, Fundamentals of experimental design for cDNA microarrays, Nat Genet 32 Suppl (2002) 490-495.

[129] H.M. Fathallah-Shaykh, Microarrays: applications and pitfalls, Arch Neurol 62 (2005) 1669-1672.

[130] H.M. Fathallah-Shaykh, B. He, L.J. Zhao, A. Badruddin, Mathematical algorithm for discovering states of expression from direct genetic comparison by microarrays, Nucleic Acids Res 32 (2004) 3807-3814.

17

High Content and Throughput Drug Discovery

Quin Wills
SimuGen, London and Kuala Lumpur
United Kingdom and Malaysia

1. Introduction

1.1 The marriage of 'high throughput' and 'high content'.

While the pharmaceutical industry innovation crisis draws much debate (Kaitin & DiMasi, 2011; Macarron et al., 2011; Munos, 2009; Paul et al., 2010), there remains little consensus on how to cohesively deliver value throughout the drug development pipeline (Fig. 1). This chapter considers some of these issues in the context of a growing field for computational biology: drug discovery high throughput screening (HTS). HTS is the approach of rapidly studying physical, chemical, biological and genetic perturbations on the scale of of tens of thousands per day. Today we are faced with ultra-HTS daily screen rates of hundreds of thousands, in part thanks to the continued development of technologies such as micro-fluidics (Agresti et al., 2010). As a discovery tool, it traces it roots back over twenty years (An & Tolliday, 2010), however it is the more recent improvements in cell culture technique - with the potential for multivariate output such as gene expression - that brings it into the domain of high content computational biology. With this maturation of cell-based assays we also notice an increased focus on statistical rigour, analytical integration, and the apparent user-driven plateau in miniaturisation (Mayr & Bojanic, 2009). Rather than being faced with a continued improvement in simple assay throughput, these suggest a growing role for more data-rich high content HTS (hcHTS)[1].

Despite the implicit gains, there exists a notable and growing antipathy towards many 'big data' approaches as discovery tools. Much publication has refocused on data quality versus quantity, with some doubting the impact of high throughput science altogether (Douglas et al., 2010; Macarron et al., 2011; Mayr & Bojanic, 2009). There persists the very real hurdle of experimentalists and team leaders struggling with the interpretation, integration, and decision making based on such data. As a concern routinely witnessed in post-genome era science, it is doubtful that the blame rests primarily with problem-specific methodology. In this chapter, the need for screens to be more decision-centric and transparent across disciplines is proposed. The aim here is not just to provide the reader with specific tools that are likely to rapidy become dated, but introduce the scope and opportunities in drug screening science.

---

[1] In this chapter 'high content' is used interchangeably with 'high dimension', as applied to multiplexed technologies that produce many descriptors per sample, well or observation. A common example is gene expression microarrays. 'High content screening' is commonly used in the literature to indicate high content imaging. To avoid confusion here, the high content approach to HTS is referred to as hcHTS, and high content imaging is referred to as HCI.

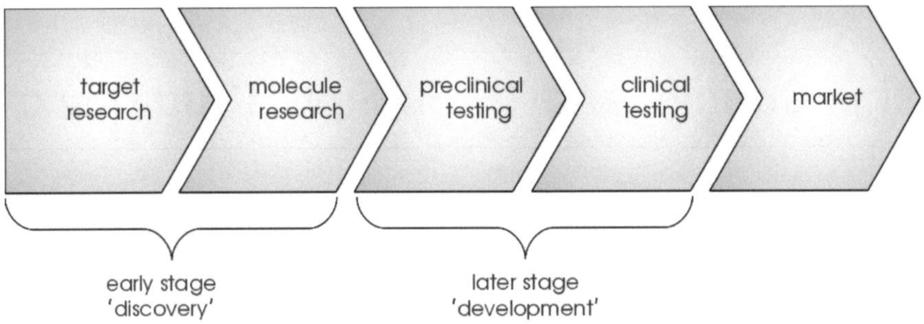

Fig. 1. No single approach is prototypical of drug development. Particularly as a growing number of therapeutic programmes focus on 'biologics' - such as proteins and RNA inhibitors - versus 'small molecule' chemical therapies. However, in general it remains an arduous, failure prone process of 10-15 years, costing hundreds of millions to billions of US$ (Adams & Brantner, 2006). The pipeline can be described as a task in managing attrition rates; a process with a very low success rate sometimes beginning with many hundreds of thousands of chemicals to launch a single successful therapy. The research and development life of a potential drug might be considered in five phases. The first being the identification, development and validation of a target for the drug; the most common targets being G-protein coupled receptors and kinases. The second phase involves the discovery of 'hit' chemicals affecting the target, and development of the hits into leads. Hit through to lead research often begins as high throughput assays, where large libraries of chemicals are screened for effect and sometimes side-effect. What remains are the development phases of animal (preclinical) and human (clinical) testing prior to market release and surveillance (pharmacovigilance).

1.2 The vital role of computational biology.

hcHTS provides a unique challenge to the computational biologist more familiar with high dimension analysis. It increases the analytical demand from the 'few observations, many descriptors' paradigm of small sample multiplex genomics to that of 'many observations, many descriptors'. Drug discovery is also gradually devolving its chemo-centric dominance into an increasingly bio-centric approach. This positions computational biology as a crucial bridge between complex science and technology, and the challenging decisions that need be made from the data produced. The melting pot of *in vitro* (cell-based and biochemical) biology, cheminformatics, bioinformatics, systems biology, and 'big data' analysis requires broad inter-disciplinary scientific and computational strengths. It affords the computational biologist the opportunity to become part of a wide ranging science. A practice where hypotheses and data iteratively refine screens and studies, converging on greater scientific understanding and defined solutions.

This chapter is divided into two main parts. Section 2 contextualises some of the challenges and considerations to guide the choice of modelling strategy, whilst section 3 provides a simple predictive toxicology example that builds on these suggestions. Two traditionally medium throughput multiplex approaches - now increasingly being used in higher throughput settings - will be discussed: gene expression and high content imaging (Bickle, 2010; Zanella et al., 2010). For other promising hcHTS technologies, such as flow cytometry (Edwards et al., 2009) and label-free methods for real time living cell assessment (Xie et al., 2009), the reader is referred to the provided citations. High content imaging (HCI) utilises high resolution multiplex fluorescence microscopy - typically immunofluoresence - to study cellular architecture and health (Karol Kozak, 2009; Zanella et al., 2010). Its strength as a tool is the single cell resolution of physiologically relevant endpoints. HCI together with transcriptomics might be thought of as high content cell and molecular phenotyping. While gene expression analysis is not typically considered part of phenotypic assays, in the context of hcHTS where perturbed pathways and their reporter genes are are studied as indicators of biological process and cell state, it should very much be seen as a proxy of the cell's molecular phenotype. An example of where the two approaches have become inextricably linked is RNA inhibition screens (Karol Kozak, 2009).

2. Modern high throughput drug discovery

2.1 'Big data' analysis paralysis.

Not without its critics (Douglas et al., 2010), the ongoing drug discovery mantra has been one of managing attrition rates by 'failing early, failing often'. However, the biological and drugability knowledge around validated targets has remained poor. An often cited FDA white paper of the early post-genome years (FDA, 2004) drew widespread attention by calling for the greater use of biomarkers and computational approaches to improve this knowledge. With the strong political willpower to modernise drug discovery, HTS has continued to gain popularity as a brute force innovation tool, entering the public domain with resources such as ChemBank (http://chembank.broadinstitute.org), PubChem (http://pubchem.ncbi.nlm.nih.gov) and ChEMBL (https://www.ebi.ac.uk/chembldb). Progress has however faced persistent concerns, with common complaints being poor chemical library design (Gillet, 2008) and that of decision-makers drowning in data. While chemical library design is beyond the scope of this chapter, the data concern is one familiar to every high content computational biologist.

A contrasting argument to the suggested deluge of data as the core concern is that the principal challenge lies with modelling strategy; not the data per se. A case in point might be made of the much publicised Large Hadron Collider with its daily data quota exceeding 40 terabytes. This represents more data than that managed by a typical computational biology team and - while still requiring considerable computing resources - remains a manageable data flow. This is arguably due to well developed theoretical models around which physics expects the data to behave. In contrast, theoretical and systems biology still suffer from a paucity of rigorous quantitative models relevant to disease and chemical biology. The ship may be sinking not because the ocean is large, but because the water is bailed with teaspoons. What then are the most appropriate strategies? To answer this we first need to appreciate that the challenge with screening science is less that of providing narrowly focused yes/no answers. Rather it is more a task of iteratively triaging the optimal options, while managing the decision-making risk across heterogeneous studies spanning months to years. Though not a style of research unaccustomed to statisticians and decision analysts, this challenges computational biology culture with its often data-centric rather than decision-centric and translational mandates.

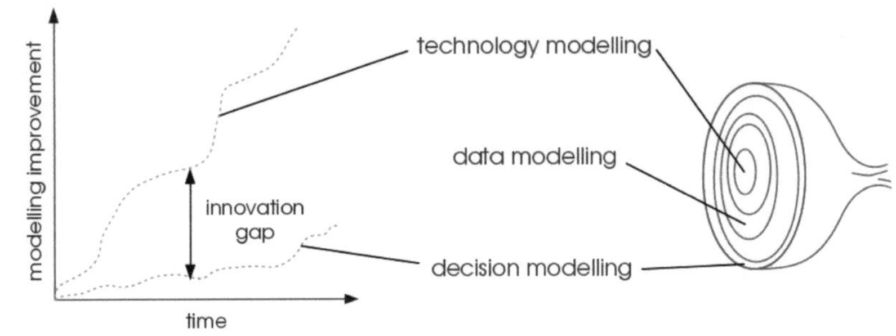

Fig. 2. A useful paradigm for HTS is that of the layers in a 'modelling onion', which emphasises the crucial role of the computational biologist, bridging technology and scientific decision making. Initial research is often driven by technology modelling: choosing the optimal biological models, experimental protocols and technologies to provide good data signal. Data modelling is the remit of computational biology, which might be divided up as a spectrum of low level data clean-up through to higher level theoretical and systems modelling. The screening computational biologist needs to balance the merits of providing detailed results versus fast results; the latter proving useful if they enable the research team to make real-time decisions and rapidly test hypotheses. In HTS this is often the balance of primary versus secondary screening strategies. The final, often neglected, layer is decision modelling. No matter how well the HTS technologists and informaticians consider their models to be performing, if these don't explicitly and transparently assist large discovery teams in making decisions, they are effectively of little use.

2.2 Improving your modelling IQ^2

Modern biology retains its distinctive knowledge-driven culture as a science; differentiated from more mature sciences as being heavily dependent on phenomenological 'stamp collecting'. Similar divides manifest in computational biology as low level data collection, clean-up and mining of bioinformatics versus computational biology modelling. In research with direct translational and economic goals - such as drug screening - it is helpful to remember that:

- Science exists to create explanatory and/or predictive models. Cohesive and comprehensive modelling practice along the entire drug development pipeline is the mandate of all researchers from *in vitro* to *in silico* to the patient.
- All models are wrong, some are useful. Particularly in HTS drug discovery, the development and use of models should be driven by their utility as transparent, triaging decision tools, not narrowly focused technological arguments.
- HTS combines three levels of modelling. Technology, data and decision models should be seen as essential layers in a 'modelling onion' (Fig. 2).

There is, of course, no silver bullet to address data rich problems in drug screening. Notwithstanding, there are general considerations before deciding on methods to optimise the screening model (Fig. 3). A few overlapping rules of thumb are suggested here as a measure of a screen's IQ^2. The test for IQ^2 summarises the need for better *integration*, assay *quantitative*

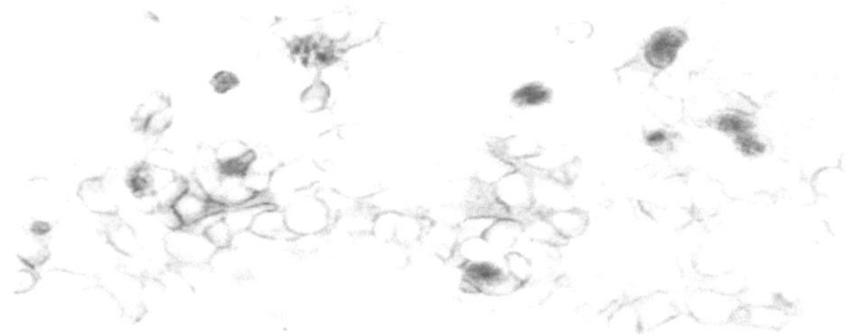

Fig. 3. The above immuno-stained cells - after timed exposure to a toxic chemical - provide a simple example of the screening model in three parts: technology, data and decision modelling. Here technology modelling involves the choice of an informative fluorescent biomarker in an appropriate cell model after an optimal perturbation duration. The data model might be to infer the lowest concentration at which 10% of cells are statistically significantly brighter than twice the average baseline fluorescence. This type of model is a 'lowest dose with an effect' model, where the combination of statistical and biological significance define the concentration output. The utility as a decision tool might be to provide a reproducible relative measure of cytotoxicity across collaborating laboratories interested in a simple ranking of cytotoxic effect, viz. a robust measure best suited as an ordinal triage of effect. The aim being hazard identification, with little explicit attention to risk management and translational or economic impact.

performance, and the decision-making *synergies* (the squared exponent) which present with the action-enabling results.

How well does your approach integrate? Integration entails more than just the use of all available data, but includes the effective integration along the entire flow of data through to knowledge and scientific wisdom (Fig. 4). This is the central tenet of translational bioinformatics, which aims to promote free flow of data between the lab and patient (Buchan et al., 2011). Still in the early stages, translational informatics projects such as Informatics for Integrating Biology and the Bedside (http://www.i2b2.org) hold much promise for feeding back into HTS.

How quantitative is your approach? The quality of the inference is limited by the quantitative performance of the screen. Too often it would seem that post hoc analysis attempts to stretch the assertions made by screening models not fit for purpose. In HTS the primary measures of interest are dose and time. If, for example, a screening programme is required to predict a new drug's safety concerns ('how toxic?'), these might be framed as one or more of many dose and time relevant questions. A few translational toxicity concerns are listed below:

- The concentration at which a percentage of the population begin to experience an effect.
- The concentration at which the risk of rare (unpredictable) adverse effects becomes too great.
- The extent of pathology after a set dose and time exposure.
- The optimal dosing schedule to minimise toxicity without significant loss of efficacy.
- Chronic affects - such as bioaccumulation - less easily extrapolated from acute and sub-acute testing.

Fig. 4. The flow of data into results and decisions reflects the well described flow of information into knowledge and wisdom. Bioinformatics began, in part, as a field to address data integration concerns (Searls, 2010). Today the integration of technologies, laboratories and heterogeneous databases is common practice, and remains vibrant with emerging resources such as cloud computing (Mak, 2011; Schadt et al., 2010). Less well practised is the routine and formal integration of results beyond simple score-based meta-analyses. Bayesian computation promises more formal approaches to update results and incorporate prior information, yet advanced statistical treatment remains underutilised in modern HCS (Malo et al., 2006). Again, the importance of assisting with decision making deserves greater attention. Modelling approaches need be transparent enough to allow a diverse community of scientists to easily communicate and understand the analytical assumptions and limitations.

A role of the screening computational biologist should be seen as providing reliable quantitative measures of concentration and time to hypotheses/questions; not just the provision of IC_{50} or EC_{50} values per variable. The concern, for example, is not the reliable measure of gene expression and the confidence around these measures per se. Rather, it's the transformation of these values into measures of concentration and time, and the confidence around these measures.
Three quantitative concerns often deserving better consideration are suggested:

- The first is the signal-to-noise ratio (Fig. 5A), commonly measured as $Z = 1 - \frac{3(\hat{\sigma}_p + \hat{\sigma}_n)}{|\hat{\mu}_p - \hat{\mu}_n|}$, for positive and negative controls p and n respectively. A Z-score greater than 0.5 is typically accepted to suggest a good assay. The Z-score is a narrowly focused measurement aimed at single-plex assays, which is not robust and assumes data normality. It also does not take into account the performance impact on decision making.
- A second concern is dynamic range. Not all cell models or technologies provide an adequate dynamic range in which drug effects over wide concentration ranges can reliably be measured by a broad spectrum of markers exhibiting near-linear correlation with the effects they proxy. A well known example of this is the poor dynamic performance of gene expression microarrays demonstrated by the Microarray Quality Control project (MAQC Consortium et al., 2006). The screening computational biologist needs to clearly demonstrate the adequate dynamic performance of their data prior to establishing any routine screening.
- The third and final consideration can be broadly defined as that of information resolution. Screens and their follow-up secondary screens/studies need to define clear goals of improving concentration and time sampling density to ensure accurate and sufficiently precise quantitative assertions. The results also need to be presented to the decision-making team at an optimal resolution to be informative without being overwhelming (Fig. 5B - 5D).

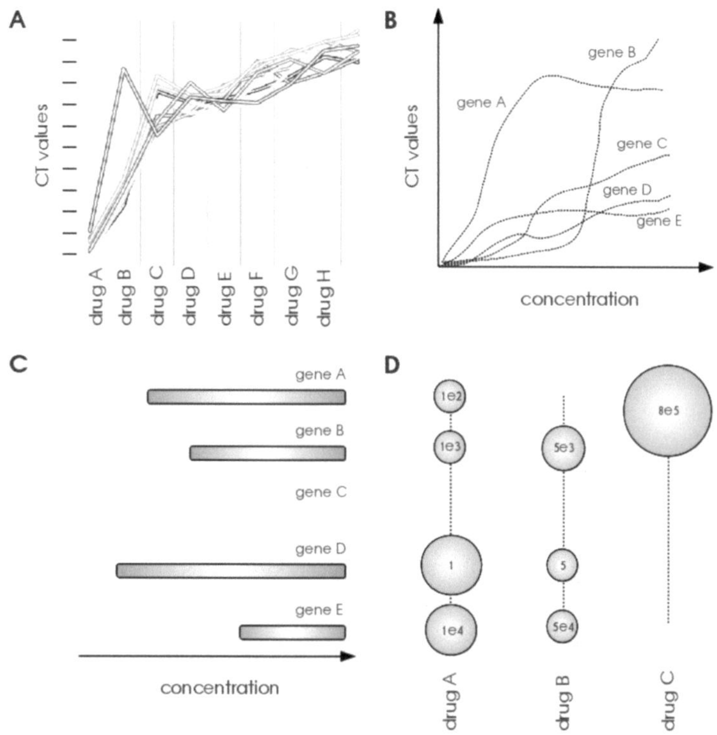

Fig. 5. **(A)** quantitative PCR measurement of gene expression remains the most common gold standard for assessing gene expression without the dynamic limitations of nucleic acid hybridisation technologies, such as gene expression microarrays. However, as can be seen in this well controlled example of highly replicated measurements for a single gene across several compounds, achieving repeated *in vitro* measurement within 1 CT unit remains a challenge. Assuming near optimal reaction efficiency, the CT scale approaches $\log_2$, indicating the cost and time challenge of adequate replication to confidently discern the doubling of a gene's expression under screening conditions. **(B-D)** The resolution at which it is optimal to present results affects the design and/or execution of the data modelling. Figure B provides a detailed trace of five genes perturbed by a compound in a secondary screen. While being detailed, it is ineffective at answering 'at what dose?' and quickly becomes intractable in terms of technical cost and analysis when comparing multiple compounds. Figure C partly resolves this by presenting the results as bars beginning at the lowest concentration at which the answer to the question becomes true. Figure D presents this information for the same genes, comparing tested compounds, numerically providing the concentrations and displaying the statistical confidence in the results as being proportional to the bubble size. Here we see that drug B behaves most similarly to drug A, but at a 5-fold higher concentration (lower potency). It would seem from the screen that we can be fairly confident that drug C behaves differently from drugs A and B.

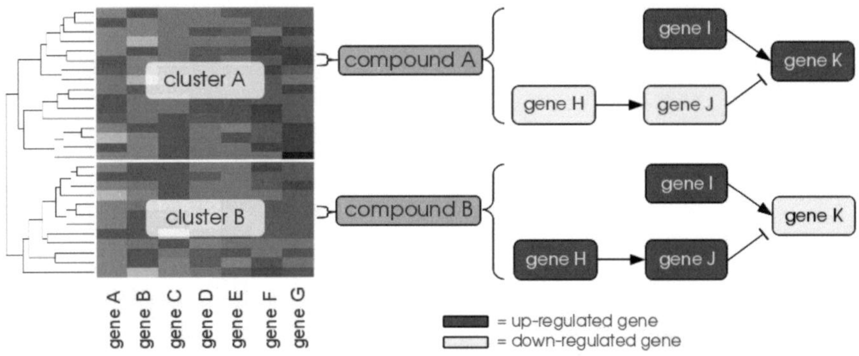

Fig. 6. Careful consideration need always be given to the actionability of screening methods used. If, for example, compound A clusters (or classifies) together with a prototypical compound in cluster A, while compound B clusters together with another prototype in cluster B, does this provide sufficient information to prioritise compound A with a defined dose and time dependent confidence? Similarly with pathway-driven approaches. If compound A does not up-regulate genes H and J at low concentrations, how does this translate into dose and time dependent effects for the purposes of screen prioritisation?

How actionable is your approach? The most important consideration is how effective the screening strategy is at enabling the team to make informed decisions that lead to clear actions where the utility, cost and risk attached to those actions are understood. These can again be considered at the technology, data and decision modelling levels.

Technologist bias will routinely be towards increasing technology complexity within time and cost restraints. However, increased complexity needs to translate into improved actionability. The debate on simple cell culture techniques replaced by the earlier use of lower throughput three-dimensional approaches (Fernandes et al., 2009) highlights this concern. Complex *in vitro* approaches run the risk of compromised data reproducibility. If reduced reproducibility and cost of technology complexity outweigh potential gains in insight, the technological improvements and necessary data modelling changes need be questioned.

The over-reliance on exploratory bioinformatics without clear quantitative questions, hypotheses and follow up is arguably core to current innovation failures (Fig. 6). Three pillars of result significance enable the rational implementation of screens:

- The first is statistical significance, which has traditionally played a minor role within single-plex HCS (Karol Kozak, 2009; Malo et al., 2006).
- Not to be confused with statistical significance is biological significance. A differentially expressed gene defined purely in terms of statistical confidence above baseline needn't represent its biological relevance as a useful marker to elucidate mechanism or enable clear actions based on screening questions.
- The economic argument forms the final pillar, where cost is considered together with utility (Swamidass et al., 2010).

Bayesian methods provide a useful framework to formally work with these different notions of importance, whilst also enabling the use of external data - such as cheminformatics and

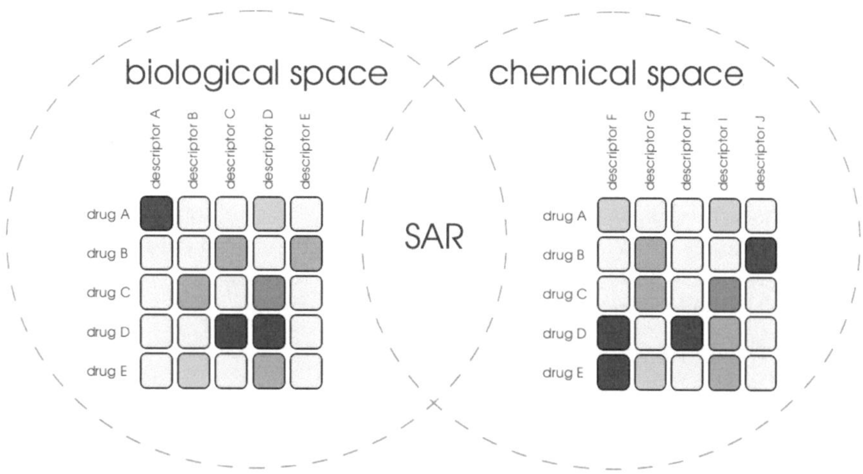

Fig. 7. Common cheminformatics practice is to define descriptors of structure and physiochemical properties in order to position a compound in chemical space. If the purpose of bioinformatics is not only to define molecular and cellular phenotype patterns and mechanisms, then the role of the screening computational biologist effectively becomes a collaborative counterpart to the cheminformatician, defining biological descriptors that are not merely useful as biologically predictive or mechanistic markers, but can be mapped to chemical descriptors in order to define structure-activity relationships (SAR) for rational drug design.

pharmacoeconomics - to establish meaningful priors. However Bayesian analysis appears notably absent in routine published practice (Klon, 2009; Nidhi et al., 2006). Computational limitations seem less likely than the poor understanding surrounding the use of these methods. The need for model understanding and transparency by the entire decision making team is paramount. So until formal frameworks can meet this need, we are left to rely on simpler approaches, some of which are discussed later in this chapter.

Finally, it might be argued that screening methods should ultimately strive not to provide a 'post-mortem' of results but actively assist the discovery team to design better therapies. If, for example, a screen has been developed to predict a spectrum of toxicities from tested chemicals, it should not only accurately identify the correct toxicities, but also their dose-time properties while guiding the chemists on how to alter the compound structures to improve their safety profiles. This re-emphasises the argument for generalist computational biologists in HTS who are able to collaborate with the chemical design and cheminformatics teams, identifying actionable structure-activity relationships (Fig. 7).

3. 'Next generation' drug screening

The title above has deliberately been borrowed from the same description applied to second generation nucleic acid sequencing technologies. It is used in part to stress the increasingly high content flavour of HCS, but also the need for a new screening paradigm focused on a cohesive, transparent and actionable modelling practice. Drawing from real data, this section demonstrates how a screen for genotoxicity might be created using currently available software. The aim is to present how simple rules, weights and thresholds can be used as one approach to create screens not only with good performance characteristics but which can be easily understood and acted on by all team members.

3.1 Rules, weights and thresholds

Cheminformatics has routinely utilised machine-driven pattern recognition to distil large data sets into rule-based models as a form of 'human readable' modelling, or rules of thumb, to predict drug properties. A well known example applicable to ADME is Lipinski's Rule of Five, which assesses how likely it is that a chemical will be orally active. To pass Lipinski's rule, a chemical is limited to violating no more than one criterion:

- Less than or equal to five hydrogen bond donors.
- Less than or equal to ten hydrogen bond acceptors.
- Less than or equal to 500 daltons in molecular weight.
- An octanol-water partition coefficient log P less than or equal to 5

In a similar vein, standard bioinformatics methods can be distilled into combinations of rules, creating such models. These can be tested, refined and understood by non-specialists across the drug discovery team (Fig. 8), with biological and decision-relevant significance better ensured by applying transparent weights and thresholds (Fig. 9). As a consequence of increased computing power and data set size, it seems likely that rule-based approaches will grow in popularity as a tractable modelling strategy. Rule-based modelling has already proven popular in systems biology, where unmanageable lists of differential equations have yielded to agent-based rules of interaction used to drive simulations (Barnes, 2010; Krakauer et al., 2011; Yoav Shoham, 2009). As a methodological approach, rule-based modelling provides a natural bridge for team-driven hypothesis generation and the maturation of generalities for mechanistic and screening biology treated as information science. A particular benefit of rule-based models in screening is that it also allows for the seamless integration of multiple data types. A model might be a collection of rules from multiple cell culture models, multiple time scales, and multiple technologies such as quantitative PCR, HCI, and classical cytometry. Collectively, all of these benefits ensure a high IQ^2 for rule-based models.

3.2 Screening with HT-Stream™

In vitro drug safety screening currently falls within the domain of what are typically medium throughput models aimed at predicting drug absorption, distribution, metabolism and excretion (Ekins et al., 2005). These models are collectively referred to as predictive ADME. Combinatorial chemistry and the shift of ADME to early stage discovery have both significantly improved our ability to design efficacious pharmaceuticals. This has left drug toxicity as an important bottleneck contributing to the innovation crisis, and has prompted its shift to earlier 'off target' cell screens. Whilst born out of lower throughput toxicogenomics (Van Hummelen & Sasaki, 2010), this shift of high content application to high throughput screening requires new methodology and software.

Fig. 10 provides an example of two established DNA damage and stress markers tested on the hepaRG® human liver cell co-culture (Guguen-Guillouzo & Guillouzo, 2010) using quantitative PCR. Microfluidic 'lab-on-a-chip' improvements have enabled cost-effective, high throughput quantitative PCR (Stedtfeld et al., 2008); dramatically improving the scalability of this gold-standard technology as a stand-alone tool or in conjunction with

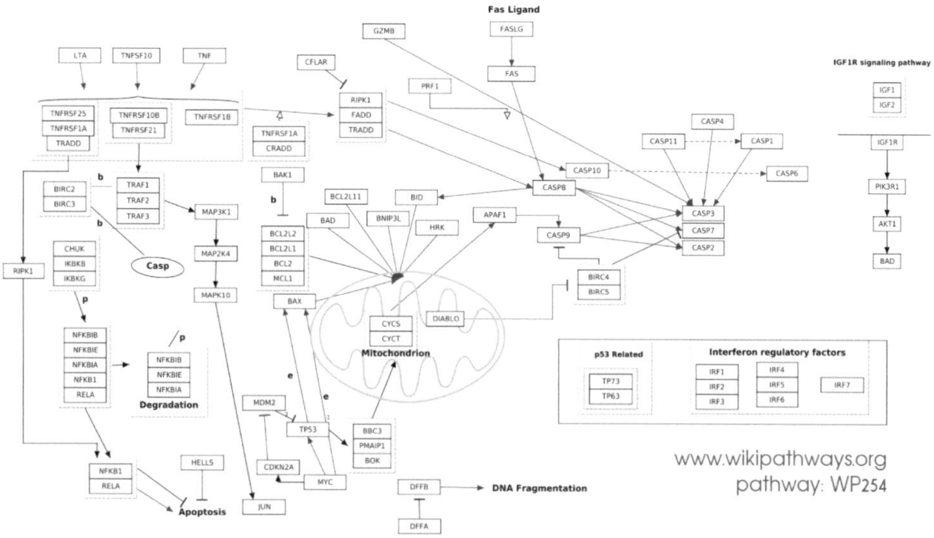

Fig. 8. The apoptosis pathway, as represented above, is commonly used to predict drug safety concerns. A typical bioinformatics approach in a high dimension setting might be to search for apoptosis gene enrichment. Gene set enrichment based on established pathways or ontologies is a rough exploratory tool that translates poorly into a setting where a precise dose-response relationship is required. Using unguided machine learning, literature mining and expert knowledge, complex pathways can be stripped down to collections of rules able to be refined over time and combined to form rule-based models. An example of a rule-based model might be 'clinically significant human apoptosis when at least one, but no greater than four of the following gene ratios hold true. . . '. In a dose-response setting, the lowest concentration at which the rule holds true is called, as a 'lowest dose with an effect' model.

other high content methods such as HCI. While it remains traditional to begin testing gene expression at cell cytotoxicity IC_{50} concentrations, these do not represent physiologically appropriate dosings. As suggested in the data correlations of Fig. 11, HCI allows for mechanistically relevant concentrations to refine classic viability assays in the absence of reliable human data; and discover transcriptomic biomarkers for use in rule-based models.
Once the models have been developed, HT-Stream^{TM} (www.ht-stream.com, www.simugen-global.com) proves useful as online collaborative software that presents the results in a decision-focused manner (Fig. 12). All submitted screening data undergoes automated quality control (Fig. 10A), prior to the inference of the lowest concentration at which each rule becomes true. HT-Stream^{TM} uses the derived concentrations, together with weights and thresholds to help prioritise compounds, visualise results, and compare models (Fig. 13). Easy-to-use software such as this helps make it possible for teams to create 'ecosystems' of applications, rules and models; continuously refining them as collective interdisciplinary knowledge grows with transparent, decision-centric screens.

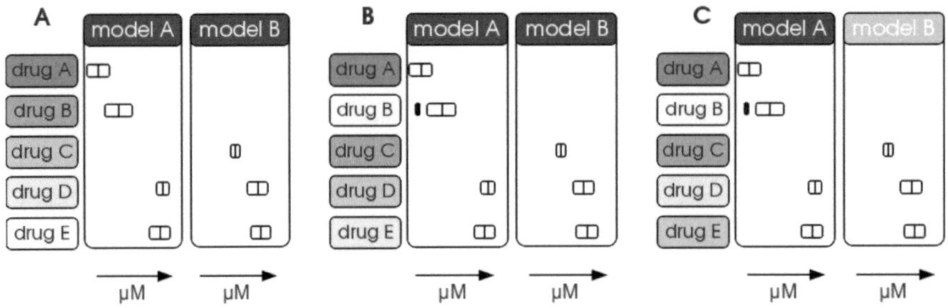

Fig. 9. The above plots represents the output for two 'lowest dose with an effect' models for five tested compounds. The aim here is to rank compounds from highest to lowest toxicity, viz. prioritise those presenting with toxicity at lower concentrations. The darker a compound's label, the higher it is prioritised. The x-axes represent increasing microMolar concentrations of the compounds, with bars representing 95% confidence intervals. **(A)** Here we observe the simplest prioritisation: drug A demonstrates high potency (from model A) and so is ranked first. The benefit of presenting data as concentration values, and the statistical confidence around those values, is evident. While drug A is prioritised over drug B, their confidence intervals overlap, suggesting insufficient statistical evidence to support the ranking. **(B)** The same results are plotted, but with a threshold added to drug B. In this example, drug B has prior information regarding its therapeutic efficacy. The discovery team have decided that any toxicity called above a certain threshold will be of little consequence, as it is unlikely to be reached at therapeutic concentrations. By including this threshold, drug B is de-prioritised. **(C)** The previous rankings assume an equal weighting of the two models. In reality this is rarely the case. If model A represents the drug's carcinogenic potential, whilst model B represents a low-grade safety concern, then model A requires a greater weighting in the global prioritisation. Here drugs D and E switch positions as model B is assigned a low weighting. Thresholds and weights ensure transparent assumptions of biological relevance. With the inclusion of prototypical compounds in the test rankings, transparent weights, thresholds, and statistical significance enable the team to collectively make informed, defensible decisions.

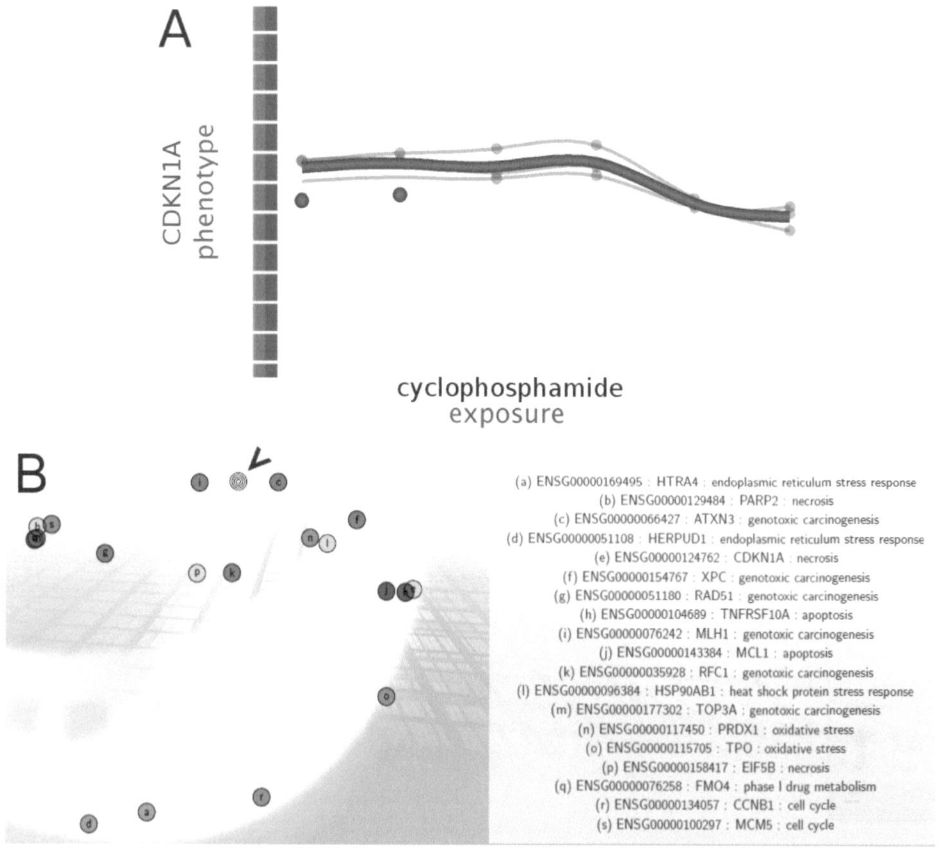

Fig. 10. Genotoxicity biomarkers in hepaRG® viewed using the online tools provided by SimuGen. **(A)** The doubling of CDKN1A's expression with high dose cyclophosphamide. The x-axis represents increasing compound concentration, while each tick in the y-axis represents a CT unit; a drop in one unit thus representing a doubling in gene expression. The analysis tools provide robust automated quality control, in this case identifying two measurements believed to be outliers in bold. **(B)** SimuGen's biomarker discovery tools provide a reference database for over 22,000 genes tested across multiple chemical perturbations in hepaRG®. The above result for GADD45A has identified its most strongly correlated toxic biomarkers, and plotted their first two principal components. GADD45A is highlighted with an arrow and can be seen to be closely clustered with, and enriching for, known genotoxic biomarkers.

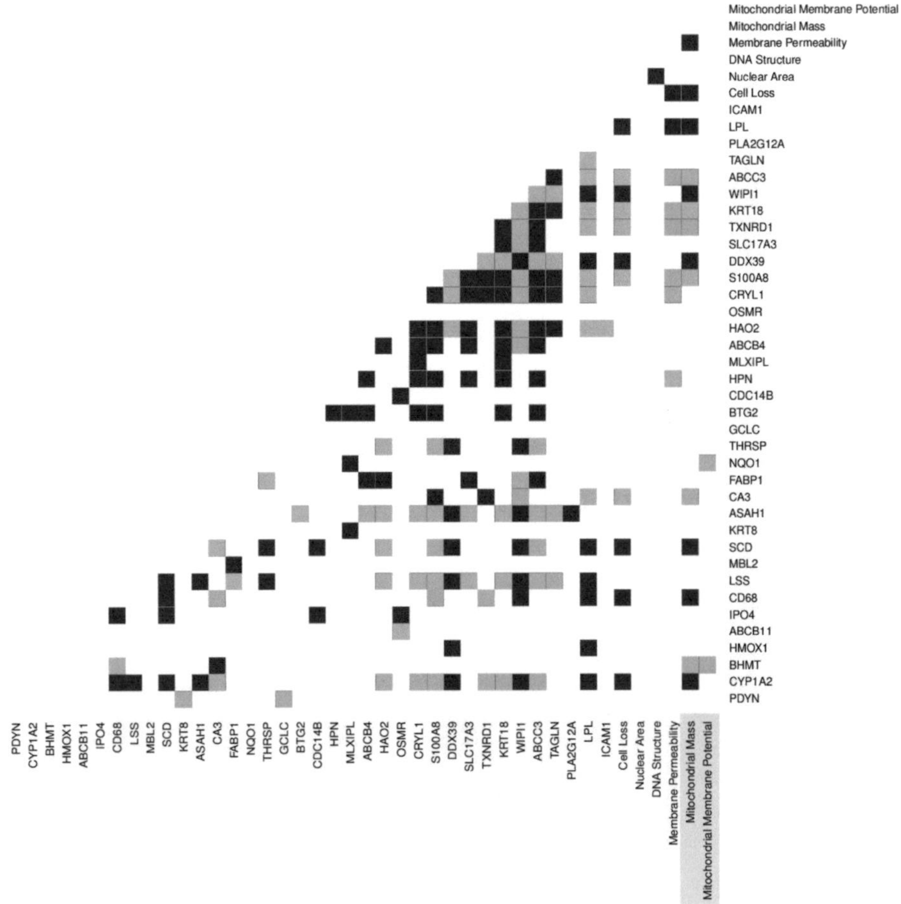

Fig. 11. HCI allows the standardisation of compound concentrations using mechanistic criteria. This correlation plot demonstrates strong correlation (dark squares: Pearson >0.8) and anti-correlation (light squares: Pearson<-08) between known hepatotoxic biomarkers and microscopic phenotypes. The gene expression profile for each compound is measured at the lowest concentration at which any HCI phenotype emerges. Considering drop in mitochondrial mass and potential as a joint phenotype (highlighted) shows strong association with with stress, metabolic and cirrhosis markers. Strong correlations such as these indicate the compatibility of the approaches and the ability to used joint HCI and gene expression data in rule-based models.

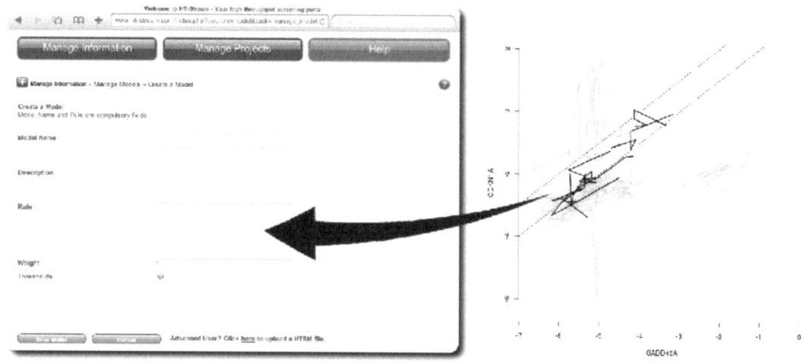

Fig. 12. The right-hand side plot traces the paths of GADD45A and CDKN1A over almost 50 chemicals as their concentrations increase. The black paths represent known genotoxic drugs. It can be seen that there is a 'golden ratio' for the two genes between the dotted lines. Most compounds fall below, whilst non-genotoxic compounds typically present above. HT-Stream™ allows such rules to be entered.

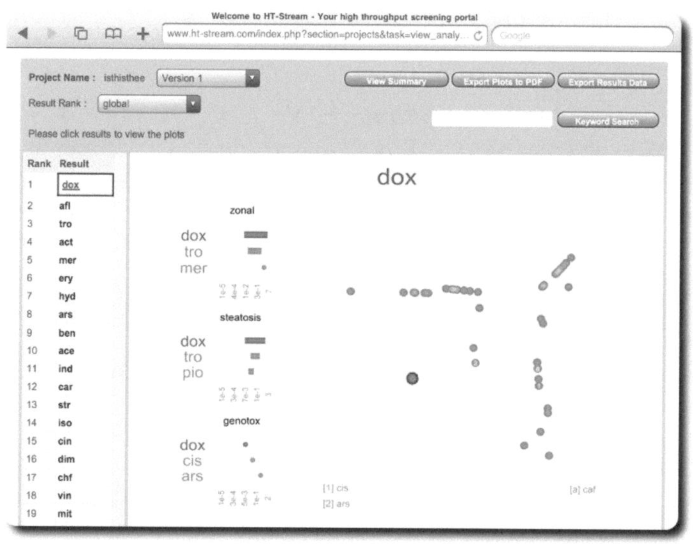

Fig. 13. Using the weights and thresholds, all tested compounds are ranked in HT-Stream™. Any model with a positive result has its results plotted, and contrasted to similar behaving compounds, as described in Fig. 9. A principal components plot, using all models, is also provided to allow the computational biologist and chemist to identify overall patterns that might be related back to chemical structure.

4. References

Adams, C. P. & Brantner, V. V. (2006). Estimating the cost of new drug development: is it really 802 million dollars?, *Health affairs (Project Hope)* 25(2): 420–428.
URL: *http://dx.doi.org/10.1377/hlthaff.25.2.420*

Agresti, J. J., Antipov, E., Abate, A. R., Ahn, K., Rowat, A. C., Baret, J.-C. C., Marquez, M., Klibanov, A. M., Griffiths, A. D. & Weitz, D. A. (2010). Ultrahigh-throughput screening in drop-based microfluidics for directed evolution., *Proceedings of the National Academy of Sciences of the United States of America* 107(9): 4004–4009.
URL: *http://dx.doi.org/10.1073/pnas.0910781107*

An, W. F. & Tolliday, N. (2010). Cell-based assays for high-throughput screening., *Molecular biotechnology* 45(2): 180–186.
URL: *http://dx.doi.org/10.1007/s12033-010-9251-z*

Barnes, D. J. (2010). *Agent-based modeling*, 1 edn, Springer.
URL: *http://www.springer.com/computer/theoretical+computer+science/book/978-1-84996-325-1*

Bickle, M. (2010). The beautiful cell: high-content screening in drug discovery., *Analytical and bioanalytical chemistry* 398(1): 219–226.
URL: *http://dx.doi.org/10.1007/s00216-010-3788-3*

Buchan, N. S., Rajpal, D. K., Webster, Y., Alatorre, C., Gudivada, R. C., Zheng, C., Sanseau, P. & Koehler, J. (2011). The role of translational bioinformatics in drug discovery, *Drug Discovery Today* .
URL: *http://dx.doi.org/10.1016/j.drudis.2011.03.002*

Douglas, F. L., Narayanan, V. K., Mitchell, L. & Litan, R. E. (2010). The case for entrepreneurship in R&D in the pharmaceutical industry., *Nature reviews. Drug discovery* 9(9): 683–689.
URL: *http://dx.doi.org/10.1038/nrd3230*

Edwards, B. S., Young, S. M., Ivnitsky-Steele, I., Ye, R. D., Prossnitz, E. R. & Sklar, L. A. (2009). High-content screening: flow cytometry analysis., *Methods in molecular biology (Clifton, N.J.)* 486: 151–165.
URL: *http://dx.doi.org/10.1007/978-1-60327-545-3_11*

Ekins, S., Nikolsky, Y. & Nikolskaya, T. (2005). Techniques: application of systems biology to absorption, distribution, metabolism, excretion and toxicity., *Trends in pharmacological sciences* 26(4): 202–209.
URL: *http://dx.doi.org/10.1016/j.tips.2005.02.006*

FDA (2004). Innovation or Stagnation: Challenge and Opportunity on the Critical Path to New Medical Products.

Fernandes, T. G., Diogo, M. M., Clark, D. S., Dordick, J. S. & Cabral, J. M. S. (2009). High-throughput cellular microarray platforms: applications in drug discovery, toxicology and stem cell research, *Trends in Biotechnology* 27(6): 342–349.
URL: *http://dx.doi.org/10.1016/j.tibtech.2009.02.009*

Gillet, V. J. (2008). New directions in library design and analysis, *Current Opinion in Chemical Biology* 12(3): 372–378.
URL: *http://dx.doi.org/10.1016/j.cbpa.2008.02.015*

Guguen-Guillouzo, C. & Guillouzo, A. (2010). General review on in vitro hepatocyte models and their applications., *Methods in molecular biology (Clifton, N.J.)* 640: 1–40.
URL: *http://dx.doi.org/10.1007/978-1-60761-688-7_1*

Kaitin, K. I. & DiMasi, J. A. (2011). Pharmaceutical innovation in the 21st century: new drug approvals in the first decade, 2000-2009., *Clinical pharmacology and therapeutics*

89(2): 183–188.
URL: *http://dx.doi.org/10.1038/clpt.2010.286*

Karol Kozak, A. A. (2009). Data Mining Techniques in High Content Screening: A Survey, *J Comput Sci Syst Biol* 2: 219–239.
URL: *http://www.omicsonline.com/ArchiveJCSB/2009/August/04/JCSB2.219.xml*

Klon, A. E. (2009). Bayesian Modeling in Virtual High Throughput Screening, *Combinatorial Chemistry & High Throughput Screening* 12(5): 469–483.
URL: *http://dx.doi.org/10.2174/138620709788489046*

Krakauer, D. C., Collins, J. P., Erwin, D., Flack, J. C., Fontana, W., Laubichler, M. D., Prohaska, S. J., West, G. B. & Stadler, P. F. (2011). The challenges and scope of theoretical biology, *Journal of Theoretical Biology* .
URL: *http://dx.doi.org/10.1016/j.jtbi.2011.01.051*

Macarron, R., Banks, M. N., Bojanic, D., Burns, D. J., Cirovic, D. A., Garyantes, T., Green, D. V. S., Hertzberg, R. P., Janzen, W. P., Paslay, J. W., Schopfer, U. & Sittampalam, G. S. (2011). Impact of high-throughput screening in biomedical research, *Nature Reviews Drug Discovery* 10(3): 188–195.
URL: *http://dx.doi.org/10.1038/nrd3368*

Mak, H. C. (2011). Trends in computational biology[mdash]2010, *Nature Biotechnology* 29(1): 45.
URL: *http://dx.doi.org/10.1038/nbt.1747*

Malo, N., Hanley, J. A., Cerquozzi, S., Pelletier, J. & Nadon, R. (2006). Statistical practice in high-throughput screening data analysis., *Nature biotechnology* 24(2): 167–175.
URL: *http://dx.doi.org/10.1038/nbt1186*

MAQC Consortium, Shi, L., Reid, L. H., Jones, W. D., Shippy, R., Warrington, J. A., Baker, S. C., Collins, P. J., de Longueville, F., Kawasaki, E. S., Lee, K. Y., Luo, Y., Sun, Y. A. A., Willey, J. C., Setterquist, R. A., Fischer, G. M., Tong, W., Dragan, Y. P., Dix, D. J., Frueh, F. W., Goodsaid, F. M., Herman, D., Jensen, R. V., Johnson, C. D., Lobenhofer, E. K., Puri, R. K., Schrf, U., Thierry-Mieg, J., Wang, C., Wilson, M., Wolber, P. K., Zhang, L., Amur, S., Bao, W., Barbacioru, C. C., Lucas, A. B. B., Bertholet, V., Boysen, C., Bromley, B., Brown, D., Brunner, A., Canales, R., Cao, X. M. M., Cebula, T. A., Chen, J. J., Cheng, J., Chu, T.-M. M., Chudin, E., Corson, J., Corton, J. C., Croner, L. J., Davies, C., Davison, T. S., Delenstarr, G., Deng, X., Dorris, D., Eklund, A. C. & Fan, X.-h. H. (2006). The MicroArray Quality Control (MAQC) project shows inter- and intraplatform reproducibility of gene expression measurements., *Nature biotechnology* 24(9): 1151–1161.
URL: *http://dx.doi.org/10.1038/nbt1239*

Mayr, L. M. & Bojanic, D. (2009). Novel trends in high-throughput screening, *Current Opinion in Pharmacology* 9(5): 580–588.
URL: *http://dx.doi.org/10.1016/j.coph.2009.08.004*

Munos, B. (2009). Lessons from 60 years of pharmaceutical innovation., *Nature reviews. Drug discovery* 8(12): 959–968.
URL: *http://dx.doi.org/10.1038/nrd2961*

Nidhi, Glick, M., Davies, J. W. & Jenkins, J. L. (2006). Prediction of Biological Targets for Compounds Using Multiple-Category Bayesian Models Trained on Chemogenomics Databases, *Journal of Chemical Information and Modeling* 46(3): 1124–1133.
URL: *http://dx.doi.org/10.1021/ci060003g*

Paul, S. M., Mytelka, D. S., Dunwiddie, C. T., Persinger, C. C., Munos, B. H., Lindborg, S. R. & Schacht, A. L. (2010). How to improve R&D productivity: the pharmaceutical industry's grand challenge, *Nature Reviews Drug Discovery* 9(3): 203–214.
URL: *http://dx.doi.org/10.1038/nrd3078*

Schadt, E. E., Linderman, M. D., Sorenson, J., Lee, L. & Nolan, G. P. (2010). Computational solutions to large-scale data management and analysis, *Nature Reviews Genetics* 11(9): 647–657.
URL: *http://dx.doi.org/10.1038/nrg2857*

Searls, D. B. (2010). The Roots of Bioinformatics, *PLoS Comput Biol* 6(6): e1000809+.
URL: *http://dx.doi.org/10.1371/journal.pcbi.1000809*

Stedtfeld, R. D., Baushke, S. W., Tourlousse, D. M., Miller, S. M., Stedtfeld, T. M., Gulari, E., Tiedje, J. M. & Hashsham, S. A. (2008). Development and experimental validation of a predictive threshold cycle equation for quantification of virulence and marker genes by high-throughput nanoliter-volume PCR on the OpenArray platform., *Applied and environmental microbiology* 74(12): 3831–3838.
URL: *http://dx.doi.org/10.1128/AEM.02743-07*

Swamidass, S. J., Bittker, J. A., Bodycombe, N. E., Ryder, S. P. & Clemons, P. A. (2010). An Economic Framework to Prioritize Confirmatory Tests after a High-Throughput Screen, *Journal of Biomolecular Screening* 15(6): 680–686.
URL: *http://dx.doi.org/10.1177/1087057110372803*

Van Hummelen, P. & Sasaki, J. (2010). State-of-the-art genomics approaches in toxicology, *Mutation Research/Reviews in Mutation Research* 705(3): 165–171.
URL: *http://dx.doi.org/10.1016/j.mrrev.2010.04.007*

Xie, J., Thapa, R., Reverdatto, S., Burz, D. S. & Shekhtman, A. (2009). Screening of Small Molecule Interactor Library by Using In-Cell NMR Spectroscopy (SMILI-NMR), *Journal of Medicinal Chemistry* 52(11): 3516–3522.
URL: *http://dx.doi.org/10.1021/jm9000743*

Yoav Shoham, K. L. (2009). *Multiagent Systems - Algorithmic, Game-Theoretic, and Logical Foundations*, Cambridge University Press.
URL: *http://www.cambridge.org/gb/knowledge/isbn/item1175725/?site_locale=en_GB*

Zanella, F., Lorens, J. B. & Link, W. (2010). High content screening: seeing is believing, *Trends in Biotechnology* 28(5): 237–245.
URL: *http://dx.doi.org/10.1016/j.tibtech.2010.02.005*